Proceedings in Technology Transfer

The book series "Proceedings in Technology Transfer" provides a valuable platform for researchers, practitioners, and policymakers to share their experiences and insights related to technology transfer. Our aim is to facilitate knowledge exchange between academic researchers and industry professionals, accelerating the introduction of new technologies and driving economic growth.

Our series publishes books and conference proceedings that focus on original research findings, best practice recommendations, emerging trends, and technology evaluations. The series covers all topics which explore the transfer of technology in various industry sectors that require innovations and economic growth.

Join us in advancing the technology transfer and its impact on innovation. Publish your proceedings with us and offer insights and expertise to professionals in your field.

Ranjith Dissanayake · Ajith De Alwis ·
Maksud Bekchanov · Pradeep Gajanayake ·
Sujith Gunawardhana
Editors

Proceedings of the International Conference on Resource Efficiency Towards Sustainability

ICRES 2025, 06 March, Colombo, Sri Lanka

Springer

Editors
Ranjith Dissanayake
University of Peradeniya
Peradeniya, Sri Lanka

Maksud Bekchanov
Ca Foscari University of Venice
Venice, Italy

Sujith Gunawardhana
INSEE Ecocycle Lanka (Private) Limited
Colombo, Sri Lanka

Ajith De Alwis
Department of Chemical and Process Engineering
University of Moratuwa
Moratuwa, Sri Lanka

Pradeep Gajanayake
Department of Biosystems Technology
University of Sri Jayewardenepura
Homagama, Sri Lanka

ISSN 2948-2321 ISSN 2948-233X (electronic)
Proceedings in Technology Transfer
ISBN 978-981-96-9550-8 ISBN 978-981-96-9551-5 (eBook)
https://doi.org/10.1007/978-981-96-9551-5

This Springer imprint is published by the registered company Springer Nature Singapore Pte Ltd.
The registered company address is: 152 Beach Road, #21-01/04 Gateway East, Singapore 189721, Singapore

About the Book

This book showcases the latest knowledge and innovations in resource efficiency and sustainability, addressing critical challenges and opportunities in achieving a sustainable future. It features top-quality and cutting-edge research findings selected from the proceedings of the International Conference on Resource Efficiency Towards Sustainability (ICRES 2025). The themes of this book include Circular Economy, Renewable Energy Solutions, Sustainable Agriculture, Waste Management and Resource Recovery, Climate Change Adaptation and Mitigation, Green Supply Chains, Life Cycle Assessment, and Community Engagement for Sustainability. The ICRES 2025 conference is powered by INSEE Ecocycle, Sri Lanka, whose commitment to advancing sustainable practices has significantly supported the success of this event. For more information, please visit the conference website at https://icres.lk/.

Contents

Circularity in Urban Planning and Infrastructure

Driving Sustainability through AI, Circular Economy, and Governance for Resource Efficiency and Climate Resilience

About the Editors

Prof. Ranjith Dissanayake is a Emeritus Professor in University of Peradeniya and Professor in the Curtin Colombo. He is a Chartered Engineer and a Fellow of the Institution of Engineers, Sri Lanka (IESL) and holds membership from several other prestigious institutions. He was a Fulbright Scholar in the USA, Endeavour Fellow in Australia, and JASSO Research Fellow in Japan. He was awarded the Young Scientist Award in 2007 for Excellence in Scientific Research by the National Science and Technology Commission (NASTEC) of Sri Lanka. He received the Overseas Prize of the Institution of Civil Engineers (ICE) in the UK in 2007 and the Australia Alumni Excellence Award in 2012. He has published over 150 papers in reputed journals. He has also chaired more than fifteen international conferences. He is the founder of several companies, including Lego International (Pvt.) Ltd, ICB (Pvt.) Ltd, Lanka AAC (Pvt) Ltd, GAP (Pvt) Ltd and WSP Pvt. Ltd. E-Mail: ranjith@fulbrightmail.org

Prof. Ajith de Alwis is a Senior Professor in the Department of Chemical and Process Engineering at the University of Moratuwa, Sri Lanka. He earned his BSc (Eng) with First Class Honors from the University of Moratuwa and completed his PhD in Chemical Engineering at the University of Cambridge, UK, focusing on "Ohmic heating of foods." Prof. de Alwis has held significant administrative roles, including serving as the Head of the Department of Chemical and Process Engineering and the Dean of the Faculty of Graduate Studies at the University of Moratuwa. He has also contributed to national science and technology initiatives as the Acting Chief Innovation Officer at the National Innovation Agency. His research interests include process optimization, renewable energy systems, and sustainable industrial practices, with numerous publications to his credit. Prof. de Alwis has been recognized with prestigious awards, including the Senior Moulton Medal from the Institution of Chemical Engineers, UK, and the Danckwerts-Maxwell Prize for the best Chemical Engineering PhD thesis at the University of Cambridge. E-Mail: ajith@uom.lk

Dr. Maksud Bekchanov is an environment and development economist holding a PhD degree from Bonn University, Germany. He worked for International Water Management Institute in Sri Lanka, Universities of Bonn and Hamburg in Germany as a researcher after PhD studies. He received prestigious Marie-Curie Fellowship and joined to Ca' Foscari University of Venice in Italy in 2022, for conducting research on strategic investments in adaptation and mitigation in water, energy and food systems. Dr. Bekchanov has broad experience in developing integrated environmental-economic assessment models to analyze environmental policies and technologies. He effectively contributed to water-economic modeling fields and addressing zero-hunger and zero-carbon challenges. His research on circular economy of composting and biogas in Sri Lanka became impactful. Overall, he led writing more than 20 peer-reviewed articles and more than 40 book chapters and discussion papers. E-Mail: maksud.bekchanov@cmcc.it

Dr. Pradeep Gajanayake is a Senior Lecturer in the Department of Biosystems Technology, Faculty of Technology, University of Sri Jayewardenepura, Sri Lanka. He earned his Bachelor's and M.Phil. degrees in Agricultural Engineering from the University of Peradeniya and his Ph.D. in Environmental Engineering from Saitama University, Japan. An Accredited Green Professional with the GBCSL and a Chartered Environmental Professional with the IEPSL, Dr. Gajanayake has been recognized for his exemplary contributions to sustainability, earning the prestigious Outstanding Green Professional of the Year Award from the GBCSL in 2024. He is also a Governing Council Member of the Sri Lanka Institute of Agriculture. Dr. Gajanayake's extensive expertise in environmental engineering, agricultural innovation, and sustainable technology has significantly influenced both local and international practices, reinforcing his dedication to fostering sustainable development and green education. E-Mail: pradeepgajanayake@sjp.ac.lk

Eng. Sujith Gunawardhana is a Director at INSEE Ecocycle Lanka (Private) Limited, Sri Lanka's leading provider of waste management services. He holds a B.Sc. in Chemical and Process Engineering from the University of Peradeniya, an MBA from Cardiff Metropolitan University, and is currently pursuing a Doctor of Business Administration (DBA) at the University of Kelaniya. His expertise spans business strategy, circular economy, waste management, renewable energy, waste-to-energy, project management, change management, and business transformation. Prior to his current role, he held positions such as Head of Projects and Project Department Manager at Siam City Cement (Lanka) Limited, and Platform Manager - Geocycle and Project Engineer/Technical Support Engineer at Holcim (Lanka) Limited. He is also a Certified Project Management Professional (PMP), a Chartered Professional Manager (MCPM) and a Council Member of the Institute of Chartered Environmental Professionals Sri Lanka (IEPSL) E-Mail: sujith.gunawardhana@siamcitycement.com

Advanced Waste Management and Recycling Technologies

Enhancing Properties of Compressed Stabilized Earth Blocks Using Pet Fibres

L. Sivaganesalingam[1(✉)], S. N. Malkanthi[1], P. D. Dharmaratne[2], and Harsha Galabada[3]

[1] Department of Civil and Environmental Engineering, University of Ruhuna, Galle, Sri Lanka
lakshminisiva21@gmail.com

[2] Department of Civil Engineering, Sri Lanka Institute of Information Technology, Malabe, Sri Lanka

[3] Esoft Metro Campus, Colombo, Sri Lanka

Abstract. Compressed stabilized earth blocks (CSEBs) are an earthen building material made from soil, cement, and water. CSEB can be considered as an environmentally friendly and sustainable alternative to traditional bricks and cement blocks. Since the main raw material of CSEB is soil, the clay content in the soil greatly affects its strength and durability. Hence fibre can be used to improve ductility and toughness. This study provides an experimental analysis of the properties of CSEBs reinforced with PET (Polyethylene Terephthalate) fibres. Further controlling of soil gradation was carried out by using construction waste and applying the particle packing optimization concept. Furthermore, the study highlights soil gradation was adjusted to match the optimization curve for producing CSEBs. To compare dry and wet compressive strength, water absorption, and dry density with SLS 1382:PART 1 requirement, test cubes (150mmx150mmx150mm) were cast with 20% finer content and 10% cement for different PET fibre lengths (1cm, 2cm, 3cm,4cm) and different PET fibre percentages by weight (0.1%,0.2%,0.3%,0.4%). Out of the test cubes, 0.4 weight percentage and 2cm length PET-reinforced CSEB were selected as optimum. So, with those results optimum industrial blocks(356mmx178mmx102mm), were cast and industrial scale blocks results satisfied Grade 1 SLS 1382: PART 1 requirement. This study also concerns the cost-effectiveness of the blocks.

Keywords: Compressed stabilized earth blocks · Compressive strength · Construction waste · Particle packing optimization · Soil modification · PET fibres

1 Introduction

Earth is the oldest building material known. For centuries, earth has been used in different forms such as mud, adobe, rammed earth and bricks. With the growing concern of awareness regarding environmental and ecological issues, the possible alternatives for traditional building materials such as burnt bricks, and cement blocks are compressed earth cement blocks or cement-stabilized earth blocks (CSEB) (Thanushan and

R. Dissanayake et al. (Eds.): ICRES 2025, PTT, pp. 3–12, 2025.
https://doi.org/10.1007/978-981-96-9551-5_1

Sathiparan, 2022). CSEBs possess high compressive strength, similar to conventional masonry materials, and can be used for load-bearing walls. Since the main raw material for CSEB production is soil, the clay content in the soil greatly affects its strength and durability (Malkanthi and Perera, 2018). Moisture causes for reducing the strength of CSEB. However, past research stated that compressive strength increased with decreasing clay content and higher cement content. These studies set a minimum limit of 15% for clay content (Heathcote, 1991). According to their experiments, compressive strength increases with decreasing finer content for varying amounts of cement. (Walker, 1995) concluded that high compressive strength can be achieved with a low plasticity index and 10% cement content, however only when the soil has a minimum clay content of 15%. A compressive strength of over 10 N/mm2 has also been found when the clay content is between 10% and 15%, even with a low plasticity index.

On the other hand, PET contributes a more significant percentage to the plastic waste stream. The concept of using fibres as reinforcement is not a new concept for CSEB. The incorporation of fibres into the soil can significantly enhance its resistance against non-elastic shrinkage, cracking, and impacts, thereby increasing its rigidity (Jesudass et al., 2021). Reduced clay and silt content in the soil is a critical factor in enhancing the durability issues associated with CSEB. Hence, adjusting soil gradation to match the optimization curve for producing CSEB is an advantageous concept (Malkanthi and Perera, 2019), (Malkanthi, Wickramasinghe and Perera, 2021). However, the reduction of clay and silt content has limitations. Hence, adding natural or synthetic fibers enhances the properties of CSEB (Malkanthi et al., 2024). It was likely that further research was needed to investigate the specific use of PET fibres in combination with the modification of soil gradation for CSEB production.

The main aim of this research is to investigate the suitability of using construction waste and waste fibre to enhance the properties of the CSEB with the concern of cost. Further, the study focused on applying the particle packing optimization concept for soil grade modification. Construction waste mainly with crushed concrete was used to modify the soil gradation while PET fibre was used as reinforcement to enhance the properties of CSEB. With that background, the research objectives are, to identify the optimum PET fibre content to enhance the properties of CSEB and to determine the cost-effectiveness of CSEB with PET fibre.

2 Research Methodology

2.1 Soil Selection

The soil samples were taken from a borrow pit in Hapugala, Galle, Southern Province of Sri Lanka. Dry and wet sieve analysis and Atterberg tests were performed to determine the finer percentage, total particle size distribution, specific gravity, and plasticity index (PI) of soil. The selected soil showed 30.7% clay and silt content. Furthermore, it had a Plasticity Index of 8. According to the SLS1382:PART 1 requirements (Sri Lanka Standard Institution, 2010), Soil should be selected with a finer percentage of less than 35% and PI of less than 12. Past studies also mentioned that soil with a plasticity index of less than 15 is suitable for CSEB production (Walker, 1995), (Raj et al., 2017). So, the selected soil is suitable for CSEB production. Many previous studies have proven that the clay and slit content in the soil should be less than 35% (Nagaraj, Rajesh and Sravan, 2016), (Venkatarama Reddy and Latha, 2014), (Jayasinghe and Perera, 1999).

2.2 Soil Modification by Particle Packing Optimization

Since the low clay and silt content in the soil is preferred for CSEB production, past researchers have suggested different methods to modify the soil gradation. (Fitzmaurice, 1958) says, there is a possibility that unsuitable soil could be modified by adding sand when there is an excess of clay or silt. However, there could be economic issues since sand is comparatively expensive. Some researchers have studied alternative materials such as coal ash and cassava peels (Villamizar et al., 2012), sugarcane bagasse and lime (Alavéz-Ramírez et al., 2012), cork granules (Guettala, 2016) that can be used to lower the clay and silt content instead of adding sand to reduce the finer content. Preliminary testing carried out by (Malkanthi and Perera, 2018) has shown that soil washing can be used to reduce the clay and silt content in the soil. They showed that soil having up to a minimum of 5% clay and silt content could be used for CSEB production and it gives significant strength and durability properties.

Other than the amount of clay and silt content, the soil gradation is also an influential factor for the properties of CSEB. Even if the soil properties are dominant, different particle combinations are available here. For the appropriate sizes and shapes of aggregates, particle packing technology is vital in concrete technology. This particle packing technology considers optimizing the correct sizes and amounts of various particles to increase particle density (Fenis and Walraven 2010). So this concept which is shown in Figure is used to adjust the particle size distribution of soil to the ideal grading curve which indicates grading with the greatest density (Senthil Kumar V and Manu Santhanam, 2003) (Fig. 1).

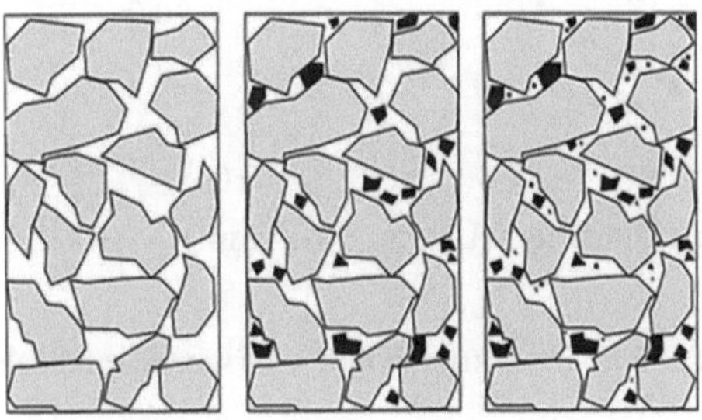

Fig. 1. The concept of particle packing

When considering optimization curves, optimum packing grading curves were tried by many researchers for different purposes (Wong and Chan, 2013), (Hettiarachchi and Mampearachchi, 2019). So, Fuller's curve is based on a correlation between the small and large particle distribution in the soil (Fennis and Walraven, 2012), (Borges et al., 2014). The Fuller's curve theory is shown in Eq. 1 and its modification performed by Funk and Dinger is shown in Eq. 2. (Malkanthi and Perera, 2019) have proved this concept is applicable to CSEB production too.

$$\mathrm{P}(d) = \left(\frac{d}{d_{\max}}\right)^{q} \tag{1}$$

$$P(d) = \frac{d_{\max}^{q} - d_{\min}^{q}}{d^{q} - d_{\min}^{q}} \tag{2}$$

P(d) = size cumulative distribution function.
d = considered particle diameter.
d_{max}= maximum particle diameter in the mixture.
q = parameter (0.33–0.5) which adjusts the curve for fineness or coarseness.
$d_{\min}$=minimum particle diameter in the mixture (m).

By using these equations, optimization curves were drawn and the particle size distribution of the selected soil sample was also plotted in the same plot which is shown in Fig. 2.

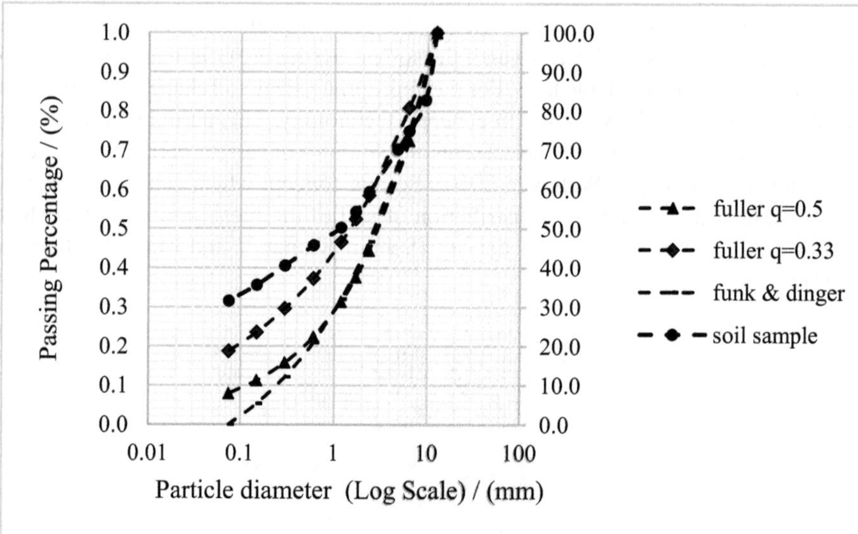

Fig. 2. Particle size distribution of soils with theoretical curves

By comparing these optimization curves, soil modification was carried out by adding construction waste to reduce the clay and silt content from 30.7% to 20%.

2.3 CSEBs Casting

CSEB cube was cast for the selected mix design in the size of $150mm \times 150mm \times 150mm$ mold, For preliminary laboratory testing, the control cube was cast only by mixing construction waste with soil and cement. So, sixteen mix proportions were designed with 20% clay and silt contents in the modified soil with a cement content of 10%.

PET fibres were used with different lengths (1cm, 2cm, 3cm, 4cm) and different fibre percentages by weight (0.1%,0.2%,0.3%,0.4%). 10% of water of total weight was used for mixing. Initial curing was carried out within 7 days by using gunny bags and final curing was performed within 28 days by exposing them to the environment.

2.4 Testing of Test Blocks

Cast blocks were tested for 28 days of dry and wet compressive strength, dry density and water absorption. The cubes were placed in a water bath for 24 h before wet compressive strength testing. The results were compared with the requirements in SLS 1382 Part 1.

The optimal mix proportion of the CSEB was used to cast blocks on an industrial scale. The cost estimation was then carried out for these industrial-scale blocks. Finally, it was compared to the market-available cement sand block and brick. The proposed final dimensions for the CSEB block are 356 $mm\times$ 178 $mm\times$ 356 mm. Moreover, the cost estimation for wall construction was calculated for a unit area for CSEB, cement block and burnt brick.

3 Results and Discussion

Table 1 shows the results of the control blocks. Figure 3 and Fig. 4 show the dry and wet compressive strengths of blocks for different weight percentages and different lengths of PET fibre. It has been realized that all the results are greater than the control sample's results.

Table 1. Properties of the control samples

Dry compressive strength (MPa)	Wet compressive strength (MPa)	Dry weight (g)	Dry density (kg/m^3)	Water absorption (%)
3.5	2.0	5,354	1,586	6.75

The dry and wet compressive strengths exhibit similar trends. The results indicate that cubes reinforced with a low weight percentage of PET fibers have lower dry and wet compressive strengths. However, as the fiber weight percentage increases, both strengths improve. This trend is particularly noticeable for shorter PET fibers. For 1 cm PET fibers, the dry compressive strength values for 0.1 wt.%, 0.2 wt.%, 0.3 wt.%, and 0.4 wt.% are 4 MPa, 7.75 MPa, 10.67 MPa, and 13.20 MPa, respectively. A similar relationship is observed for 2 cm fibers. Likewise, the wet compressive strength values for 1 cm PET fibers at the same weight percentages are 3.3 MPa, 4.7 MPa, 7.47 MPa, and 9.6 MPa, with 2 cm fibers following the same pattern. The highest dry and wet compressive strengths were achieved with 2 cm fibers for all fiber percentages except 0.1%. However, for fiber percentage optimization, the maximum strength was recorded at a fiber length of 3 cm.

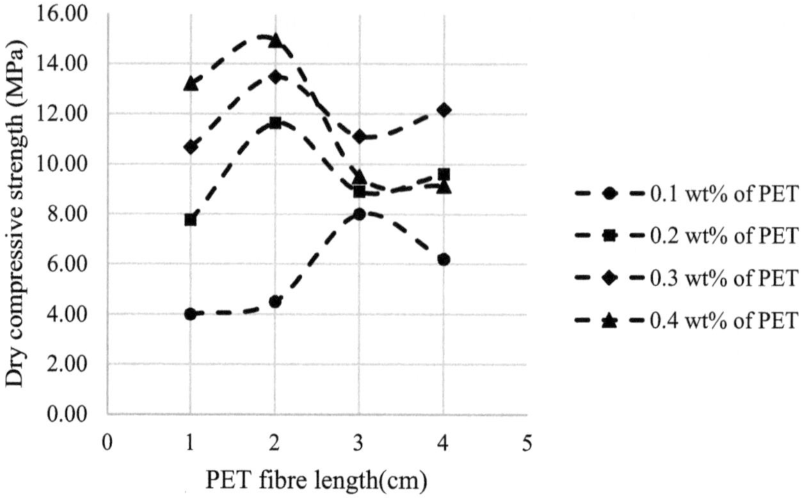

Fig. 3. Dry compressive strength of soil blocks with varying fibre length and weight

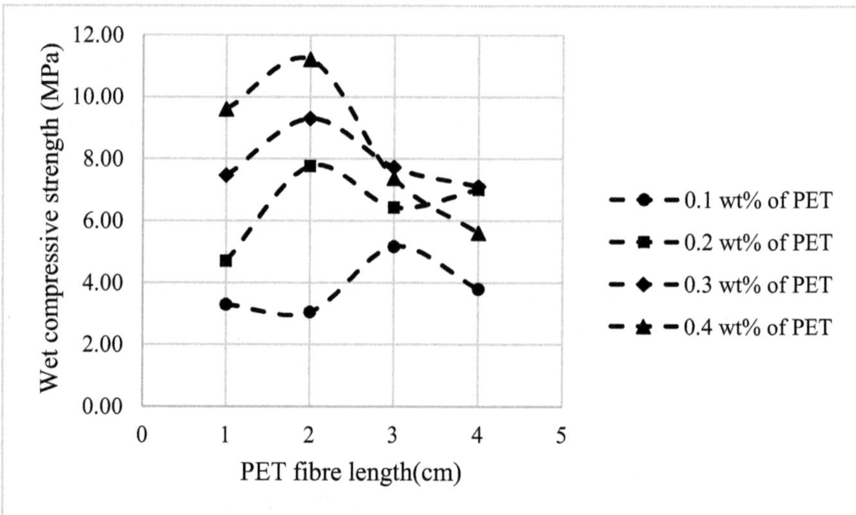

Fig. 4. Wet compressive strength of soil blocks with varying fibre length and weight

These findings suggest that, for shorter PET fibers, increasing fiber content leads to an improvement in both dry and wet compressive strengths. That is because shorter PET fibers are more evenly distributed throughout the matrix, creating a denser and more interconnected network. This improved distribution enhances the bond between the fibers and the cementitious material, allowing for better stress transfer and resistance to cracking. Further, PET fibers act as reinforcement within the matrix by bridging microcracks. As fiber content increases, more fibers are available to resist crack propagation, thereby improving the overall strength of the material. This relationship was proved by Muhit by using plastic dust (Muhit, 2013).

But in a particular length, there are some variations when the fiber content increases. When weight percentage increases from 0.1% to 0.4%, first it indicates a significant increase and then a notable drop then it completes with a marginal uptick. So the reason for this is when it comes to the 0.3% or 0.4% with a long length the PET fiber clumps together and forms a large ball (Muhit, 2013). So, 0.3% of combinations have this problem in 2cm, 3cm and 4cm CSEBs. But comparatively 1cm fibres neither formed a ball nor clamped together. The reason for the marginal uptick is when mixing it's easy to find the bigger ball which is made by 4cm PET than 3cm.

According to SLS 1382 requirements, 6.0 MPa, 4.0 MPa, and 2.8 MPa are the dry compressive strength limits for Grade 1, Grade 2, and Grade 3, respectively. 2.4 MPa, 1.6 MPa, and 1.2 MPa are the wet compressive strength limits for Grade 1, Grade 2. And Grade 3, respectively. Therefore, it can be concluded that dry and wet compressive strength values of this CSEB fall at least under Grade 3 blocks.

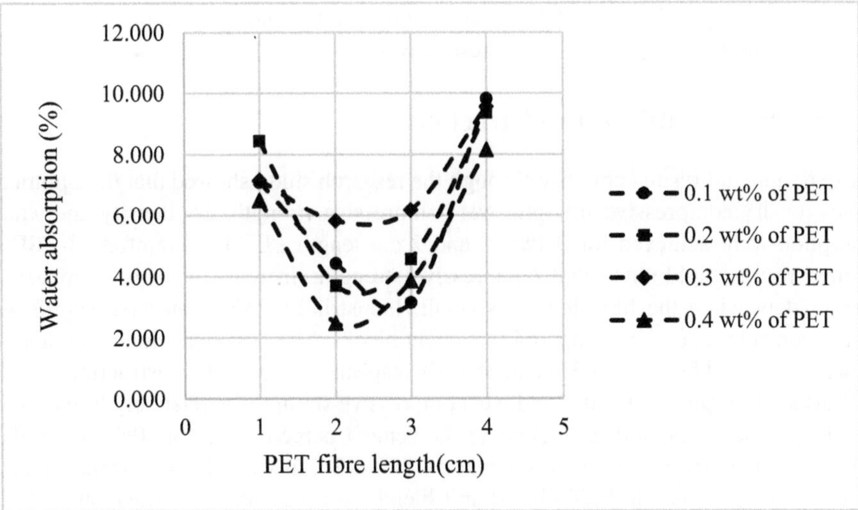

Fig. 5. Water absorption of soil blocks with varying fibre length and weight

Water absorption is the main parameter to measure the durability of blocks. The amount of water absorption depends on the type of soil used and is related to the compressive strength (Malkanthi and Perera, 2018). Figure 5 shows the water absorption variation of soil blocks with varying fibre lengths and weights. So, because of the same soil used in these blocks, it can be concluded that the water absorption changes relate to the compressive strength and it can be seen that the optimum value among 16 combinations for dry and wet strength have the least water absorption and lesser compressive strength also has higher water absorption. According to Fig. 5, all the water absorption values are less than 15% which satisfies the SLS 1382 requirements.

For all the combinations except one (2 cm length and 0.1 wt%) dry density values are between 1,887 kg/m^3 (for 4 cm and 0.1wt%) and 2,138 kg/m^3 (2cm and 0.4wt%). From these results, 0.4wt.% 2 cm combination was selected as the optimum mix proportion (design). Industrial blocks were cast with this optimum combination and the results are shown in Table 2.

Table 2. Results of the industrial-scale CSEB block

Dry compressive strength (MPa)	Wet compressive strength (MPa)	Water absorption (%)	Dry density (kg/m^3)	Grade according to SLS standard
9.13	7.10	9.8	1436.8	Grade 1

According to the cost estimation, unit CSEB manufacturing cost is Rs.89.00 and it is around 22% price reduction of the present market price of cement blocks in Sri Lanka. The unit rate calculation of 1m^2 wall panel with 4-inch thickness of CSEB, cement block and brick were Rs.2,923.00, Rs.3,000.00, and Rs.3,478.00 respectively. Comparing the cost of 1m^2 wall panel from CSEB, the cost reduction percentages of cement block wall and brick wall are 2.63% and 18.96% respectively.

4 Conclusion and Future Directions

The experimental results obtained through the research study showed that the optimum values for dry compressive strength, wet compressive strength, dry density and water absorption were achieved for 0.4wt % and 2 cm length PET fibre-reinforced CSEB. Further, results highlighted that compressive strengths increase with the increase of fibre content when the fibre length is small. Industrially CSEB can have significant production cost reduction compared to cement blocks. The cost reduction compared to cement block and brick is 2.63% and 18.96% respectively for wall construction.

Further, the value of the dry and wet compressive strengths is relatively higher than the SLS Grade 1 standard. So, reducing the cement percentage from 10% to 8% also can achieve the grade 1 SLS recommended CSEBs. So, this will help to further reduce the cost. In this research, PET fiber didn't blend with the mix. So, Investigating PET fiber blending techniques and blending effect to strength variation will be conducted as future research.

Because of the cement addition for stabilization purposes PET fibers have the opportunity to degrade in the long term. Therefore, investigations of degradation and its impact on the CSEBs can be carried out in the future. The PET fiber-reinforced CSEBs have more strength than other fiber-reinforced CSEBs compared to the past literature. Thus, it is crucial to conduct long-term durability and water erosion tests before introducing it to the industry. Additionally, some difficulties came across during the casting process, particularly in maintaining the uniform distribution of PET fibers and generation of dust during production. So, research addressing these issues will be welcomed.

Finally, for a sustainable future, social awareness should be spread to increase confidence in CSEBs presenting successful projects.

Acknowledgments. This work was immensely supported by the Department of Civil and Environmental Engineering, University of Ruhuna, Sri Lanka with the provision of well-equipped laboratory facilities and technical officers of the laboratories. We would like to express our deep gratitude to this organization. Further, the authors would like to acknowledge the support given by Ms. M.W.P. Sandamali Nandasena, Technical officer of Geotechnical Laboratory and Mr. T.G.P Wasantha Kumara, Technical officer of Building Materials Laboratory, Faculty of Engineering, University of Ruhuna, Sri Lanka.

References

Alavéz-Ramírez, R., Montes-García, P., Martínez-Reyes, J., Altamirano-Juárez, D.C., Gochi-Ponce, Y.: The use of sugarcane bagasse ash and lime to improve the durability and mechanical properties of compacted soil blocks. Constr. Build. Mater. **34**, 296–305 (2012). https://doi.org/10.1016/j.conbuildmat.2012.02.072

Borges, P.H.R., Fonseca, L.F., Nunes, V.A., Panzera, T.H., Martuscelli, C.C.: Andreasen Particle Packing Method on the Development of Geopolymer Concrete for Civil Engineering (2014). https://doi.org/10.1061/(ASCE)MT.1943-5533.0000838

Fennis, S.A.A.M., Walraven, J.C.: Using particle packing technology for sustainable concrete mixture design. Heron **57**(2), 73–101 (2012)

Fitzmaurice, R.: Manual on stabilized soil construction for housing. United Nations Technical Assistance Programme, NewY ork (1958)

Guettala, S.: 'Properties of the compressed-stabilized earth brick containing cork granules journal of earth science and climatic properties of the compressed-stabilized earth brick containing cork granules. **9**(May) (2016). https://doi.org/10.4172/2157-7617

Heathcote, K.: Earthwall construction compressive strength of cement stabilized pressed earth blocks. Building Res. Inf. **19**(2), 101–105 (1991). https://doi.org/10.1080/096132191087 27106

Hettiarachchi, H.A.C.K., Mampearachchi, W.K.: Validity of aggregate packing models in mixture design of interlocking concrete block pavers (ICBP). Road Mater. Pave. Design **20**(2), 462–474 (2019). https://doi.org/10.1080/14680629.2017.1393001

Jayasinghe, C., Perera, A.: Stabilized Soil Block Technology for Sri Lanka. In: Research for Industry, pp. 147–160. University of Moratuwa, Sri Lanka, Engineering Research Unit (1999)

Jesudass, A., Gayathri, K., Harish, S., Sree Ram, S., Mohammed, R.: Earthen blocks with synthetic fibres – a review. In: IOP Conference Series: Materials Science and Engineering **1145**(1), p. 012039 (2021). https://doi.org/10.1088/1757-899x/1145/1/012039

Malkanthi, S.N., Sathsara, K.G.K., Dharmaratne, P.D., Galabada, H.: Proposed mix design improvements of compressed stabilized earth blocks (CSEB) with particle packing optimization and coir reinforcement. J. National Sci. Found. **52**(2), 159–167 (2024)

Malkanthi, S.N., Perera, A.A.D. A.J.: 'Particle packing application for improvement in the properties of compressed stabilized earth blocks with reduced clay and silt. Technol. Appl. Sci. Res. **9**(4), pp. 4538–4542 (2019). https://doi.org/10.48084/etasr.3002

Malkanthi, S.N., Perera, A.A.D.A.J.: Durability of compressed stabilized earth blocks with reduced clay and silt. In: IOP Conference Series: Materials Science and Engineering (2018). https://doi.org/10.1088/1757-899X/431/8/082010

Malkanthi, S.N., Wickramasinghe, W.G.S., Perera, A.A.D.A.J.: Use of construction waste to modify soil grading for compressed stabilized earth blocks (CSEB) production. Case Stud. Constru. Mater. **15**(July), e00717 (2021). https://doi.org/10.1016/j.cscm.2021.e00717

Muhit, I.B.: Evaluation of stabilized soil blocks with the inclusion of " Plastic Fibre " as sustainable building material : a complete review 2013(November 2013), pp. 1–3 (2013)

Nagaraj, H.B., Rajesh, A., Sravan, M.V.: Influence of soil gradation, proportion and combination of admixtures on the properties and durability of CSEBs. Constr. Build. Mater. **110**, 135–144 (2016). https://doi.org/10.1016/j.conbuildmat.2016.02.023

Raj, S. Mohammad S., Das R., Saha, S.: Coconut fibre reinforced cement stabilized rammed earth blocks. World J. Eng. **14**(3), 208–216 (2017). https://doi.org/10.1108/WJE-10-2016-0101

Senthil Kumar, V., Manu, S.: Particle packing theories and their application in concrete mixture proportioning: a review effect of acid attack on concrete view project national centre for safety of heritage structures view project (September 2003) (2003)

Sri Lanka Standard Institution (2010) Specification for Compressed Stabilized Earth Blocks:Part 1 Requirements. Colombo

Thanushan, K., Sathiparan, N.: (2022) 'Mechanical performance and durability of banana fibre and coconut coir reinforced cement stabilized soil blocks.' Materialia **21**, 101309 (2021). https://doi.org/10.1016/j.mtla.2021.101309

Venkatarama Reddy, B.V., Latha, M.S.: Influence of soil grading on the characteristics of cement stabilised soil compacts. Mater. Struct. Mater. Construct. **47**(10), 1633–1645 (2014). https://doi.org/10.1617/s11527-013-0142-1

Villamizar, M.C.N., Araque, V.S., Reyes, C.A.R., Silva, R.S.: Effect of the addition of coal-ash and cassava peels on the engineering properties of compressed earth blocks. Constr. Build. Mater. **36**, 276–286 (2012). https://doi.org/10.1016/j.conbuildmat.2012.04.056

Walker, P.J.: Strength, durability and shrinkage characteristics of cement stabilised soil blocks. Cement Concr. Compos.Concr. Compos. **17**(4), 301–310 (1995). https://doi.org/10.1016/0958-9465(95)00019-9

Wong, V., Chan, K.W.: Applying theories of particle packing and rheology to concrete for sustainable development **5**(2) (2013). https://doi.org/10.5592/otmcj.2013.2.3

Removal of Heavy Metals from Aqueous Solution Using Locally Available Low Cost Biosorbents

J. A. L. W. Gunathilaka(✉), E. K. G. G. Jith, and A. M. W. Menike

Department of Chemical and Process Engineering, Faculty of Engineering, University of Peradeniya, Peradeniya 20400, Sri Lanka
e18122@eng.pdn.ac.lk

Abstract. Heavy metal contamination in aquatic ecosystems presents a severe threat to both environmental and human health. Biosorption has proven to be an effective and eco-friendly alternative to conventional methods for heavy metal removal. This study investigates the biosorption of Lead (Pb) and Cadmium (Cd) from aqueous solutions using modified coco peat, jackfruit rags, and mango seeds under various experimental conditions. Fourier-transform infrared (FTIR) spectroscopy was utilized to analyse the functional groups present in the biosorbents. Among the tested materials, coco peat exhibited superior adsorption capabilities, achieving a Pb biosorption capacity of 20.70 mg/g. The adsorption behaviour of Pb onto Coco peat followed the Freundlich isotherm, indicating favourable multilayer adsorption. Conversely, Cd exhibited poor fitting to both Freundlich and Langmuir isotherms. It was selected for process optimization due to its high performance, availability, and cost-effectiveness. Optimal conditions for Pb adsorption onto coco peat were identified as a pH of 6, a contact time of 150 min, an adsorbent dosage of 1 g/L, and a maximum adsorption capacity of 73.03 mg/g at an initial metal concentration of 100 ppm, achieving a removal efficiency of 73%. For Cd adsorption, the optimal conditions included a pH of 6, a contact time of 20 min, an adsorbent dosage of 1 g/L, and a maximum adsorption capacity of 81.81 mg/g at the same initial concentration, resulting in a removal efficiency of 81%. These findings highlight potential of modified coco peat as an effective biosorbent for heavy metal removal, offering a sustainable and affordable solution for mitigating contamination in water.

Keywords: Coco peat · Jackfruit Rags · Mango seeds · Biosorption · Lead · Cadmium

1 Introduction

Heavy metals are a group of metallic chemical elements that have a high atomic weight and density (Koller and Saleh 2018). Some of the most common heavy metals include lead (Pb), cadmium (Cd), mercury (Hg), arsenic (As), chromium (Cr), and nickel (Ni). According to the World Health Organization (WHO), Cd^{2+}, Pb^{2+}, Hg^{2+}, and As^{3+} ions

R. Dissanayake et al. (Eds.): ICRES 2025, PTT, pp. 13–24, 2025.
https://doi.org/10.1007/978-981-96-9551-5_2

are described as the most hazardous heavy metal ions (Sen 2023). Heavy metal contamination in aquatic ecosystems is a growing problem that is closely linked to urbanization, climate change, and growing industrial activity (Briffa et al. 2020a). Sources such as landfill leachates, industrial and municipal wastes, and urban runoff, coupled with natural phenomena such as weathering, rock abrasion, and volcanic eruptions, contribute to the complexity of pollution (Hama Aziz et al. 2023). Mining and irrigation of crop fields with industrial water are two of the most common human activities that pollute water bodies with heavy metals. The wide-reaching impact of industrialization further compounds the problem, spanning processes such as electroplating, electronic device production, metallurgy, smelting, fertilizer production, nuclear fuel processing, paper manufacturing, power plant emissions, and chemical etching (Sud et al. 2008). Nowadays, heavy metal pollution is a major issue, especially in Sri Lankan lagoon ecosystems (Nanayakkara et al. 2022). According to the studies, the levels of Cd and Pb in these ecosystems exceed the recommended limits (Nanayakkara et al. 2022; Chandrasekara et al. 2014; SR 2020). This poses a significant concern for the environment and human health, leading to various health issues such as headaches, fatigue, appetite loss, and hypertension (Zaynab et al. 2022). The extended existence of heavy metal pollution in the environment leads to significant alterations in soil characteristics, such as pH, color, porosity, and overall chemistry, making the land unfit for cultivation (Briffa et al. 2020c). The tenacity of soil-borne metals spreads to water sources, causing pollution and posing a double threat to the environment as they immediately affect plant life and make their way into the food chain. The issue is further worsened by air pollution, as heavy metals can be introduced into the atmosphere through various means, contributing to phenomena such as eutrophication, corrosion, and the generation of particulate matter that gives rise to serious health issues (Briffa et al. 2020c).

With differing degrees of effectiveness, a number of conventional technologies such as reverse osmosis (RO), adsorption, membrane processes, filtration and ion exchange have been used to remove heavy metal contaminants from wastewater and water (Elhafez et al. 2017). Among these mechanisms, adsorption was selected for this study due to its simplicity, efficiency, selectivity, and environmental considerations (Sud et al. 2008). In industrial applications, the most prevalent adsorbents include activated carbon, zeolite, activated alumina, and silica gel (Sen 2023). However, the drawbacks associated with these materials, particularly their high cost, have prompted a shift towards utilizing biological materials such as plant biomass and microorganisms for the adsorption of heavy metals (Ali Redha 2020).The advantages of using biosorbent materials, include their minimal pre-treatment requirements, local availability, low cost, environmental friendliness, feasibility of physical and chemical modification, and good biosorption capacity (Elhafez et al. 2017).

In the context of this study, Jackfruit rags, Mango seed and Coco peat have been selected as biosorbents. These materials have gained recognition for their use in dye and heavy metal removal. (Esther et al. 2021; Zakaria et al. 2023; Shah and Hussain 2022). However, limited research has been conducted on their heavy metal adsorption capacities when they are modified. Therefore, the present research is mainly focused on the application of modified jackfruit rags, mango seeds and coco peat for removal of Cd and Pb from aqueous solutions and aims to assess the Cadmium (Cd) and Lead

(Pb) adsorption capacity from aqueous solutions using modified biosorbents made from jackfruit rags, mango seeds, and coco peat, determine the biosorbent material with the maximum biosorption capacity, and investigate the effect of biosorbent dose and contact time on the biosorption capacity of Cd and Pb.

2 Method

This study involved the following methodological steps, 1) Preparation of modified biosorbents from selected materials; 2) Determination of biosorption capacity for each biosorbent and identifying the most effective one for removing heavy metals; 3) Optimization of process parameters for the selected biosorbent. These steps enabled a comprehensive investigation into the efficacy of modified biosorbents for heavy metal removal, ultimately leading to the identification and optimization of process conditions.

2.1 Biosorbents Preparation

Initially, all three biosorbents were washed with distilled water to remove the dust and other impurities. Subsequently, the collected materials were dried in an oven at 70 °C for 24 h. Once completely dried, the materials were finely ground and sieved through a 0.5 mm sieve to achieve a consistent particle size. The resulting materials were then carefully stored in containers, preserving their integrity for use in subsequent modifications.

2.2 Modification of Biosorbents

For the biosorbent modification, a solution of 0.3 M Nitric acid (HNO_3) was used. The ground and sieved samples were soaked in the acid solution for a period of one day, followed by a thorough rinsing procedure with distilled water. Subsequently, the treated samples were dried for 24 h using an oven dryer at a temperature of 50 °C.

2.3 Identification of Functional Groups

Fourier transform infrared spectroscopy (FTIR) (PerkinElmer Spectrum IR Version 10.6.2) was used to obtain the infrared spectrum of the biosorbents.

2.4 Metal Solution Preparation

A lead metal solution of 1000 ppm was prepared by dissolving the necessary quantity of lead nitrate in distilled water. Similarly, a 1000 ppm cadmium solution was prepared by dissolving Cd metal in 10 ml of concentrated Nitric acid and 50 ml of distilled water. For the batch experiment, metal solutions with required concentration were prepared by diluting the 1000 ppm stock solutions. pH of the experimental solutions was adjusted using 0.5 M HCl and 0.5 M NaOH.

2.5 Design of Experiment for Biosorption Optimization

According to the literature, factors, such as pH, initial metal concentration, biosorbent dosage, and contact time, affect the efficiency of heavy metal removal. Therefore, it is important to assess the combined influence of these parameters on the heavy metal adsorption capacity. For biosorption process optimization, the design of experiments was executed using Minitab 17 statistical software (Version 17.1.0). A fractional factorial design incorporating 3 center points was utilized to interpret the relationship between input and output variables. This study focused on examining the effects of biosorbent dosage, ranging from 1 to 20 g/l, and contact time, varying from 20 to 150 min, on the Pb and Cd adsorption capacity. pH and initial metal concentration were not included as variables due to experimental time constraints and the high number of required sample analyses, which would have increased the complexity and resource demand of the study (Table 1).

Table 1. Design matrix for process parameter optimization

Run	Biosorbent dosage (g/l)	Contact time (min)
1	10.5	85
2	10.5	85
3	20.0	150
4	1.0	150
5	10.5	85
6	20.0	20
7	1.0	20

2.6 Biosorption Study

For biosorbent selection, 0.2 g of coco peat was separately added to four samples containing 50 ml of metal solutions with concentrations of 10 mg/l, 30 mg/l, 50 mg/l, and 70 mg/l, and two samples with a concentration of 100 mg/l, respectively. pH of the metal solutions was maintained at 6. The mixture was then stirred for 150 min at 150 rpm using an incubated shaker. Then, the biosorbents were filtered, and the filtrate was subjected to analysis using Atomic Absorption Spectrometry (AAS). The same procedure was followed for the Jackfruit rags and Mango seeds to test the biosorption capacity. The biosorption capacity was calculated using Eq. (1), while under the optimized parameters, the percentage of heavy metal removal was determined using Eq. (2).

$$q = \frac{(c_i - c_e)v}{m} \tag{1}$$

$$\%\text{Removal} = \frac{(\mathrm{c_i} - \mathrm{c_e})}{\mathrm{c_i}} \times 100\% \tag{2}$$

where q is the biosorption capacity (mg of metal ion/ g of biosorbent), C_i is the initial metal concentration in solution (mg/l), C_e is the equilibrium metal concentration in solution (mg/l), V is the solution volume (l), and m is the dry weight of the biosorbent (g) (Pérez-Marín et al. 2007).

To fit the experimental biosorption data, the Langmuir model, characterized by Eq. (3), and the Freundlich model, expressed by Eq. (5), were used. The separation factor, also known as the equilibrium parameter, which determines the favorability of adsorption according to the Langmuir isotherm, was calculated using Eq. (4).

$$\frac{c_e}{q_e} = \frac{c_e}{Q_m} + \frac{1}{Q_m K_L} \tag{3}$$

where, C_e is the equilibrium concentration of metal (mg/l), q_e is the amount of metal adsorbed at equilibrium (mg/g), Q_m is the saturated monolayer adsorption capacity (mg/g), and K_L is the Langmuir bonding energy coefficient (l/mg) (Garg et al. 2007).

$$R_L = \frac{1}{1 + K_L C_i} \tag{4}$$

where R_L is the separation factor, K_L is the Langmuir bonding energy coefficient and C_i is the initial concentration of metal. The separation factor can take on different values, and these values provide information about the adsorption behaviour: (i) unfavourable adsorption, $R_L > 1$, (ii) favourable adsorption, $0 < R_L < 1$, (iii) linear adsorption, $R_L = 1$, and (iv) irreversible adsorption, $R_L = 0$ (Farasati et al. 2016).

$$\ln(q_e) = \ln\left(K_F + \frac{1}{n}\ln(C_e)\right) \tag{5}$$

where, C_e is the equilibrium concentration of metal (mg/l), q_e is the amount of metal adsorbed at equilibrium (mg/g), K_F is the equilibrium constant indicative of adsorption capacity and n is the equilibrium constant indicative of adsorption intensity. The constant n can indicate the favourability of adsorption according to Freundlich isotherm. If $n > 1$, it indicated that the adsorption is favourable (Hashtroudi 2018).

3 Results and Discussion

3.1 FTIR Analysis

The FTIR spectra of modified coco peat, jackfruit rags and mango seeds were used to determine the functional groups available in the biosorbent. Upon comparing the results obtained from the FTIR analysis of the modified biosorbents with the FTIR analysis conducted on the raw coco peat documented in the literature (Lazim et al. 2015; Dzigbor and Chimphango 2019), notable shifts in peaks, changes in the intensities of peaks and formation and absence of some functional groups were observed. These changes signify significant modifications in the chemical structure and surface characteristics of the biosorbents induced by the acid treatment.

In modified coco peat (Fig. 1), the adsorption peak around 3307.5 cm^{-1} indicated the existence of hydroxyl groups. The peaks observed at 1625 cm^{-1} can be assigned to stretching vibration C = C group. The peak around 1516.5 cm^{-1} corresponds to the N-O stretching. The peak observed at 1444.5 cm^{-1} indicated the existence of carboxylic acid group. The additional peaks at 1372.5 cm^{-1}, 1262.5 cm^{-1}, and 1104.5 cm^{-1} can be assigned to bending modes of phenol, stretching modes of alkyl aryl ether, and stretching mode of secondary alcohol respectively.

In modified jackfruit rags (Fig. 2), the broad adsorption band is observed at 3301.5 cm^{-1} which can be attributed to the bonded -OH groups in the structure. The peak around 1734 cm^{-1} corresponds to the C = O stretching. The peak at 1437 cm^{-1} indicated the presence of carboxylic acid group.

In modified mango seeds (Fig. 3), the broad adsorption peak around 3285 cm^{-1} can be assigned to the stretching vibration of O-H group. The peaks observed around 2917.5 cm^{-1} and 1336.5 cm^{-1} correspond to the stretching modes of alcohol and bending modes of alcohol respectively.

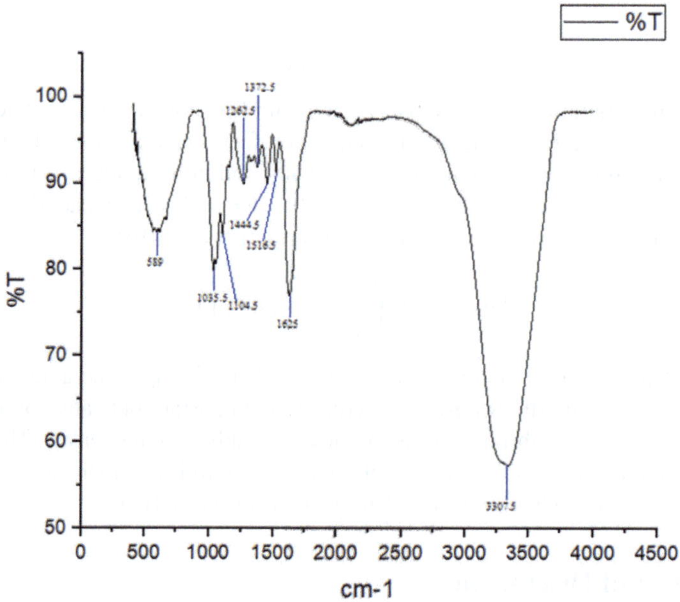

Fig. 1. FTIR spectra for modified Coco peat

3.2 Adsorption Isotherms

The results obtained of Lead (Pb) and Cadmium (Cd) removal by the coco peat, mango seeds and jackfruit rags were analyzed by the Langmuir and Freundlich models. Langmuir and Freundlich parameters (Table 2) for both metals were obtained by plotting C_e versus $\frac{C_e}{Q_e}$ and ln (C_e) versus ln (Q_e) respectively. According to the results shown in Table 2, it can be seen that the biosorption data for Pd is best fitted to Freundlich isotherm due to the highest coefficient of determination (R^2) value compared to other models. So,

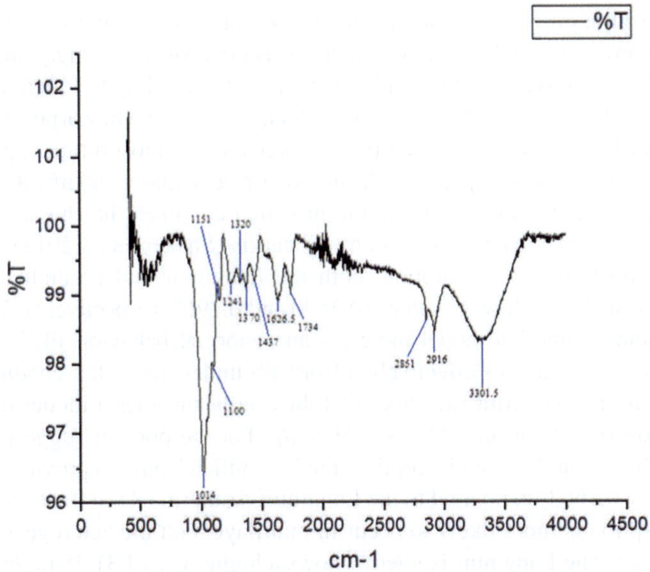

Fig. 2. FTIR spectra for modified Jackfruit rags

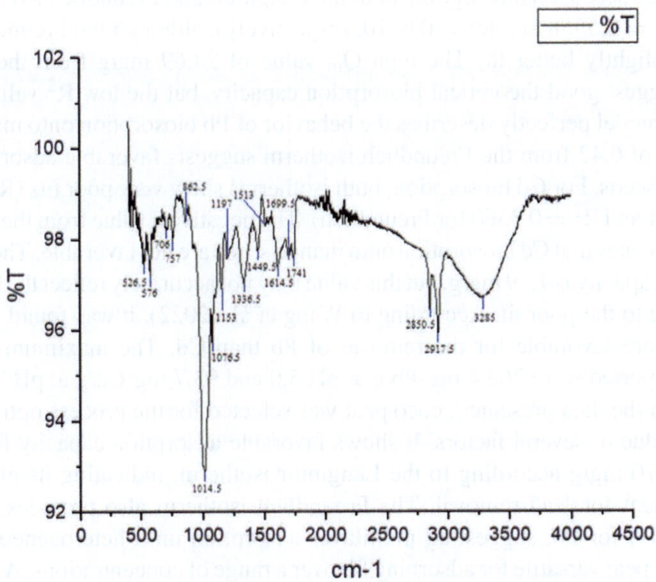

Fig. 3. FTIR spectra for modified Mango seeds

for coco peat, Pb biosorption was found to follow Freundlich isotherms compared to Langmuir isotherm. Thus, it can be concluded that there is a multilayer coverage of Pb over the heterogeneous surface of coco peat. Also, the Langmuir isotherm can represent the Cd biosorption onto coco peat but shown poor fit. The biosorption capacity of the coco peat according to the Langmuir isotherm was observed as 20 mg/g and 4.14 mg/g for Pb and Cd respectively. In this study, $0 < R_L < 1$ according to Langmuir isotherm and $n > 1$ according to Freundlich isotherm which says that the biosorption is favorable for both metal biosorption onto coco peat. Moreover, the biosorption capacity of the modified coco peat was compared with the coco peat without modification available in the literature. Results indicated that the modified coco peat has higher biosorption capacity compared to coco peat without modification (Shakeri et al. 2012).

For jackfruit rags, Pb biosorption, both the Langmuir and Freundlich isotherms show poor fits, with very low R^2 values (0.2690 and 0.1969, respectively). The negative 'n' value in the Freundlich model indicates an abnormal behavior, likely pointing to inability of jackfruit rags to effectively adsorb Pb under the tested conditions. Thus, Pb biosorption onto jackfruit rags doesn't follow either isotherm model reliably. The Langmuir biosorption capacity Q_m is 9.81 mg/g, but the poor fit suggests caution in interpreting this result. For Cd biosorption, the Freundlich isotherm provides a relatively better fit ($R^2 = 0.7663$) compared to the Langmuir isotherm ($R^2 = 0.0941$), suggesting that Cd adsorption is more likely to occur in multilayers on the heterogeneous surface of jackfruit rags. The Langmuir isotherm shows a higher Q_m of 31.25 mg/g for Cd, but again, the poor fit indicates that this model does not accurately describe the adsorption process.

For mango seeds, Pb biosorption, both the Langmuir and Freundlich isotherms show poor fits ($R^2 = 0.5558$ and $R^2 = 0.6210$, respectively), although the Freundlich model provides a slightly better fit. The high Q_m value of 24.69 mg/g from the Langmuir isotherm suggest good theoretical biosorption capacity, but the low R^2 values indicate that neither model perfectly describes the behavior of Pb biosorption onto mango seeds. The n value of 6.42 from the Freundlich isotherm suggests favorable adsorption of Pb onto mango seeds. For Cd biosorption, both isotherms show very poor fits ($R^2 = 0.1477$ for Langmuir and $R^2 = 0.3360$ for Freundlich). The negative n value from the Freundlich isotherm indicates that Cd biosorption onto mango seeds are not favorable. The Langmuir biosorption capacity is 7.79 mg/g, but this value may not accurately reflect the biosorption behavior due to the poor fit. According to Wang et al. (2022), it was found that mango seeds are more favorable for the removal of Pb than Cd. The maximum adsorption capacities reported were 263.4 mg-Pb/g at pH 5.0 and 93.7 mg-Cd/g at pH 7.5.

Based on the data presented, coco peat was selected for the process optimization in biosorption due to several factors. It shows favorable adsorption capacity for Pb, with a Q_m of 20.70 mg/g according to the Langmuir isotherm, indicating its effectiveness as a biosorbent for lead removal. The Freundlich isotherm also provides a better fit ($R^2 = 0.7864$) for Pb, suggesting multilayer adsorption on a heterogeneous surface, making coco peat versatile for adsorbing Pb over a range of concentrations. Additionally, the Freundlich exponent n is greater than 1 ($n = 1.39$ for Pb and $n = 2.07$ for Cd), indicating favorable adsorption for both metals. Although Cd adsorption has a lower Q_m of 4.14 mg/g, there is potential for improvement through optimization, especially given the moderate Freundlich fit for Cd biosorption ($R^2 = 0.3107$). Moreover, coco peat is inexpensive and widely available, ensuring cost-effectiveness and scalability in commercial applications (Sireesha and Sreedhar 2023).

Table 2. Isotherm parameters for biosorption of Pb and Cd

Adsorbent	Heavy Metal	Langmuir Isotherm			Freundlich Isotherm		
		Q_m (mg/g)	K_L (l/mg)	R^2	K_F	n	R^2
Coco Peat	Pb	20.70	0.07	0.6766	1.42	1.39	0.7864
	Cd	4.14	0.31	0.3115	0.82	2.07	0.3107
Mango Seeds	Pb	24.69	405	0.5558	23.49	6.42	0.6210
	Cd	7.79	0.51	0.1477	13.09	-1.23	0.3360
Jackfruit Rags	Pb	9.81	5.86	0.2690	7.21	-8.34	0.1969
	Cd	31.25	0.01	0.0941	0.57	1.28	0.7663

3.3 Process Parameter Optimization

The process parameter optimization experiment was conducted by varying the contact time and adsorbent dosage, while keeping the initial metal concentration and pH constant at 100 ppm and 6, respectively. Samples were prepared and allowed to interact based on the values from the design matrix generated using Minitab 17 statistical software. The adsorption capacity corresponding to each sample is shown in Table 3.

Table 3. Adsorption Capacities of Lead (Pb) and Cadmium (Cd) on to Coco Peat at various Biosorbent dosages and Contact times

Run	Biosorbent dosage (g/l)	Contact time (min)	Adsorption Capacity of Pb (mg/g)	Adsorption Capacity of Cd (mg/g)
1	10.5	85	9.25	7.75
2	10.5	85	8.99	7.74
3	20.0	150	4.85	4.12
4	1.0	150	73.03	80.75
5	10.5	85	9.18	7.72
6	20.0	20	4.88	4.05
7	1.0	20	68.76	81.81

When modelling the adsorption capacity of Pb onto coco peat, it was observed that the p-values for all variables were less than 0.05 (Fig. 4). This indicates that the adsorbent dosage, contact time, and the 2-way interaction between adsorbent dosage and contact time are significant for Pb adsorption capacity with 95% confidence. Based on the results, a p-value less than 0.05 for the curvature test shows that the model has significant curvature, indicating that the relationship between the factors and the response is properly described by the curvature model.

Factorial Regression: Q - Pb versus Adsorbent dose, contact time, CenterPt

Analysis of Variance

Source	DF	Adj SS	Adj MS	F-Value	P-Value
Model	4	5785.06	1446.26	79904.10	0.000
Linear	2	4364.46	2182.23	120565.06	0.000
Adsorbent dose	1	4359.96	4359.96	240881.82	0.000
contact time	1	4.49	4.49	248.31	0.004
2-Way Interactions	1	4.62	4.62	255.39	0.004
Adsorbent dose*contact time	1	4.62	4.62	255.39	0.004
Curvature	1	1415.98	1415.98	78230.87	0.000
Error	2	0.04	0.02		
Total	6	5785.09			

Model Summary

S	R-sq	R-sq(adj)	R-sq(pred)
0.134536	100.00%	100.00%	*

Fig. 4. Factorial Regression for Adsorption of Pb onto Coco Peat

It was observed that the regression equation for Pb adsorption onto coco peat is:

$$\begin{aligned} Q_{(Pb)} &= 71.430 - 3.3273\,\text{Adsorbent dose} + \\ &0.03459\,\text{Contact time} - \\ &0.001741\,\text{Adsorbent dose} \times \text{Contact time} - \\ &28.740\,C_tP_t \end{aligned} \tag{6}$$

When the initial metal concentration is 100 ppm and the pH is 6, the optimum conditions for Pb adsorption were observed to be a contact time of 150 min, an adsorbent dosage of 1 g/L, and an adsorption capacity of 73.03 mg/g, with a removal efficiency of 73%.

When modelling the adsorption capacity of Cd onto coco peat, it was observed that the p-values for all variables were also less than 0.05 (Fig. 5). This indicates that the adsorbent dosage, contact time, and the 2-way interaction between adsorbent dosage and contact time are significant for Pb adsorption capacity with 95% confidence. Similarly, for Pb adsorption, the p-value for the curvature test was also less than 0.05, indicating that there is a curvature relationship between the factors and the response.

Factorial Regression: Q-Cd versus Adsorbent dose, contact time, CenterPt

Analysis of Variance

Source	DF	Adj SS	Adj MS	F-Value	P-Value
Model	4	8053.14	2013.28	8628361.40	0.000
Linear	2	5959.31	2979.66	12769956.54	0.000
Adsorbent dose	1	5959.07	5959.07	25538862.96	0.000
contact time	1	0.25	0.25	1050.11	0.001
2-Way Interactions	1	0.32	0.32	1368.11	0.001
Adsorbent dose*contact time	1	0.32	0.32	1368.11	0.001
Curvature	1	2093.51	2093.51	8972164.41	0.000
Error	2	0.00	0.00		
Total	6	8053.14			

Model Summary

S	R-sq	R-sq(adj)	R-sq(pred)
0.0152753	100.00%	100.00%	*

Fig. 5. Factorial Regression for Adsorption of Cd onto Coco Peat

The regression equation for Cd adsorption onto coco peat was observed as:

$$\begin{aligned} Q_{(C_d)} &= 86.0749 - 4.10178\,\text{Adsorbent dose} + \\ &0.008611\,\text{Contact time} - \\ &0.000457\,\text{Adsorbent dose} \times \text{Contact time} - \\ &34.9458\,C_t P_t \end{aligned} \tag{7}$$

When the initial metal concentration is 100 ppm and the pH is 6, the optimum conditions for Cd adsorption were observed to be a contact time of 20 min, an adsorbent dosage of 1 g/L, and an adsorption capacity of 81.81 mg/g, with a removal efficiency of 81%.

4 Conclusions

The study evaluated the removal of lead (Pb) and cadmium (Cd) from aqueous solutions using acid-modified coco peat, jackfruit rags, and mango seeds. Significant changes in functional groups resulting from acid modification were detected by FTIR analysis, especially in modified coco peat, which had improved adsorption properties with functional groups such as hydroxyl, carboxylic acid, and phenolic groups. It outperformed mango seeds and jackfruit rags in terms of biosorption capacity, exhibiting 20.70 mg/g of Pb. Optimization revealed that the best conditions for Pb removal using coco peat were an adsorbent dosage of 1.0 g/L and a contact time of 150 min, achieving 73.03 mg/g adsorption capacity and a removal efficiency of 73%. For Cd, the optimal conditions included an adsorbent dosage of 1.0 g/L and a contact time of 20 min, resulting in an adsorption capacity of 81.81 mg/g and a removal efficiency of 81%. Significant curvature relationships were observed, indicating that simple linear models do not fully capture the complex interactions between factors. Future research should focus on optimizing all factors (adsorbent dose, contact time, initial metal concentration, pH) for improved adsorption. Additionally, exploring alternative modification techniques for jackfruit rags and mango seeds and studying desorption methods for metal recovery and adsorbent regeneration will be important for enhancing the practical application of these biosorbents in water treatment.

References

Ali Redha, A.: Removal of heavy metals from aqueous media by biosorption. Arab J. Basic Appl. Sci. **27**(1), 183–193 (2020). https://doi.org/10.1080/25765299.2020.1756177

Briffa, J., Sinagra, E., Blundell, R.: Heavy metal pollution in the environment and their toxicological effects on humans. Heliyon **6**(9), e04691 (2020)

Chandrasekara, C.K., et al.: Heavy metal contamination of water in negombo lagoon and interconnected water sources. Lakes, Reservoirs and Ponds **8**(2), 96–110 (2014)

Dzigbor, A., Chimphango, A.: Production and optimization of NaCl-activated carbon from mango seed using response surface methodology. Biomass Conver. Biorefinery **9**(2), 421–431 (2019)

Elhafez, S.E.A., et al.: Management of agricultural waste for removal of heavy metals from aqueous solution: adsorption behaviors, adsorption mechanisms, environmental protection, and techno-economic analysis. Environ. Sci. Pollut. Res. **24**(2), 1397–1415 (2017). https://doi.org/10.1007/s11356-016-7891-7

Esther, N.W., Ruth, W.N., James, N.: 'Adsorption studies of lead (II) ions from a synthetic media using jackfruit (Artocarpus heterophyllus L.) rags: kinetics, equilibrium and thermodynamic studies. Int. J. Sci. Res. Res. Paper. Chemic. Sci. **8**(4), 5–12 (2021). <www.isroset.org>

Farasati, M., Haghighi, S., Boroun, S.: Cd removal from aqueous solution using agricultural wastes. Desalin. Water Treat. **57**(24), 11162–11172 (2016)

Garg, U.K., et al.: Removal of hexavalent chromium from aqueous solution by agricultural waste biomass. J. Hazard. Mater. **140**(1–2), 60–68 (2007)

Hama Aziz, K.H., et al.: Heavy metal pollution in the aquatic environment: efficient and low-cost removal approaches to eliminate their toxicity: a review. RSC Adv. **13**(26), 17595–17610 (2023)

Hashtroudi, H.: Using agricultural wastes to treat lead-contaminated water in Using agricultural wastes to treat lead-contaminated water in Western Australia. The Grants Register **2024**, 470–472 (2018). https://doi.org/10.1016/j.heliyon.2020.e04691

Koller, M., Saleh, H.M.: introductory chapter: introducing heavy metals. Heavy Metals, pp. 3–12 (2018)

Lazim, Z.M., et al.: Adsorption characteristics of bisphenol a onto low-cost modified phyto-waste material in aqueous solution', Water, Air, and Soil Pollution, **226**(3) (2015)

Nanayakkara, C., et al.: Spatial distribution of heavy metals in sediments of the negombo lagoon, Sri Lanka', pp. 62–69 (2022)

Pérez-Marín, A.B., et al.: Removal of cadmium from aqueous solutions by adsorption onto orange waste. J. Hazard. Mater. **139**(1), 122–131 (2007)

Sen, T.K.: Agricultural solid wastes based adsorbent materials in the remediation of heavy metal ions from water and wastewater by adsorption: a review. Molecules **28**(14) 2023

Shakeri, A., et al.: Removal of lead (II) from aqueous solution using cocopeat: An investigation on the isotherm and kinetic. Iran. J. Chem. Chem. Eng. **31**(3), 45–50 (2012)

Sud, D., Mahajan, G., Kaur, M.P.: Agricultural waste material as potential adsorbent for sequestering heavy metal ions from aqueous solutions - A review. Biores. Technol. **99**(14), 6017–6027 (2008)

Zakaria, A, Aynuddin, Djasmasari, W.: Exploring the potential of mango seed as a bioadsorbent for Pb(II) removal in aqueous solution. J. Kimia Valensi **9**(1), 155–162 (2023)

Zaynab, M., et al.: Health and environmental effects of heavy metals. J. King Saud Univ. (2022)

Sustainable Biofuel Generation from Natural Rubber Sludge Waste Using Pyrolysis Process

E. P. R. H. H. W. Nilmalgoda[1,4,5], E. A. P. Divyanjali[1], I. B. Wijethunga[1], A. D. Ampitiyawatta[3], and K. R. Koswattage[2,4](✉)

[1] Department of Biosystems Technology, Faculty of Technology, Sabaragamuwa University of Sri Lanka, Belihuloya 70140, Sri Lanka

[2] Department of Engineering Technology, Faculty of Technology, Sabaragamuwa University of Sri Lanka, Belihuloya 70140, Sri Lanka
koswattagekr@appsc.sab.ac.lk

[3] Department of Export Agriculture, Faculty of Agricultural Sciences, Sabaragamuwa University of Sri Lanka, Belihuloya 70140, Sri Lanka

[4] Centre for Nanodevices Fabrication and Characterization, Faculty of Technology, Sabaragamuwa, University of Sri Lanka, Belihuloya, Sri Lanka

[5] Faculty of Graduate Studies, Sabaragamuwa University of Sri Lanka, P.O. Box, Belihuloya 70140, Sri Lanka

Abstract. The pyrolysis of natural rubber sludge, a significant byproduct of the rubber industry, was investigated to produce valuable energy products, including bio-oil, syngas, and biochar. This study aimed to explore the potential of converting rubber sludge into renewable fuels and chemicals using a lab-scale pyrolysis process. The natural rubber sludge was pre-treated to remove non-biodegradable fillers such as calcium carbonate ($CaCO_3$) using hydrochloric acid (HCl) washing and water washing. The pyrolysis process was conducted at 450 °C for 30 min in an anaerobic environment. The resulting bio-oil was analyzed using gas chromatography-mass spectrometry (GC-MS) to identify the key aromatic and phenolic compounds formed during pyrolysis. Major compounds detected included syringol (7.651 min), creosol (6.392 min), vanillin (8.457 min), syringyl acetone (10.772 min), eugenol derivatives (Z-isoeugenol: 9.450 min, E-isoeugenol: 12.009 min), and guaiacol (5.548 min). These compounds are typical byproducts of lignin degradation and indicate the significant presence of lignin in the rubber sludge. The bio-oil produced exhibited a high calorific value (44.47–44.68 MJ/kg), making it a promising renewable fuel, though its low flash point and high acidity suggest the need for further processing. Syngas analysis revealed high hydrogen (32.8%) and carbon monoxide (24.3%) yields, which were significantly improved by the pre-treatment process. This study demonstrates that pyrolysis of natural rubber sludge, especially with optimized pre-treatment, offers a sustainable approach to waste management while generating high-quality bio-oil and syngas for renewable energy applications. The findings highlight the viability of integrating pyrolysis into a circular economy model, offering an alternative energy solution with environmental benefits.

Keywords: Pyrolysis · Natural Rubber Sludge · Bio-oil · Sustainable Energy · Waste Management

R. Dissanayake et al. (Eds.): ICRES 2025, PTT, pp. 25–39, 2025.
https://doi.org/10.1007/978-981-96-9551-5_3

1 Introduction

Sri Lanka's rubber industry is a significant pillar of its economy, contributing to both the domestic and export markets. Rubber plantations, mainly concentrated in the central and southern regions of the country, thrive due to the favorable climatic conditions, with an average rainfall of over 2,000 mm per year and temperatures ranging between 23 °C and 28 °C (Gamage and Udya, 2023). The industry supports rural livelihoods, with small-scale growers managing approximately 70% of the cultivated area (Andooz et al., 2023). These smallholders, often operating on limited land sizes, play a vital role in maintaining the rubber supply chain, making the sector essential for rural economic development and poverty alleviation.

Natural Rubber (NR), composed primarily of cis-1,4-polyisoprene, is known for its unique properties, including elasticity, durability, and resistance to abrasion. These properties make NR indispensable for a wide range of products, including tires, industrial seals, gloves, footwear, and medical devices (Arabiourrutia et al., 2020; Payungwong, Wu and Sakdapipanich, 2024). NR is used globally in various industrial applications, with the automotive industry being one of the largest consumers of rubber. However, despite its widespread use, the industrial production of rubber generates substantial waste, particularly in the form of sludge from the glove manufacturing process (Czajczyńska et al., 2017). This sludge contains high concentrations of rubber residues and calcium carbonate ($CaCO_3$), which are difficult to dispose of safely and contribute to environmental pollution (Mendis et al., 2020; Chittella et al., 2021).

Traditional waste management methods, such as incineration and landfilling, are commonly employed to handle rubber waste, but these approaches exacerbate environmental problems. Incineration, for instance, can release harmful pollutants, including greenhouse gases, particulate matter, and volatile organic compounds, into the atmosphere, contributing to air pollution and climate change. Landfilling, on the other hand, leads to the accumulation of non-biodegradable waste in landfills, further depleting natural resources and occupying valuable land space (Dhanapala et al., 2023). The disposal of rubber sludge in landfills is particularly problematic, as the high content of calcium carbonate in the sludge can result in leachate contamination, potentially polluting nearby water sources (Pathirana et al., 2024). These challenges highlight the urgent need for innovative, sustainable waste management solutions that can reduce the environmental impact of rubber sludge and convert it into valuable resources.

In recent years, pyrolysis has emerged as a promising alternative for the disposal of rubber waste (Pathiraja et al., 2025). Pyrolysis is a thermochemical process that involves the decomposition of organic materials in the absence of oxygen, typically at elevated temperatures of 350 °C to 600 °C (Andooz et al., 2023). During pyrolysis, organic materials such as rubber sludge break down into several valuable byproducts, including bio-oil, biochar, and syngas. Bio-oil, a liquid fraction, contains a complex mixture of organic compounds, including phenolic compounds, aldehydes, and acids, and can be used as a renewable fuel or as a chemical feedstock for further refinement. Syngas, a gaseous byproduct, primarily contains hydrogen (H_2) and carbon monoxide (CO), and can be used for electricity generation or as a fuel for industrial processes. Biochar, a solid byproduct, has potential applications in carbon sequestration, soil enhancement, and even as an adsorbent in wastewater treatment (Pahnila et al., 2023; Tin et al., 2024).

Pyrolysis has gained significant attention in recent years due to its potential to convert waste materials into high-value energy products while addressing waste management challenges. In the context of Sri Lanka's energy strategy, pyrolysis could contribute to the country's efforts to diversify its energy mix by incorporating more renewable energy sources (Dharmasiri, 2019). Sri Lanka has long been reliant on imported fossil fuels, which constitute a major portion of its energy consumption. The government has set ambitious targets for increasing the share of renewable energy in the national energy mix, with a particular emphasis on wind, solar, and biomass energy (Koswatte et al., 2024). Bio-oil produced through the pyrolysis of natural rubber sludge could serve as a viable substitute for fossil fuels, helping to reduce dependence on imports while also providing a sustainable energy solution for industrial and domestic applications. The high calorific value and chemical properties of bio-oil make it suitable for use in electricity generation, heating, and other industrial processes (Han, Han and Chen, 2023; Lachos-Perez et al., 2023).

Moreover, pyrolysis aligns with global efforts to promote circular economies, which seek to minimize waste and maximize the value of resources. Circular economy practices focus on keeping products, materials, and resources in use for as long as possible, recovering and regenerating them at the end of each service life. Pyrolysis is an ideal process for implementing circular economy principles in the rubber industry, as it converts rubber waste into valuable resources such as renewable fuel and chemical feedstocks while reducing environmental impact (Tsipa et al., 2024).

This study focuses on the pyrolysis of natural rubber sludge, a byproduct of glove manufacturing, as an effective and sustainable waste management strategy. Specifically, the research investigates the effectiveness of pre-treatment methods, including HCl washing and water washing, in removing calcium carbonate and other impurities from the sludge. By improving the quality of the bio-oil produced, these pre-treatment methods can enhance the overall performance of the pyrolysis process (Harshani and Shantha, 2021). Laboratory-scale pyrolysis experiments are carried out to evaluate key properties of the bio-oil, including its calorific value, density, viscosity, and pH (Roy et al., 2021)These characteristics are critical for determining the bio-oil's potential as a renewable fuel and its suitability for various applications.

The findings of this research are of particular relevance to Sri Lanka's waste-to-energy initiatives, as they demonstrate the feasibility of using pyrolysis to convert industrial rubber waste into high-value bio-oil. This study supports the broader transition to renewable energy and circular economy practices by showing that pyrolysis can be an effective tool for managing rubber waste while providing a renewable source of energy. Additionally, integrating pyrolysis into Sri Lanka's rubber industry could play a key role in achieving both environmental and economic sustainability by addressing waste management challenges and energy demands.

2 Methodology

Figure 1 shows the overall methodology of the study as a summary.

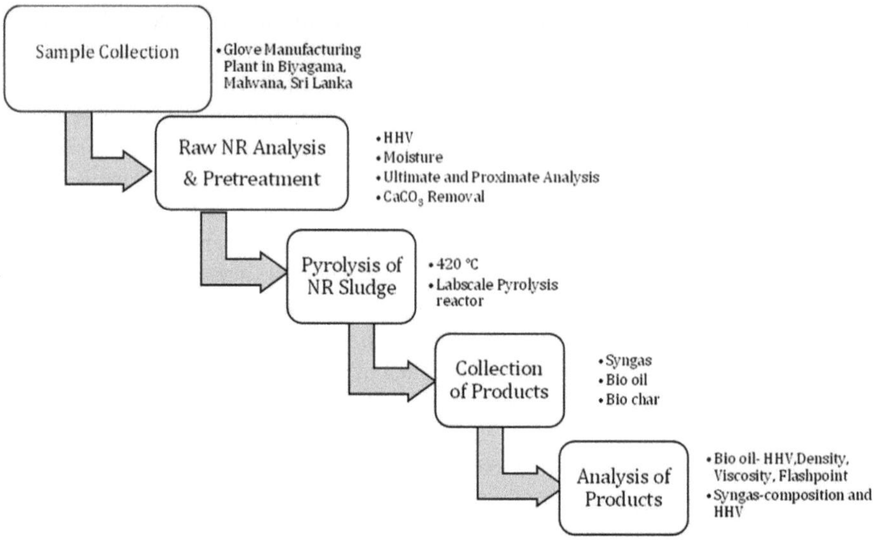

Fig. 1. Summary flowchart of the methodology

2.1 Study Area and Sample Collection

The study utilized natural rubber (NR) sludge waste collected from a glove manufacturing company in the Biyagama Export Processing Zone, Malwana, Sri Lanka. The sludge was sourced from the effluent treatment plant (ETP), specifically from the equalization tank and filter press. The composition of the sludge consisted of approximately 40% rubber, 40% Calcium Carbonate ($CaCO_3$), and 20% other materials, including corn starch and cotton flock. Sufficient quantities of the sludge were collected to facilitate subsequent experimental analysis (Fig. 2).

2.2 Analysis of Raw Rubber Sludge

The moisture content and calorific value of the raw rubber sludge were assessed before pyrolysis.

Moisture Content: The moisture content was determined using the oven-dry method, where samples were heated at 105 °C until a consistent weight was achieved. This method ensured precise measurement of moisture loss, which was calculated using Eq. 1.

$$Moisture\ Content\ (\%) = \left(\frac{Initial\ mass - Dry\ mass}{Initial\ mass}\right) \times 100\% \tag{1}$$

Calorific Value (CV): An automatic bomb calorimeter (CKIC Model 5E-C5508) was used to measure the calorific value of 1 g samples of raw feedstock. The calorimeter setup included a pressurized oxygen bomb that enclosed the fuel sample. ASTM D5865 standard for determining the calorific value of NR sludge was used in the experiment.

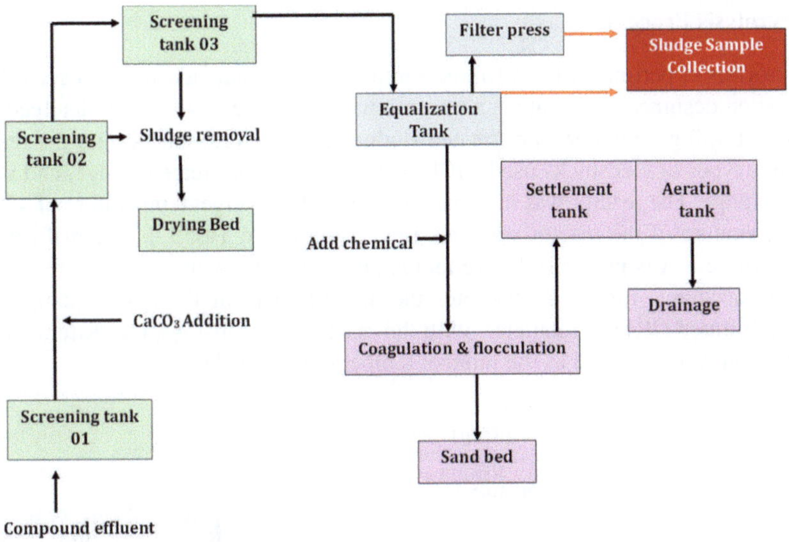

Fig. 2. NR sludge collection from the effluent treatment plant and the standard treatment process.

2.3 Pre-Treatment of Natural Rubber Sludge

To enhance the pyrolysis process's efficiency, pre-treatment was carried out to remove ($CaCO_3$). Two methods were employed for this purpose: HCl washing and water washing.

HCl Washing Method (Sample 1): 1 M (HCl) solution was prepared by diluting 8.33 mL of concentrated HCl (12.1 M) with distilled water to make 100 mL of solution. For each batch, 250 g of sludge was treated with the HCl solution, thoroughly stirred to ensure a complete reaction between $CaCO_3$ and HCl, which produced Calcium Chloride ($CaCl_2$) and released carbon dioxide (CO_2) gas. The reaction progress was monitored until the pH stabilized, indicating its completion. The treated sludge was filtered to separate the solid from the liquid phase, washed with distilled water to remove residual acid, and dried in an oven at 105–110 °C for 24 h.

Non-Treated Sludge (Sample 2): This sludge has not undergone any pre-treatment. It retains all the impurities, including calcium carbonate, which may affect its efficiency and yield in pyrolysis. Its properties may be less optimal for biofuel production compared to the treated samples.

Water Washing Method (Sample 3): NR sludge 500 g was mixed with 5 L of water and agitated the mixture for two hours. The suspension was filtered through muslin cloth, and the retained sludge was further washed under running water to remove impurities. Finally, the washed sludge was dried in an oven at the same temperature for 24 h to reduce its moisture content.

2.4 Pyrolysis Process

The pyrolysis experiments for all three samples were conducted in a lab-scale batch type reactor, designed to operate at a maximum temperature of 450 °C with a feedstock capacity of 450 g. Dried and pre-treated sludge samples were loaded into the reactor, with three types of feedstocks used: HCl-treated sludge, non-treated sludge, and water-washed sludge. The reactor was incrementally heated to a target temperature of 450 °C. The temperature of the reactor was measured using Fluke TiS55 + Thermal imaging camera where it was possible to take temperature measurements from a safe distance at a high accuracy. Figure 4 illustrates the measurement of the reactor temperature using an infrared (IR) thermometer. With the reactors' dimensions, the IR temperature measurement was assumed as the internal temperature (Fig. 3).

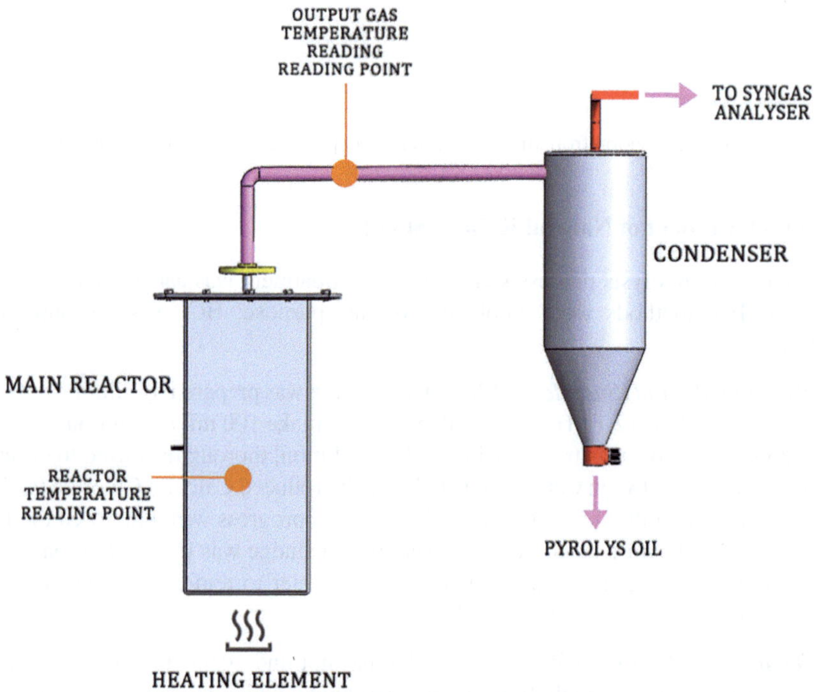

Fig. 3. Lab-scale pyrolysis reactor setup

During the process, the organic compounds in the sludge underwent thermal decomposition in the absence of oxygen, yielding three primary products: bio-oil, biochar, and syngas. The vapors produced during pyrolysis were condensed to collect bio-oil in a separate container, while solid residues, referred to as biochar, were retrieved from the reactor chamber. Non-condensable gases, or syngas, were captured and stored for potential use as a fuel source.

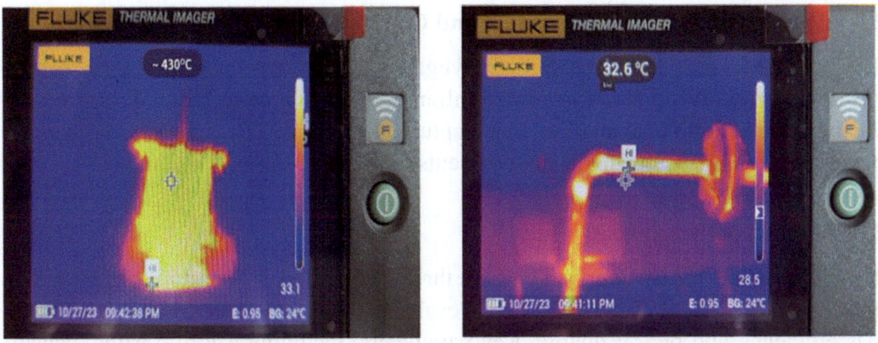

(a) **(b)**

Fig. 4. Measurement of reactor temperature: **(a)** reactor chamber; **(b)** exhaust tube.

2.5 Analysis of Biooil Properties

The bio-oil produced through pyrolysis was analyzed to assess its potential as a renewable energy source.

Calorific Value (CV): Measured using a CKIC 5E-C5508 bomb calorimeter in accordance with ASTM-D5865 standards, calculating the energy content based on the heat released during the complete combustion of 1 g of bio-oil.

Flash Point: Measured using an Anton Paar PMA 500 Pensky-Marten's flash point tester, adhering to ASTM-D93B standards, to measure the lowest temperature at which the vapors from the bio-oil could ignite.

Density and Viscosity: Measured using an Anton Paar SVM 3001 viscometer, which recorded kinematic viscosity, dynamic viscosity, and density, enabling the evaluation of flow properties and fuel performance.

***pH*:** Measured using an ST 3100 pH meter to determine its acidity or alkalinity. Before testing, the pH electrode was calibrated, and the results were documented for comparison among the different samples.

2.6 GC-MS Analysis of Biooil

The bio-oil sample was analyzed using an Agilent 7890B gas chromatograph coupled with an Agilent 5977B mass-selective detector (MSD). The GC column used was HP-5MS (30 m × 0.25 mm × 0.25 μm). Helium was used as the carrier gas at a flow rate of 1.5 mL/min. The injection was conducted in splitless mode, with a 5 μL injection volume. The oven temperature program was as follows: starting at 50 °C for 1.5 min, ramping at 30 °C/min to 180 °C, and then at 20 °C/min to a final temperature of 280 °C, which was held for 20 min. The MS detector was operated in Electron Ionization (EI) mode at 70 eV, with a mass scan range of 50–450 m/z and a sampling rate of 3.6 scans/s.

2.7 Analysis of Syngas Composition and Content

The Gasboard-3100GP portable infrared syngas analyzer was utilized to analyze the syngas composition using Non-Dispersive Infrared Spectroscopy (NDIR), Thermal Conductivity Detector (TCD) and Electron Capture Detector (ECD) methods. It is capable of simultaneously measuring key components such as CO, CO_2, CH_4, and H_2.

2.8 Comparative Analysis

The bio-oil characteristics derived from the three feedstocks (HCl-treated, water-washed, and non-treated sludge) were compared to evaluate the impact of pre-treatment on pyrolysis efficiency and bio-oil quality. Key parameters, including calorific value, density, flash point, viscosity, and pH, were analyzed to assess the suitability of bio-oil for industrial applications and energy generation. This comparative analysis provided insights into optimizing pre-treatment methods to enhance bio-oil yield and quality, ensuring its feasibility as a sustainable energy source.

2.9 Cost-Benefit Analysis (CBA) of Traditional NR Sludge Waste Management vs Sludge Pyrolysis

The Cost-Benefit Analysis (CBA) compares traditional NR sludge waste management (landfilling or incineration) with sludge pyrolysis. For traditional waste management, costs include waste handling and environmental impacts such as emissions and contamination. For pyrolysis, costs cover operational expenses, pre-treatment, and capital investment for the facility. Revenue from pyrolysis is generated through the sale of byproducts like bio-oil, syngas, and biochar. Environmental benefits of pyrolysis, such as reduced pollution and landfill use, are also factored in. The analysis compares the costs, revenues, and environmental impact of both methods to assess their economic and environmental feasibility.

3 Results and Discussion

3.1 Raw NR Waste Analysis

Table 1 shows the proximate and ultimate analysis of NR waste sludge.

Water-washed sludge (Sample 3) is likely the best candidate for biofuel production due to its higher volatile matter, fixed carbon, and lower ash content, suggesting a higher calorific value. Non-treated sludge (Sample 2) has more ash and may be less efficient, while HCl-treated sludge (Sample 1) falls in between. Further studies on pyrolysis behaviour and energy output are needed to confirm their biofuel potential.

The calorific values (CV) of the natural rubber (NR) feedstock samples were measured, with Sample 01 and Sample 02 recording 27.78MJ/kg and 27.45 MJ/kg, respectively, using a bomb calorimeter. In contrast, Sample 03, the water-washed NR sludge, demonstrated a higher CV of 30.25 MJ/kg, making it the most suitable feedstock for pyrolysis. This finding highlights the superior energy potential of water-washed NR sludge and its importance in optimizing feedstock for efficient pyrolysis. This suggests that the removal of impurities, such as calcium carbonate, through water washing enhances the energy density of the feedstock, making it more suitable for pyrolysis.

Table 1. Proximate and Ultimate analysis of NR waste

	Proximate Analysis (wt.%)				Ultimate Analysis (wt.%)				
Sample	Moisture	Volatile	Ash	Fixed Carbon	C	H	O	N	S
1	22.26	60	15	17	45.2	6.2	38.5	0.8	0.3
2	22.27	58.5	20	15.3	43.8	6	40	0.9	0.4
3	26.51	62	12.5	18	46.5	6.5	37	0.7	0.3

1: HCl-Treated Sludge, 2: Non-Treated Sludge, and 3: Water-Washed Sludge.

3.2 Pre-Treatment of Natural Rubber Sludge

Table 2. Comparison of HCl-Treated Sludge (Sample 1) and Water-Washed Sludge (Sample 3)

Parameter	Sample 1: HCl-Treated Sludge	Sample 3: Water-Washed Sludge
Initial Sludge Weight	1250 g	1000.33 g
Sludge Weight After Removing $CaCO_3$	525 g	586.36 g
% of Sludge Weight Without $CaCO_3$	42%	58.61%

The comparison between HCl-treated sludge (sample 1) and water-washed sludge (sample 3) in Table 2 reveals significant differences in their composition after the removal of $CaCO_3$. Case 1, with a lower percentage of sludge remaining after $CaCO_3$ removal (42%), suggests a more aggressive treatment that effectively removes a larger proportion of impurities. This results in a cleaner sludge, which could be more suitable for processes like pyrolysis, as it contains fewer contaminants that might otherwise interfere with energy production. On the other hand, Case 3 retains a higher proportion of its original material (58.61%) after the removal of $CaCO_3$, indicating that the water-washing process is milder. This means that Case 3 retains more of its organic and inorganic components, potentially affecting its efficiency and performance in energy conversion processes. Overall, HCl-treated sludge may offer higher purity and more efficient energy production, while water-washed sludge could be a viable option where a less aggressive treatment is preferred, but it may yield lower efficiency due to the higher content of remaining impurities.

3.3 Pyrolysis Product Distribution and Material Balance

At the end of the waste NR sludge pyrolysis process, three main products were obtained. These include pyrolysis oil, syngas, and solid char illustrated in Fig. 5.

(a) (b) (c)

Fig. 5. Pyrolysis products of the trials: (a) Pyrolysis oil; (b) Char sample; (c) Syn gas

Product distribution of different products is illustrated in Fig. 6.

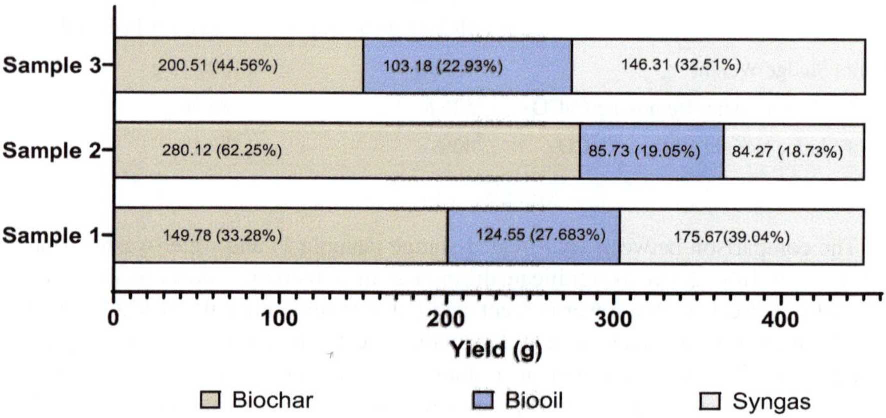

Fig. 6. Pyrolysis product distribution with material balance.

The pyrolysis product distribution across the three samples highlights the impact of pre-treatment and decomposition efficiency on biochar, bio-oil, and syngas yields. Sample 2, which produced the highest biochar (62.25%) and lowest syngas (18.73%), suggests incomplete thermal breakdown, likely due to the presence of $CaCO_3$, which restricted volatile release. In contrast, Sample 3 exhibited the highest syngas yield (39.04%) and the lowest biochar content (33.28%), indicating more efficient thermal cracking, possibly due to complete $CaCO_3$ removal via HCl washing. Sample 1 showed a balanced distribution (biochar: 44.56%, bio-oil: 22.93%, syngas: 32.51%), representing moderate decomposition, likely corresponding to partial $CaCO_3$ removal (water washing). These findings demonstrate that pre-treatment significantly influences pyrolysis efficiency, with HCl washing favouring higher syngas production, while untreated sludge retains more solid carbon. Optimizing pre-treatment and pyrolysis conditions could further enhance gas yields and energy recovery, making this process more viable for sustainable waste valorization.

3.4 Biooil Analysis

The pyrolysis oil sample underwent analysis to determine its density, pH, Higher Heating Value (HHV), flash point, and viscosity. The density and viscosity measurements were conducted using an SVM 3001 instrument (Anton Paar) following the ASTM D7042 standard methods. A summary of the measured properties of the pyrolysis oil, along with a comparison to other biomass feedstocks, is provided in Table 3.

Table 3. The comparison of properties of NR sludge waste pyrolysis oil with various biomass feedstocks

Biomass origin	Calorific Value (MJ/kg)	Flash Point (°C)	Density (g/cm^3)	pH	Viscosity	
					Dynamic (mPa s)	Kinematic (mm^2/s)
Sample 1	44.47	47.5	0.859	2.42	1.3465	1.566
Sample 2	39.46	52.3	0.830	1.94	0.8681	1.0455
Sample 3	44.68	46.2	0.857	2.33	1.7267	1.4803
Rice Husk [a]	45.75	119	1.14	3.69	5–15	4–12
Sugarcane Bagasse [b]	20–30	70–90	1.1–1.4	3–4	20–40	15–35
Coconut Shell [b]	20–30	70–100	1.1–1.3	2–3	40–80	30–70
Rice Straw [b]	16–18	< 100	1.1–1.3	2–3	10–30	8–25

[a] (Mashayekh-Salehi et al., 2021); [b] (Sydney et al., 2019)

The analysis of bio-oil properties highlights that Samples 1 and 3, with high calorific values (44.47–44.68 MJ/kg), moderate viscosities (dynamic: 1.34–1.72 mPa·s, kinematic: 1.56–1.48 mm^2/s), and manageable densities (~0.85 g/cm^3), are promising candidates for energy production. However, their low flash points (46.2–47.5 °C) and high acidity (pH 1.94–2.42) require careful handling and potential mitigation for storage and industrial use. Compared to biomass sources like rice husk, sugarcane bagasse, and coconut shell, which exhibit higher viscosities and densities but lower energy efficiency, the tested samples are more suitable for applications requiring easy flow and high energy output, provided their storage challenges are addressed.

3.5 GC-MS Identification of Compounds

The GC-MS analysis of the bio-oil revealed a complex mixture of aromatic and phenolic compounds. The major compounds identified are given in Table 4.

Table 4. GC Results of Pyrolysis Oil from Natural Rubber Sludge, Showing the Identified Compounds and Their Corresponding Retention Times.

Compound	Retention Time (min)
Guaiacol	5.548
Creosol	6.392
Syringol	7.651
Vanillin	8.457
Eugenol Derivative (Z-Isoeugenol)	9.45
Syringyl Acetone	10.772
Eugenol Derivative (E-Isoeugenol)	12.009

These compounds are primarily phenolic compounds typically derived from the lignin degradation during the pyrolysis of organic matter. The presence of these compounds suggests that the natural rubber sludge contains significant lignin content, which breaks down into aromatic molecules upon pyrolysis.

3.6 Syngas Analysis

Gas Composition (%) from Pyrolysis of Natural Rubber (NR) Sludge with Different Pre-Treatments Data is Summarized in Fig. 7. The results indicate that pre-treatment of NR sludge significantly affects the pyrolysis gas yield and quality. The HCl-washed NR sludge produces the highest H_2 (32.8%) and CO (24.3%), with the lowest CO_2 (26.5%), making it the best option for syngas production. This is due to the complete removal of $CaCO_3$, preventing its decomposition into CO_2 and Allowing for more efficient thermal cracking of hydrocarbons. In contrast, the non-treated NR sludge shows the highest CO_2 (38.7%) and the lowest H_2 (21.9%), as the presence of $CaCO_3$ leads to additional CO_2 formation and CaO influences gas-phase reactions. The water-washed NR sludge has intermediate values, with CO_2 (31.12%), H_2 (28.03%), and CO (20.4%), showing partial improvement due to some $CaCO_3$ removal. These findings suggest that HCl washing enhances gas quality by increasing H_2 and CO while reducing CO_2, making it a preferred method for high-energy syngas production.

3.7 Cost-Benefit Analysis

The **Cost-Benefit Analysis** (CBA) conducted between traditional NR sludge waste management and sludge pyrolysis reveals distinct differences in the economic and environmental outcomes of each method as shown in Table 5.

The Cost-Benefit Analysis clearly indicates that sludge pyrolysis is not only a more economically profitable method compared to traditional waste management but also offers significant environmental benefits. Pyrolysis provides a sustainable solution for NR sludge disposal while generating valuable products and reducing the environmental footprint. The findings highlight the potential for large-scale adoption of pyrolysis in

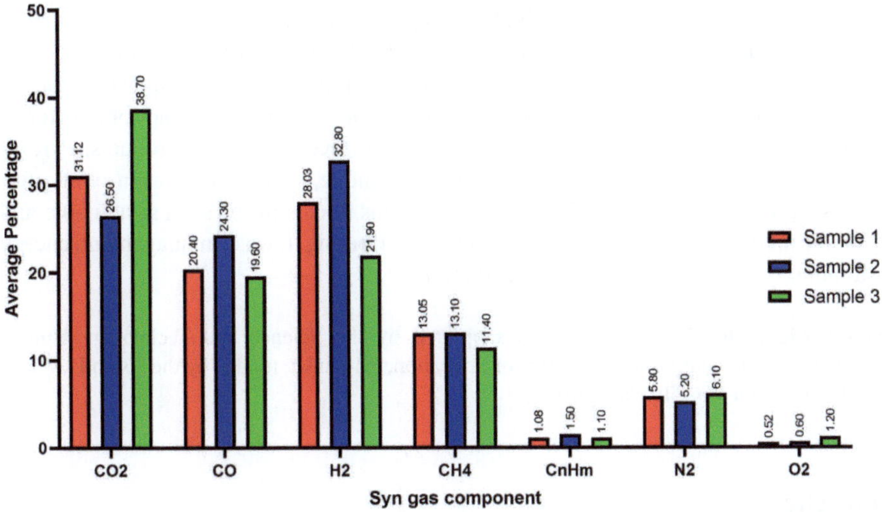

Fig. 7. Gas composition (%) from pyrolysis of NR sludge with different pre-treatments

Table 5. Summary of the cost-benefit analysis

Parameter	Traditional Waste Management	Sludge Pyrolysis
Cost per Ton	$150	$100 (Operational Cost)
Revenue per Ton (from byproducts)	$0	$275 (Bio-oil, Syngas, Biochar)
Environmental Benefit	$0	-$150 (Environmental Saving)
Net Profit per Ton	N/A	$175 (Revenue - Operational Cost)
Break-even Point	N/A	2,857 tons of waste
Total Cost (1,000 Tons)	$150,000	$100,000 (Operational)
Total Revenue (1,000 Tons)	$0	$275,000
Net Profit (1,000 Tons)	N/A	$175,000

managing rubber waste, contributing to both environmental sustainability and economic growth.

4 Conclusion

This study demonstrated the potential of converting natural rubber (NR) sludge into valuable energy products through pyrolysis. Among the feedstock samples, the water-washed NR sludge proved most efficient, exhibiting superior calorific value and enhanced energy potential due to the removal of impurities. GC-MS analysis identified valuable phenolic compounds in the bio-oil, highlighting its high energy density and potential as a renewable fuel, though further optimization is needed for handling and storage.

Syngas, with a calorific value of 11.7 MJ/m^3, and biochar also showed promise as energy sources. The Cost-Benefit Analysis revealed that pyrolysis not only offers a lower operational cost but also generates significant revenue from byproducts. Pyrolysis produces a net profit of $175 per ton of NR sludge, with a break-even point after processing approximately 2,857 tons of waste. This makes pyrolysis a sustainable and profitable method for NR sludge valorization, contributing to waste management, energy sustainability, and environmental conservation.

Acknowledgments. This research was supported by the Science and Technology Human Resource Development Project, Ministry of Education, Sri Lanka, funded by the Asian Development Bank (Grant No. CRG/R3/SB6).

References

Andooz, A., Eqbalpour, M., Kowsari, E., Ramakrishna, S., Cheshmeh, A., Zahra: A comprehensive review on pyrolysis from the circular economy point of view and its environmental and social effects. J. Clean. Prod. **388**, 136021 (2023). https://doi.org/10.1016/j.jclepro.2023.136021

Arabiourrutia, M., Lopez, G., Artetxe, M., Alvarez, J., Bilbao, J., Olazar, M.: Waste tyre valorization by catalytic pyrolysis – a review. Renew. Sustain. Energy Rev. **129**, 109932 (2020). https://doi.org/10.1016/j.rser.2020.109932

Chittella, H., Yoon, L.W., Ramarad, S., Lai, Z.-W.: Rubber waste management: a review on methods, mechanism, and prospects. Polym. Degrad. Stab. **194**, 109761 (2021). https://doi.org/10.1016/j.polymdegradstab.2021.109761

Czajczyńska, D., et al.: Potentials of pyrolysis processes in the waste management sector. Energy Procedia **123**, 387–394 (2017). https://doi.org/10.1016/j.egypro.2017.07.275

Dhanapala, S., Nilmalgoda, H., Gunathilake, M.B., Rathnayake, U., Wimalasiri, E.M.: Energy balance assessment in agricultural systems; an approach to diversification. AgriEngineering **5**(2), 950–964 (2023). https://doi.org/10.3390/agriengineering5020059

Dharmasiri, L.M.: Waste management in sri lanka: challenges and opportunities. Sri Lanka J. Adv. Soc. Stud. **9**(1), 72 (2019). https://doi.org/10.4038/sljass.v9i1.7149

Gamage, L., Udya, K.A., Udani: Rubber plantations in sri lanka: an examination of historical and current grievances in the pahala karannagoda grama niladari division. OUSL J. **18**(2), 53–86 (2023). https://doi.org/10.4038/ouslj.v18i2.7621

Han, W., Han, D., Chen, H.: Pyrolysis of waste tires: a review. Polymers **15**(7), 1604 (2023). https://doi.org/10.3390/polym15071604

Harshani, K.A.P., Shantha, A.A.: Factors affecting the economic efficiency of small-scale rubber plantations: with special reference to kalutara district in Sri Lanka. Colombo Business Journal **12**(1), 145–165 (2021). https://doi.org/10.4038/cbj.v12i1.76

Koswatte, I., Iddawala, J., Kulasekara, R., Ranaweera, P., Dasanayaka, C.H., Abeykoon, C.: Can Sri Lanka be a net-zero nation by 2050?—current renewable energy profile, opportunities, challenges, and recommendations. Cleaner Energy Syst. **8**, 100126 (2024). https://doi.org/10.1016/j.cles.2024.100126

Lachos-Perez, D., et al.: Review on biomass pyrolysis with a focus on bio-oil upgrading techniques. Analytica **4**(2), 182–205 (2023). https://doi.org/10.3390/analytica4020015

Mashayekh-Salehi, A., et al.: Use of mine waste for H2O2-assisted heterogeneous fenton-like degradation of tetracycline by natural pyrite nanoparticles: catalyst characterization, degradation mechanism, operational parameters and cytotoxicity assessment. J. Clean. Prod. **291**, 125235 (2021). https://doi.org/10.1016/j.jclepro.2020.125235

Mendis, A.S.S., Dunuweera, S.P., Walpalage, S., Rajapakse, R.M.G.: Conversion of biological treatment plant sludge to organic fertilizer for applications in organic farming. Detritus (9) (2020). https://doi.org/10.31025/2611-4135/2020.13899

Pahnila, M., Koskela, A., Sulasalmi, P., Fabritius, T.: A review of pyrolysis technologies and the effect of process parameters on biocarbon properties. Energies **16**(19), 6936 (2023). https://doi.org/10.3390/en16196936

Pathiraja, K., Nilmalgoda, H., Wijethunga, I., Koswattage, K.: Sustainable management of nitrile butadiene rubber waste through pyrolysis. Sustainability **17**(3), 846 (2025). https://doi.org/10.3390/su17030846

Pathirana, S., Thngarasa, S., Yapa, A., Nilmalgoda, H.: Evaluation of the feasibility of pyrolyzing cinnamon waste biomass for the production of sustainable biofuel. In: Proceedings of International Conference on Innovation and Emerging Technologies (ICIET) 2024. Faculty of Technology, University of Sri Jayewardenepura Homagama 10206, Sri Lanka.: Faculty of Technology, University of Sri Jayewardenepura, p. 127 (2024)

Payungwong, N., Wu, J., Sakdapipanich, J.: Unlocking the potential of natural rubber: a review of rubber particle sizes and their impact on properties. Polymer **308**, 127419 (2024). https://doi.org/10.1016/j.polymer.2024.127419

Roy, K., Debnath, S.C., Basu, D., Pongwisuthiruchte, A., Potiyaraj, P.: Emerging advances in rubber technology by the suitable application of sol-gel science and technology. Rubber Chem. Technol. **94**(4), 601–625 (2021). https://doi.org/10.5254/rct.21.79955

Sydney, E.B., et al.: Current analysis and future perspective of reduction in worldwide greenhouse gases emissions by using first and second generation bioethanol in the transportation sector. Biores. Technol. Reports **7**, 100234 (2019). https://doi.org/10.1016/j.biteb.2019.100234

Tin, K.K., Taweepreda, W., Singh, A., Kumar, A.: Advancements in sludge-to-energy strategies for rubber processing industries: a comprehensive review. Int. J. Energy Water Res. (2024). https://doi.org/10.1007/s42108-024-00305-z

Tsipa, P.C., Phiri, M.M., Iwarere, S.A., Mkhize, N.M., Phiri, M.J., Hlangothi, S.P.: A novel chemical pre-pyrolysis treatment of waste tyre crumbs: a viable way for low temperature waste tyre pyrolysis. J. Anal. Appl. Pyrol.Pyrol. **181**, 106631 (2024). https://doi.org/10.1016/j.jaap.2024.106631

Valorisation of Fruit and Vegetable Wastes; Development of Biopolymers from Potato Peels and Lime Pomace

D. M. M. L. Delpitiya(✉), D. D. I. T. Denipitiya, and M. A. Elangasinghe

Department of Chemical and Process Engineering, Faculty of Engineering, University of Peradeniya, Peradeniya 20400, Sri Lanka
e18065@eng.pdn.ac.lk

Abstract. The waste generated from fruits and vegetables presents significant environmental, economic, and social challenges for the food industry. This study investigates the potential of utilizing potato peels and lime pomace to develop sustainable biopolymer films as a solution to food waste management. Sixteen biopolymer samples were prepared using two methods: Method 1 without ultrasound treatment (8 samples), and Method 2 with 60 min of ultrasound treatment (8 samples). Both methods used varying glycerol concentrations (30%, 50%, 70%, and 100%) and a potato peel to lime pomace ratio of 0.5:1. Ultrasound treatment reduced biopolymer particle size, facilitating film formation. Four samples from each method were immersed in $CaCl_2$ solution after peeling off to enhance film properties, while the other four samples were simply peeled off without immersion. FTIR, TGA, and SEM analyses provided insights into material composition, thermal stability, and structural characteristics. Increasing glycerol concentration elevated moisture content and solubility while reducing water absorption. Ultrasound treatment reduced the water repellency of the films. The films' thickness ranged from 0.30 ± 0.005 mm to 1 ± 0.005 mm. Regarding mechanical properties, the G30C sample showed better performance than the G100C sample, with both immersed in $CaCl_2$, with tensile strengths of 2.25 MPa and 2.05 MPa, and elongations at break of 10.62% and 5.18%, respectively. The G30 sample, not immersed in $CaCl_2$, had a lower tensile strength (0.55 MPa) and elongation (2.51%) compared to G30C, showing that $CaCl_2$ treatment improved mechanical performance. The UG30C sample demonstrated the lowest moisture content (21.46%) and solubility (49.65%) across all samples, indicating that ultrasound treatment and lower glycerol concentration (30%) significantly improved the mechanical and barrier properties. In conclusion, ultrasound treatment, reduced glycerol concentration, and $CaCl_2$ immersion effectively enhanced biopolymer film performance, offering a promising approach for sustainable packaging solutions.

Keywords: Valorization · Fruit and Vegetable Wastes · Biopolymers · Sustainability · Biofilms

R. Dissanayake et al. (Eds.): ICRES 2025, PTT, pp. 40–56, 2025.
https://doi.org/10.1007/978-981-96-9551-5_4

1 Introduction

Food waste (FW) is one of a major waste stream in municipal solid waste. Globally, 1/3 of food produced for human consumption is lost or wasted. It equal to 1.3 billion tons of food per year (Dimou et al., 2019). The main categories of FW are dairy, meat and poultry, fish, fruit and vegetable, bakery, brewing and winery industry and others (Blakeney, 2019). Fruit and Vegetables are essential components of human diet and agricultural practices. And also, fruits and vegetables play a crucial role while providing essential nutrients and supporting food security. Fresh produce is becoming more and more in demand worldwide as a result of population increase and improved eating habits. As a result, the Fruit and Vegetables (FVs) production also increase and waste generation also increasing due to these reasons (Magama, Chiyanzu and Mulopo, 2022). With the increasing demand for fruits and vegetables, a substantial amount of waste and loss are generated accounts for over 42% of the total food waste (Ganesh, Sridhar and Vishali, 2022). FVs have the highest wastage percentage (40–50%) due to the highly perishable materials (Blakeney, 2019). The FVWs mainly consist of inedible parts of fruits and vegetables such as, peels, seeds, kernels, tails and pomace. These wastes have a high potential for biodegradability (Ferdeş et al., 2022). These residues may cause high levels of environmental contamination, due to high biodegradability associated with high moisture content and microbial load. So, FVWs are indeed a serious issue in society, with significant economic, environmental and social implications. FW significantly contributes to climate change through methane and carbon dioxide emissions from landfills, causing odour issues and spreading diseases due to improper disposal. There are two management systems used for managing FW including traditional and alternative techniques (Magama, Chiyanzu and Mulopo, 2022). Traditional methods include landfilling, open dumping, incineration, and composting. Traditional methods also have some disadvantages such as; cause to the emission of greenhouse gases, odour issues, health hazards, slow degradation rates, and water pollution (Magama, Chiyanzu and Mulopo, 2022). To address these challenges, researchers have identified alternative methods that are more efficient, cost-effective, and environmentally sustainable.

FVs processing by-products which called FVWs consist of valuable compounds such as antioxidants, oils, fiber, fatty acids, isoprenoids, lipids, proteins, saponins, phytoestrogens, and valuable bioactive compounds (Dimou et al., 2019). In studies, it was found that the organic fraction of fruit and vegetable wastes consists of 75% sugars and hemicelluloses, 9% cellulose, and 5% lignin (Paraschiv et al., 2022). FVWs consist of dietary fibers that cause water-holding capacity, oil-holding capacity, and solubility (Sharma, Gupta and Issar, 2016). FVWs consists of high moisture content (80–90%), proteins, lipids, and other nutrients such as vitamins, minerals, and fibers(Magama, Chiyanzu and Mulopo, 2022). From the dry mass of FVWs, 70–90% of carbohydrates are included in the fruit and vegetable wastes (Magama, Chiyanzu and Mulopo, 2022). Valorizing fruit and vegetable wastes into valuable resources, such as additives, nutraceuticals, functional ingredients, pharmaceuticals, cosmetics, biopolymers, or innovative products, can be an effective waste management strategy. Among these various valorisation methods, this study mainly focused on developing biopolymer films from FVW.

Over 350 million tons of plastics are produced annually around the world (Merino, Uttam C Paul and Athanassiou, 2021). Among them only 9% of plastics are recycled

(Maraveas, 2020). In contrast to synthetic polymers, biopolymers are biodegradable and are derived from renewable sources, making them an essential alternative in addressing environmental concerns. Biopolymers, with their reduced environmental impact, hold significant potential to replace conventional plastics, providing a more sustainable solution for the future. Starch, protein and cellulose are actively used to produce biopolymers (Maraveas, 2020).

It was demonstrates that, the biopolymers can be developed by using Potato Peels (PP) and Lime Pomace (LP) (Borah, Das and Badwaik, 2017). Potato processing generates peels, pulps and rejects. The amount of waste and by- product of potato industry is estimated to be around 12–20% of their total production (Borah, Das and Badwaik, 2017). This is a considerable amount of wastes and researchers found that this waste has a potential to produce dietary fibre in medicine, adsorbent in water treatment, biopolymer materials. PP consists of high amount of starch, cellulose, hemicellulose, proteins, ash, fermentable sugars and small amount of lipids c. PP can be considered as the main raw material for the biopolymer film development due its high amount of starch content which is 25% (Khanal et al., 2023). However, starch-based polymers are highly sensitive to moisture, exhibiting high Water Vapor Permeability (WVP) and poor mechanical properties (Miller et al., 2020). To address these limitations, researchers often incorporate additional compounds. The study of Borah, Das, and Badwaik in 2017 reported that the inclusion of LP powder with PP powder can mitigate these challenges, improving the overall performance of the resulting biopolymers. It was reported that LP consists of 24.5% of pectin (Borah, Das and Badwaik, 2017). Pectin has an excellent gelling ability which makes it a suitable element for film formation (Ahmad et al., 2022). Pectin is used as a gelling and bonding agent, thickener, and colloid stabilizer (Roy et al., 2023), (Merino et al., 2021). Therefore, LP is adding as a binding agent (Merino, Uttam C Paul and Athanassiou, 2021).

The study investigates the potential of PP and LP as sustainable feedstocks for biopolymer production. By examining the effects of ultrasound treatment, varying glycerol concentrations, and calcium chloride ($CaCl_2$) immersion, this research aims to enhance the mechanical and barrier properties of biopolymers derived from FW. Most studies on biopolymer development have taken a broad approach, yet a significant challenge remains in enhancing their mechanical and barrier properties, which are often weaker than those of synthetic polymers. Research indicates that ultrasound treatment can improve the strength and performance of biopolymers by enhancing the bonding between polymer chains and making their structure more uniform (Borah, Das and Badwaik, 2017). Glycerol, a widely used plasticizer, enhances flexibility and processability but, at higher concentrations, can weaken the polymer network by disrupting intermolecular interactions. This study examines how ultrasound treatment affects the impact of higher glycerol concentrations on the mechanical and barrier properties of biopolymer films, with the goal of improving their performance for potential industrial applications. The paper provides a detailed analysis of the experimental approach, followed by a discussion on the findings and their implications for biopolymer development.

2 Materials and Methodology

2.1 Materials and Equipment

Potato peels (uncooked) and sweet lime pulp, after juice extraction were collected from the canteen of the University of Peradeniya. Glycerol, used as a plasticizer, was supplied by BDH Chemicals Ltd., England. Calcium chloride ($CaCl_2$), used as a bonding agent, was supplied by Sisco Research Laboratories Pvt. Ltd., Maharashtra, India. All solutions were prepared using distilled water. All sample materials were weighed using an electric balance, Model LPB 423i (420g × 0.001g). A DLAB MS-H340-S4 magnetic stirrer was used to mix and prepare the solutions. For the drying process, a tray dryer (Yamato Model DS-63) was used.

2.2 Powder Sample Preparation

The raw PP and LP were washed properly to remove unwanted foreign particles like sands. Washed PP and LP were dried using tray dryer at 50 °C for 48 h. Drying is essential step otherwise microbial spoilage can be occurred (Dimou et al., 2019). Once drying was completed, grinding was carried out by a grinder to make fine powder particles. After grinding, screening was done to get only the powdered particles of size <250 mm (Borah, Das and Badwaik, 2017). Powder particles were kept in a moisture-free desiccator until starting the development of biopolymer films.

2.3 Development of Biopolymer Film from PP and LP

PP and LP were used for the development of biopolymer films. The study by Borah, Das, and Badwaik in 2017 recommended that the powder ratio of PP and LP be maintained in the proportion of 2:1(Borah, Das and Badwaik, 2017). The polymer solutions were prepared by mixing 5 g of powder in 100 ml of water using a magnetic stirrer at 800 rpm for 15 min. All the solutions were heated to 85 °C for 45 min and then cooled to 30 °C (Ahmad et al., 2022). Glycerol was added at 30%, 50%, 70%, and 100%, based on the total powder amount (5 g), for samples G30C and G30, G50C and G50, G70C and G70, and G100C and G100, respectively. After, the suspension was poured into glass petri plates and dried at 50 °C for 48 h in a tray dryer to cast the films (Merino *et al.*, 2021). Dry films were peeled intact from the casting surface. The films were then immersed in 2% w/w $CaCl_2$ for 10 min and dried again at 45 °C for 12 h (Borah, Das and Badwaik, 2017). To investigate the effect of $CaCl_2$ solution immersion, samples G30, G50, G70, and G100 were not subjected to immersion in the $CaCl_2$ solution, while samples G30C, G50C, G70C, and G100C were immersed in $CaCl_2$. This allowed for a direct comparison between films treated with $CaCl_2$ and those without treatment, to assess the role of cross-linking on the mechanical and physical properties of the biopolymer films. The prepared films were conditioned at 55 ± 1.2% RH and 20 ± 1 °C in a desiccator containing silica gel for 72 h, or till further tests. Films were tested for tensile properties, solubility, moisture absorption, FTIR, TGA and thickness (Ahmad et al., 2022). Figure 1 illustrates the developed biopolymer samples from this method.

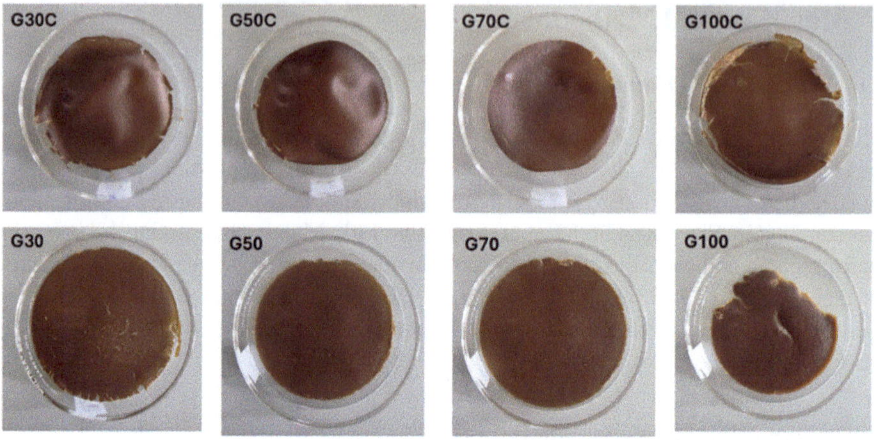

Fig. 1. Prepared biopolymer samples

2.4 Development of Biopolymer Film from PP and LP with Ultrasound Treatment

PP and LP were used for the development of biopolymer films. The study by Borah, Das, and Badwaik in 2017 recommended that the powder ratio of PP and LP be taken in the recommended proportion of 2:1(Borah, Das and Badwaik, 2017). The polymer solutions were prepared by mixing 5 g of powder in 100 ml of water using a magnetic stirrer at 800 rpm for 15 min. All the solutions were heated to 85 °C for 45 min and then cooled to 30 °C (Ahmad et al., 2022). Each solution was treated with ultrasound for 60 min in ultrasonic bath (JK-5200B, Wuxi City Bleecker Trading Co., Ltd China) (Borah, Das and Badwaik, 2017). The ultrasound frequency 40 ± 3 kHz and wattage of 50 W were used for this study. Glycerol was added at 30%, 50%, 70%, and 100%, based on the total powder amount (5 g), for samples UG30C and UG30, UG50C and UG50, UG70C and UG70, and UG100C and UG100, respectively. After, the suspension was poured into glass petri plates and dried at 50 °C for 48 h in a tray dryer to cast the films (Merino et al., 2021). Dry films were peeled intact from the casting surface. The samples UG30, UG50, UG70, and UG100 were not subjected to immersion in the $CaCl_2$ solution, while samples UG30C, UG50C, UG70C, and UG100C were immersed in 2% w/w $CaCl_2$ for 10 min and then dried again at 45 °C for 12 h (Borah, Das and Badwaik, 2017). The prepared films were conditioned at 55 ± 1.2% RH and 20 ± 1 °C in a desiccator containing silica gel for 72 h, or till further tests. Films were tested for tensile properties, solubility, moisture absorption, FTIR, TGA and thickness (Ahmad et al., 2022). Table 1 summarizes the composition of all prepared biopolymer samples, while Fig. 2 presents the ultrasound-treated samples.

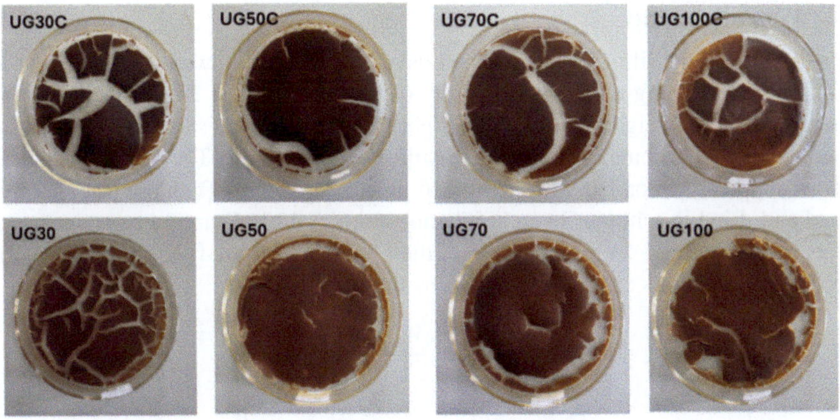

Fig. 2. Prepared biopolymer samples with ultrasound treatment

Table 1. Summary of sample composition

Sample Name	Glycerol Content (%)	$CaCl_2$	Ultrasound treatment	Sample Name	Glycerol Content (%)	$CaCl_2$	Ultrasound treatment
G30C	30	Yes	No	UG30C	30	Yes	Yes
G50C	50	Yes	No	UG50C	50	Yes	Yes
G70C	70	Yes	No	UG70C	70	Yes	Yes
G100C	100	Yes	No	UG100C	100	Yes	Yes
G30	30	No	No	UG30	30	No	Yes
G50	50	No	No	UG50	50	No	Yes
G70	70	No	No	UG70	70	No	Yes
G100	100	No	No	UG100	100	No	Yes

2.5 Characterization of Films

2.5.1 Film Thickness

Films were measured with vernier caliper having a sensitivity of 0.01 mm. The four pieces of films were stacked one above the other and average thickness was recorded.

2.5.2 Moisture Content

For the determination of moisture content, a 2 × 2 cm^2 film sample was prepared, and the initial weight (W_o) of each sample was measured. The samples were then placed in a tray dryer maintained at 105 °C for 3 h (Ahmad et al., 2022). After drying, the final weight (W_f) of the samples was measured again.

$$Moisture\ content = \frac{W_o - W_f}{W_o} \times 100\% \quad (1)$$

2.5.3 Water Solubility

The solubility of the film was determined using a method followed by (Ahmad et al., 2022). For determining the water solubility of films, about 2 × 2 cm^2 of the samples were prepared and weighed the initial weight (W_o). Then, film sample was taken and dispersed in beakers containing 50 ml of distilled water at 25 ∘C. The samples were kept at room temperature for 1 h with periodical manual agitation. Then the samples were filtered and dried in a hot air oven at a temperature of 110 ∘C until a constant weight was reached. The final dry weight (W_f) of the film was measured. The water solubility percentage of the films can be determined by the Eq. (2).

$$Water\,solubility = \frac{W_o - W_f}{W_o} \times 100\% \tag{2}$$

2.5.4 Water Absorption Capacity

The water absorption capacity of the film was determined using a method followed by (Ahmad et al., 2022). For determining the water absorption of films, about 2 × 2 cm^2 of the samples were prepared and weighed the initial weight (W_o). Then, dried film sample was taken and immersed in beakers containing 50 ml of deionized water at room temperature for 24 h. After the immersion, samples were taken out from the beaker and extra water was wiped off. The final weight (W_f) was recorded. The water absorption of the film was determined using the formula (3):

$$Waterabsorption = \frac{W_f - W_o}{W_o} \times 100\% \tag{3}$$

2.5.5 Fourier Transforms Infrared Spectroscopy (FTIR)

FTIR spectra of all the prepared biopolymer samples were recorded by using FTIR spectrum two Perkin Elmer Fourier Transform Infrared Spectrometer, USA. The FTIR measurement was carried out at room temperature (25 ± 1 °C) in the wavenumber range of 4000 cm^{-1} to 400 cm^{-1} with an averaging over 16 scans.

2.5.6 Thermo-Gravimetric Analysis (TGA)

Thermo-gravimetric analysis (TGA) is to be performed for the compound samples and the prepared thin film samples under nitrogen atmosphere from 35 °C to 800 °C by using a TGA apparatus (Model PerkinElmer TGA 4000). There the thermal weight loss (TG) curve, and its derivative weight loss (DTG) curve of the samples expected to be obtained.

2.5.7 Scanning Electron Microscope (SEM)

The morphology of the prepared selected samples was observed using SEM. And the morphology and the dispersion of biopolymer matrix can be investigated using a scanning electron microscopy (Kisku et al., 2014). To evaluate the morphologies of the sample's magnifications of 5000X, 25000X and 50000X can be used.

2.5.8 Tensile Strength

Tensile strength (TS) and elongation at break (EB) of biopolymer films were determined using the Intron 3365 Universal Testing Machine (Instron Ltd, Buckinghamshire) according to the ASTM D882–02 method at a crosshead speed of 10 mm/min. Specimens were cut from the samples ($n = 5$) and each sample had a width of 30 mm and a length of 110 mm. The TS and EB were measured during the stretching and the mechanical properties were calculated as the average value from the attained results (Borah, Das and Badwaik, 2017).

3 Results and Discussion

3.1 Film Thickness

The thickness of all the films ranged from 0.30 ± 0.005 mm to 1 ± 0.005 mm, as shown in Table 1. The average thickness of the prepared films was 0.592 ± 0.005 mm. Among all the films, the one with 100% glycerol and ultrasound treatment had the maximum thickness (0.992 ± 0.230 mm), while the film without ultrasound treatment and with 30% glycerol concentration had the least thickness (0.306 ± 0.058 mm). Compared to the biopolymers not subjected to ultrasound treatment, those treated with ultrasound exhibited greater thickness. Ultrasound treatment led to the reduction of powder particle sizes, allowing them to be uniformly distributed throughout the film (Borah, Das and Badwaik, 2017). As a result, the thickness of the biopolymers treated with ultrasound increased. Based on the results, there is no significant variation in the thickness of the samples. The thickness of the samples can vary depending on the volume poured into the petri dishes for casting the films.

3.2 Moisture Content

The moisture content results were shown in Table 2. The biopolymer sample UG30C and G100C showed of lowest moisture content (21.46%) and highest moisture content (31.46%). The lowest moisture content is observed in the ultrasound-treated biopolymer sample with the lower glycerol concentration. Ultrasound treatment further decreased the moisture content compare to the samples prepared without ultrasound treatment. This may due to the ultrasound treatment can promote more effective evaporation of moisture from the material. The vibrations generated by ultrasound can increase the rate of water removal during processing (Borah, Das and Badwaik, 2017). Further, while ultrasound can increase porosity by creating bubbles or cavities (Borah, Das and Badwaik, 2017), this enhanced porosity can also facilitate the removal of moisture from the material, leading to a lower final moisture content. Therefore, with the treatment of ultrasound the moisture content is reduced. With increasing the glycerol concentration, the moisture content also increasing. It was reported that with more glycerol concentration, there is a reduction in the intermolecular forces between polymer chains, which creates more space between them (Kang et al., 2015). Further, the glycerol has hydrophilic nature and with the higher concentration the hydrophilic nature in also increasing (Kang et al., 2015). Then, the high amount of moisture can be retained in the biofilms. According to the obtained results, the moisture content of the biopolymer films was directly related to the glycerol concentration.

3.3 Water Solubility

Water solubility patterns of films are shown in Table 2. The highest water solubility was observed in sample G100 (75.00%) for non-ultrasound-treated biopolymers and sample UG70 (76.38%) for ultrasound treated biopolymers. Neither of these samples was immersed in $CaCl_2$. Comparing the other results, it is clear that water solubility increases when the samples are not immersed in $CaCl_2$. Immersing the samples in $CaCl_2$ has a significant effect on the preparation of the biopolymer samples. Calcium chloride ($CaCl_2$), as a cross-linking agent, can adversely affect water solubility because it strengthens the molecular network within the biopolymer matrix. The cross-links formed between polymer chains reduce the ability of the material to interact with water, making the biopolymer more rigid and less prone to dissolving. This decreases its overall water solubility by limiting the penetration water molecules into the structure. Consider the ultrasound treated and non-ultrasound treated samples, generally the water solubility is reduced with the ultrasound treatment. Ultrasound treatment breaks down particles into smaller sizes, leading to a more uniform and tightly packed structure (Borah, Das and Badwaik, 2017). Then the water solubility can be decreased. Water solubility increases with glycerol concentration (30%, 50%, 70%, 100%) glycerol concentration. Glycerol is a hydrophilic substance that attracts and holds water. As its concentration increases, the overall hydrophilicity of the film or material increases, leading to greater affinity for water. Higher glycerol concentration result in weak intermolecular forces between the polymer chains (Miller et al., 2020). These weak forces create a less dense and less rigid network, which increase water infiltration ability and water solubility is increased.

3.4 Water Absorption Capacity

Based on the absorption capacity results, $CaCl_2$ immersed biopolymer films showed low water absorption capacity in both ultrasound-treated and non-ultrasound-treated biopolymers, as shown in Table 2. With the immersion of $CaCl_2$, the cross linking of the film were increased (Borah, Das and Badwaik, 2017). High cross-linking creates a tighter, denser network within the material, leaving less free space or voids for water molecules to penetrate and be absorbed. Therefore, the water absorption is reduced. Compared to non-ultrasound-treated $CaCl_2$ biopolymers, the ultrasound-treated $CaCl_2$ immersed biopolymers showed lower water absorption capacity. Ultrasound treatment breaks down particles into smaller sizes, leading to a more uniform and tightly packed structure. Then with the ultrasound treatment process, the water absorption capacity is reduced. Non-$CaCl_2$-immersed biopolymers showed no significant variation of water absorption capacity. When comparing water absorbance with glycerol concentrations, samples with higher glycerol concentrations show lower water absorbance capacity. Glycerol forms hydrogen bonds with the hydroxyl (OH) groups present in the starch compounds (Sahari et al., 2012). This is due to the forming of new hydrogen bonds between glycerol and hydroxyl groups. As a result, the plasticizing effect of the starch is enhanced, making it more difficult for water molecules to penetrate plasticized AS films.

Several studies also reported that the water absorption of carbohydrate-based films decreased with increasing plasticizer concentration (Yang, Yu and Ma, 2006), (Ma, Yu and Kennedy, 2005). In the study of Sahari in 2012 reported that the rate of water absorption of plasticized sugar palm starch films linearly decreased (20.1, 17.1, 10.3 and 7.2%) with increased glycerol (15, 20, 30 and 40% w/w) respectively (Sahari et al., 2012). The lower glycerol concentrations mean there are fewer hydrophilic sites available, reducing the material's overall ability to absorb water (Miller et al., 2020). Increased glycerol concentration enhances hydrogen bonding within the biopolymer matrix, thereby decreasing water absorption capacity.

Table 2. Moisture content (%), Water solubility (%) and Water absorption capacity (%) and Film thickness

Sample	Moisture content (%)	Water solubility (%)	Water absorption capacity (%)	Film thickness (mm)
G30C	23.16	61.79	96.88	0.306
G50C	23.71	62.86	84.09	0.367
G70C	30.52	63.47	47.06	0.408
G100C	31.46	63.75	61.63	0.350
G30	25.00	70.10	116.47	0.350
G50	25.71	70.31	146.15	0.350
G70	24.53	72.89	150.69	0.392
G100	24.88	75.00	100.61	0.539
UG30C	21.46	49.65	78.26	0.975
UG50C	25.35	56.98	73.62	0.800
UG70C	29.78	60.59	51.63	0.956
UG100C	30.83	59.15	32.76	0.992
UG30	22.30	56.89	142.31	0.550
UG50	27.81	70.34	107.53	0.550
UG70	28.33	76.38	120.41	0.708
UG100	28.37	70.22	110.29	0.617

3.5 Fourier Transforms Infrared Spectroscopy (FTIR)

Figure 3 represents the IR spectra of all the biopolymer samples according to different glycerol concentrations. Based on Fig. 3, all the films showed similar spectra. Wide bands in the range of 3000 – 3750 cm^{-1} were observed in all films, which correspond to the stretching vibration of O–H groups of starch and glycerol (Merino, Uttam C. Paul and Athanassiou, 2021a). In the study of Wathoni *et al.* in 2019, the chemical structure of pectin of FTIR analysis were reported (Wathoni et al., 2019). The major peaks of hydroxyl group were appeared at 3402.43 cm^{-1} for pectin. The aliphatic C H groups and C=O carbonyl group were shown in 2927.94 cm^{-1} and 1747.51 cm^{-1} for pectin. The peak at 1442.7 cm^{-1} describes the presence of a CH methyl group for both of extracted pectin and standard pectin. In addition, 1161.15 cm^{-1} and 1153.43 cm^{-1} indicate the presence of C–O bonds in alcohols, esters, and carboxylic acids.

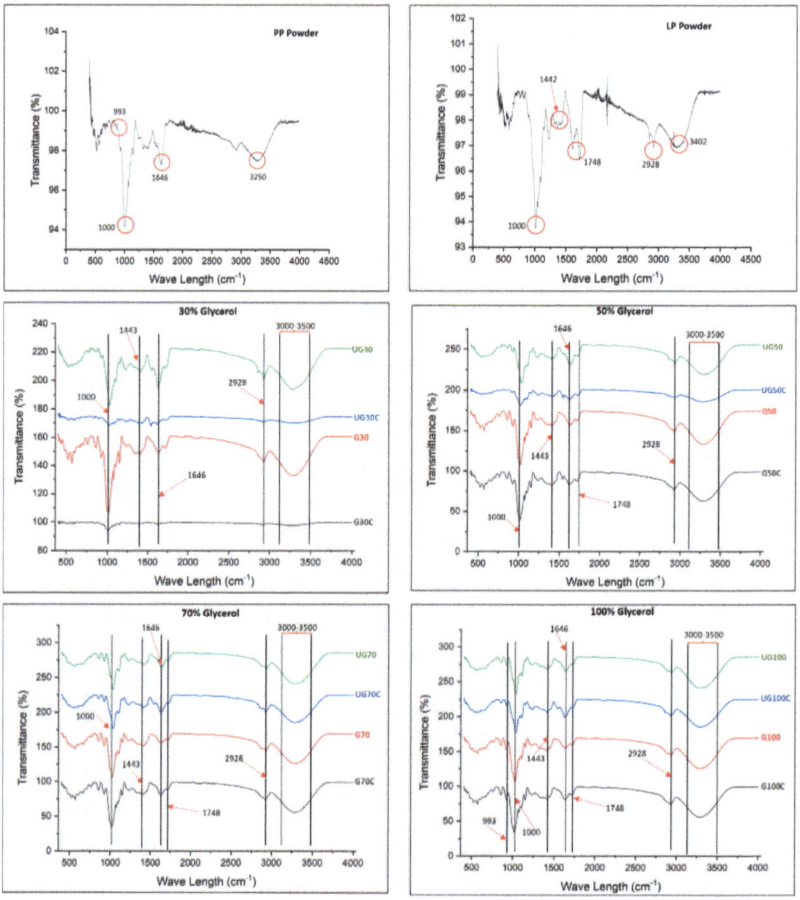

Fig. 3. FTIR spectra of samples

The peaks in the range of 2925–2930 cm^{-1} were ascribed to the (C–H) methyl group, as supported by the finding from (Merino, Uttam C. Paul and Athanassiou, 2021a), (Edhirej et al., 2017). Meanwhile, a significant absorption peak in the 1646 cm^{-1} was attributed to the hydroxyl group of tightly bound water present in starch within starch films (Merino, Uttam C. Paul and Athanassiou, 2021b), (Sahari *et al.*, 2014). The peaks at 1409 and 1433 cm^{-1} are related to the C-H bending of CH_2. This peak might also be linked with O–H stretching vibration groups of glycerol as plasticizer. The band higher than 993 cm^{-1} was associated to C–O stretching vibrations of polysaccharide compound of starch and glycerol (Syafiq, Sapuan and Zuhri, 2021), (Syafri et al., 2019). The FTIR analysis revealed that adding plasticizer at different compositions, conducting ultrasound treatment and immersing $CaCl_2$ solution to PP and LP based biofilms had no significant effect on the chemical structure of the biopolymer samples.

3.6 Thermo-Gravimetric Analysis (TGA)

TGA has been conducted to acquire the thermal degradation and stability of the developed biofilms. Figure 4 and 5 shows the decomposition of films as a function of temperature. When comparing these graphs, the thermal degradation temperature ranges appear similar, but the extent of weight loss varies. A gradual loss of weight was observed in the films till about 200 °C. This weight loss may due to loss of free and bound moisture from the films.

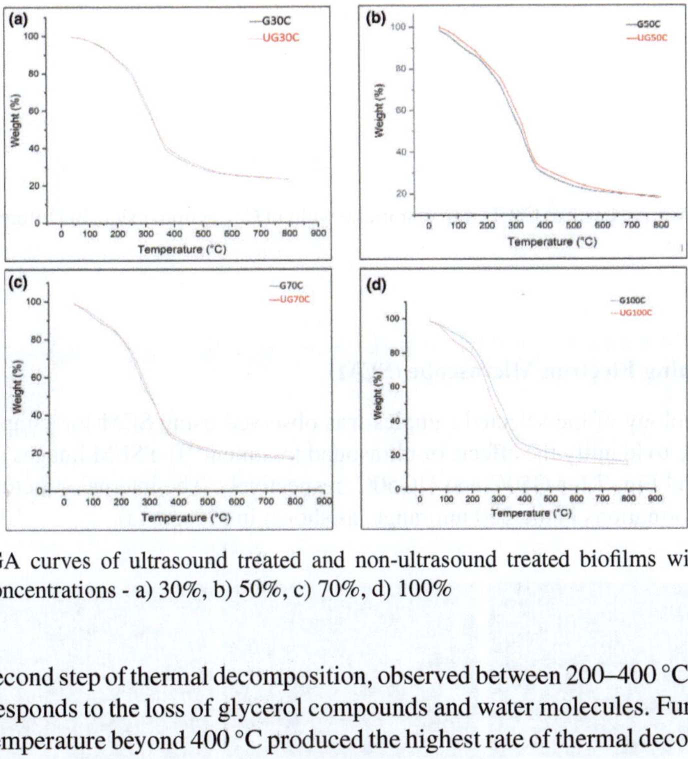

Fig. 4. TGA curves of ultrasound treated and non-ultrasound treated biofilms with different glycerol concentrations - a) 30%, b) 50%, c) 70%, d) 100%

The second step of thermal decomposition, observed between 200–400 °C in all samples, corresponds to the loss of glycerol compounds and water molecules. Further increment of temperature beyond 400 °C produced the highest rate of thermal decomposition, which was influenced by the radical weight loss of biofilms.

According to the obtained results of all the curves, the ultrasound treatment, glycerol concentration and immersion of $CaCl_2$ may not affect the thermal stability of the biopolymer films. But, the weight loss at each sample is vary with the ultrasound treatment, glycerol concentration and immersion of $CaCl_2$. According to Fig. 4, the weight loss is reduced with ultrasound treatment, except for the sample with 100% glycerol concentration. Figure 5-(b) shows the TGA curves for different glycerol concentrations, where the primary difference is in the extent of weight loss. With higher glycerol concentrations, the intermolecular bonds within the biopolymer. Matrix weaken, leading to increased weight loss.

Figure 5-(a) compares the TGA result curves of biopolymer samples with and without $CaCl_2$ immersion. The data indicate a significant impact on the thermal properties of the samples due to the $CaCl_2$ treatment. Specifically, the weight loss during the thermal degradation process is greater for the samples that were not immersed in $CaCl_2$ solution. This suggests that the $CaCl_2$-treated samples exhibit improved thermal stability, likely due to the increased cross-linking in the polymer matrix, which strengthens the material and reduces weight loss under high temperatures. Despite these differences in weight loss, the thermal degradation temperatures are consistent across all samples, indicating that the onset of degradation occurs at the same temperature regardless of $CaCl_2$ immersion.

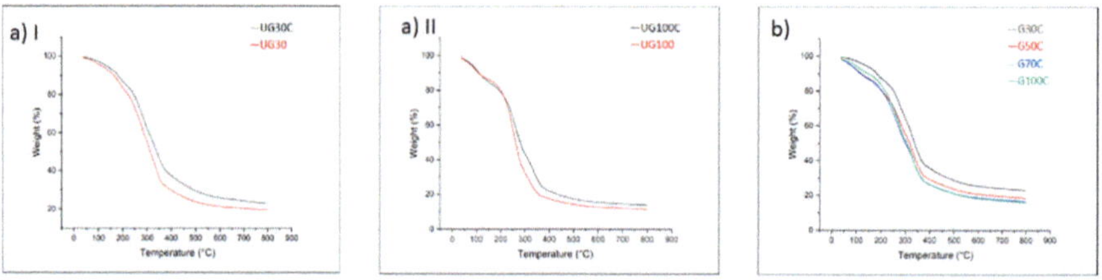

Fig. 5. TGA curves with a) $CaCl_2$ immersion and without $CaCl_2$ immersion, b) Different glycerol concentrations

3.7 Scanning Electron Microscope (SEM)

The morphology of the selected samples was observed using SEM for samples G50C and UG50C to identify the effects of ultrasound treatment. The SEM images are shown in Fig. 6 and Fig. 7 for G50C and UG50C, respectively. The internal structure reveals peak-like formations in the 200 nm range, as shown in Figures (a).

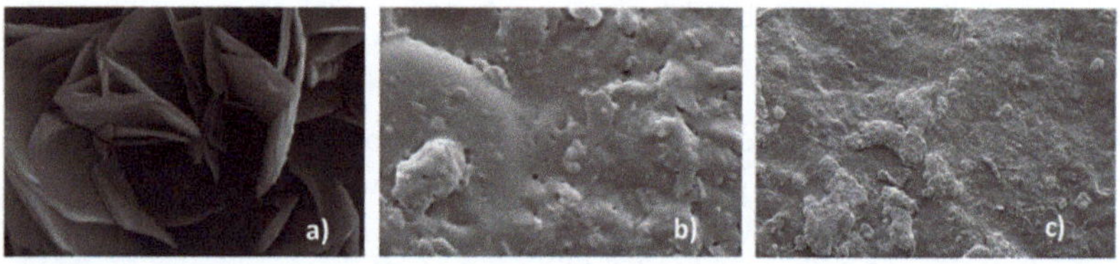

Fig. 6. SEM images of biofilms G50C a) Mag = ×50 k, 200 nm, b) Mag = ×25 k, 2 µm c) Mag = ×5k, 2 µm

The surface of the ultrasound-treated film, shown in Fig. 7-(c), was relatively smooth with fewer holes or other surface irregularities. Furthermore, the ultrasound-treated biofilms exhibited a more clustered nature compared to the non-ultrasound-treated samples, as shown in Figs. 6-(a), (b) and 7-(a), (b). This may be due to the ultrasound treatment causing stronger intermolecular bonds and a more uniform structure, as indicated in (Borah, Das and Badwaik, 2017).

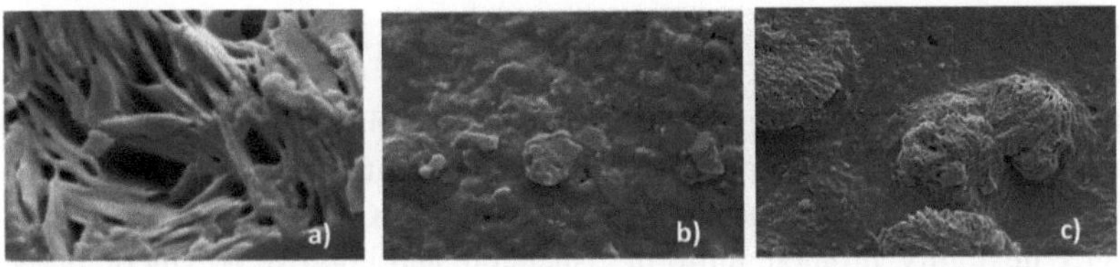

Fig. 7. SEM images of biofilms UG50C a) Mag = ×50 k, 200 nm, b) Mag = ×25 k, 2 μm c) Mag = ×5 k, 2 μm

3.8 Tensile Strength

Tensile strength (TS) and elongation at break (EB) of biopolymer films are shown in Table 3. Tensile strength was measured for three samples: G30C, G100C, and G30. Comparing the G30C and G30 samples, the G30C sample exhibited higher tensile strength (2.26 MPa) and elongation at break (10.62%) compared to the G30 sample, which had a tensile strength of 0.55 MPa and elongation at break of 2.51%.

Table 3. Tensile strength of biofilms

Sample Name	Tensile strength (MPa)	Elongation at Break (%)
G30C	2.26	10.62
G100C	2.05	5.18
G30	0.55	2.51

The G30C sample was immersed in a $CaCl_2$ solution, while the G30 sample was not. Therefore, the immersion in $CaCl_2$ solution has a significant impact on the mechanical properties of the biopolymers. As the glycerol concentration increases, the samples prepared with 100% glycerol concentration exhibit lower tensile strength (2.05 MPa) and reduced elongation at break (5.16%) compared to the samples prepared with 30% glycerol concentration. Bioplastics made from various types of starch with glycerol as a plasticizer, but without any filler materials, typically exhibit a tensile strength range between 0.22 and 18.49 MPa, as reported in the literature (Gabriel, Solikhah and Rahmawati, 2021). Therefore, the obtained results of the developed biopolymer materials are within the the range of 0.22 – 18.49 MPa. According to the biodegradable plastic standard SNI 7818:2014, the minimum required tensile strength is 1.343 MPa, as reported in the literature (Gabriel, Solikhah and Rahmawati, 2021). Samples G30C and G100C exhibited tensile strengths of 2.26 MPa and 2.05 MPa, respectively, which are both greater than the minimum standard for biodegradable plastic tensile strength. The tensile strength of the biopolymer materials developed in this study falls within this range, confirming that the mechanical properties of the fabricated films are consistent with previously reported values for starch-based bioplastics. This suggests that the biopolymer materials, despite variations in glycerol concentration and the use of $CaCl_2$, achieve tensile strength comparable to other starch-based bioplastics. The consistency of the results highlights the effectiveness of the formulation and processing methods used in the current research.

4 Conclusions

In this study, composite biofilms were prepared via solution casting, with and without ultrasound treatment, under varying glycerol concentrations (30%, 50%, 70%, and 100%) and with or without immersion in $CaCl_2$ solution. Glycerol served as a plasticizer to enhance flexibility, promote uniform dispersion, and improve polymeric properties. $CaCl_2$ acted as a cross-linking agent, strengthening the polymer matrix through ionic interactions. The biopolymer films treated with ultrasound exhibited a more compact matrix, improved tensile strength, extensibility, and enhanced water repellency compared to non-ultrasound-treated films. Analytical techniques, including FTIR spectroscopy, TGA, and SEM, revealed stronger polymer chain interactions, an amorphous intercalated structure, good thermal stability, and effective particle dispersion, despite some agglomeration. However, increasing glycerol concentrations negatively affected their mechanical strength and water repellency, even when treated with ultrasound. Therefore, the optimum glycerol concentration can be obtained as the 30%. However, $CaCl_2$ immersion facilitated additional cross-linking within the biopolymer matrix. In conclusion, biofilms treated with ultrasound and immersed in $CaCl_2$, particularly those with lower glycerol concentrations, demonstrated superior overall performance in terms of mechanical properties and water resistance.

5 Future Perspective

Future research should concentrate on the careful analysis of ultrasound treatment time and frequency, as these parameters may significantly influence the properties of the biopolymer samples. It is also crucial to investigate the potential increase in powder particle size associated with ultrasound treatment, as this may impact the performance of the biopolymers. Additionally, a thorough examination of the drying process during biopolymer development with ultrasound treatment is essential to optimize the final properties of the materials. Without proper control, the developed biopolymer samples may risk over-drying and cracking.

Acknowledgment. Department of Geology, University of Peradeniya, the Rubber Research Institute of Sri Lanka, and Joint Research and Demonstration Centre (JRDC), Peradeniya facilitated instrumental analysis.

References

Ahmad, A., et al.: Mosambi (Citrus limetta) peel and Sago based biodegradable film: development and characterization of physical, water barrier and biodegradation properties. Biores. Technol. Reports, 18(January), p. 101016 (2022). https://doi.org/10.1016/j.biteb.2022.101016

Blakeney, M.: Food loss and food waste: Causes and solutions. Food Loss and Food Waste: Causes and Solutions (2019). https://doi.org/10.4337/9781788975391

Borah, P.P., Das, P. Badwaik, L.S.: Ultrasonics sonochemistry ultrasound treated potato peel and sweet lime pomace based biopolymer film development. Ultrason. Sonochemis. **36**, 11–19 (2017). https://doi.org/10.1016/j.ultsonch.2016.11.010

Dimou, C., et al.: Valorization of fruits by-products to unconventional sources of additives, oil, biomolecules and innovative functional foods, pp. 1–11 (2019). https://doi.org/10.2174/1389201020666190405181537

Edhirej, A., et al.: Sugar palm fiber reinforced cassava starch hybrid composites : physical , thermal and structural properties. Int. J. Biolog. Macromolecules Cassava **101**, 75–83 (2017)

Ferdeş, M., et al.: Food waste management for biogas production in the context of sustainable development. Energies **15**(17) (2022). https://doi.org/10.3390/en15176268

Gabriel, A.A., Solikhah, A.F., Rahmawati, A.Y.: Tensile strength and elongation testing for starch-based bioplastics using melt intercalation method: a review. J. Phys. Conf. Seri. **1858**(1) (2021). https://doi.org/10.1088/1742-6596/1858/1/012028

Ganesh, K.S., Sridhar, A., Vishali, S.: Chemosphere utilization of fruit and vegetable waste to produce value-added products : conventional utilization and emerging opportunities-a review. Chemosphere **287**(P3), 132221 (2022). https://doi.org/10.1016/j.chemosphere.2021.132221

Kang, H.J., et al.: Plasticization and moisture sensitivity of potato peel-based biopolymer films **24**(5), pp. 1703–1710 (2015). https://doi.org/10.1007/s10068-015-0221-x

Khanal, S., et al.: Sustainable utilization and valorization of potato waste : state of the art, challenges , and perspectives. Biomass Conversion and Biorefinery [Preprint], (0123456789) (2023). https://doi.org/10.1007/s13399-023-04521-1

Ma, X., Yu, J., Kennedy, J.F.: Studies on the properties of natural fibers-reinforced thermoplastic starch composites. **62**, 19–24 (2005). https://doi.org/10.1016/j.carbpol.2005.07.015

Magama, P., Chiyanzu, I., Mulopo, J.: 'A systematic review of sustainable fruit and vegetable waste recycling alternatives and possibilities for anaerobic biorefinery. Biores. Technol. Reports **18**(February), 101031 (2022). https://doi.org/10.1016/j.biteb.2022.101031

Maraveas, C.: Production of sustainable and biodegradable polymers from agricultural waste. Polymers **12**(5) (2020). https://doi.org/10.3390/POLYM12051127

Merino, D. et al.: Avocado peels and seeds : processing strategies for the development of highly antioxidant bioplastic films (2021). https://doi.org/10.1021/acsami.1c09433

Merino, D., Uttam, P.C., Athanassiou, A.: 'Bio-based plastic films prepared from potato peels using mild acid hydrolysis followed by plasticization with a polyglycerol. Food Packaging and Shelf Life **29**(December 2020), 100707 (2021). https://doi.org/10.1016/j.fpsl.2021.100707

Merino, D., Paul, Uttam C., Athanassiou, A.: 'Bio-based plastic films prepared from potato peels using mild acid hydrolysis followed by plasticization with a polyglycerol. Food Packaging and Shelf Life **29**(December 2020), 100707 (2021a). https://doi.org/10.1016/j.fpsl.2021.100707

Merino, D., Paul, Uttam C., Athanassiou, A.: Bio-based plastic films prepared from potato peels using mild acid hydrolysis followed by plasticization with a polyglycerol. Food Packaging and Shelf Life 29(December 2020) (2021b). https://doi.org/10.1016/j.fpsl.2021.100707

Miller, K., et al.: Effects of glycerol and sorbitol on optical , mechanical , and gas barrier properties of potato peel-based films. (August), 1–13 (2020). https://doi.org/10.1002/pts.2536

Paraschiv, G., et al.: Food Waste Management for Biogas Production in the Context (2022)

Roy, S., et al.: International journal of biological macromolecules recent progress in pectin extraction, characterization , and pectin-based films for active food packaging applications : a review 239(March) (2023)

Sahari, J., et al.: 'A new approach to use arenga pinnata as sustainable biopolymer : effects of plasticizers on physical properties. **4**, 254–259 (2012). https://doi.org/10.1016/j.proche.2012.06.035

Sahari, J., et al.: Physico-chemical and thermal properties of starch derived from sugar palm tree (arenga pinnata) physico-chemical and thermal properties of starch derived from sugar palm tree (arenga pinnata). (January 2015) (2014). https://doi.org/10.14233/ajchem.2014.15652

Sharma, P.C., Gupta, A., Issar, K.: Effect of packaging and storage on dried apple pomace and fiber extracted from pomace. 00, pp. 1–10 (2016). https://doi.org/10.1111/jfpp.12913

Syafiq, R., Sapuan, S.M., Zuhri, M.R.M.: Antimicrobial activity , physical , mechanical and barrier properties of sugar palm based nanocellulose. starch biocomposite films incorporated with cinnamon essential oil. J. Mater. Res. Technol. **11**, 144–157 (2021). https://doi.org/10.1016/j.jmrt.2020.12.091

Syafri, E., et al.: Effect of sugar palm nanofibrillated cellulose concentrations on morphological, mechanical and physical properties of biodegradable films based on agro-waste sugar palm (Arenga pinnata (Wurmb.) Merr) starch. Integrative Med. Res. **8**(5), 4819–4830 (2019). https://doi.org/10.1016/j.jmrt.2019.08.028

Wathoni, N., et al.: 'Characterization and antioxidant activity of pectin from Indonesian mangosteen (garcinia mangostana l .) rind heliyon characterization and antioxidant activity of pectin from indonesian mangosteen (garcinia mangostana l .) rind. Heliyon (August), e02299 (2019). https://doi.org/10.1016/j.heliyon.2019.e02299

Yang, J., Yu, J., Ma, X.: Preparation and properties of ethylenebisformamide plasticized potato starch (EPTPS). **63**, 218–223 (2006). https://doi.org/10.1016/j.carbpol.2005.08.059

A Review and Analysis of Current Hard Plastics Recycling in Sri Lanka

H. K. T. S. Kuruppuge(✉) and K. G. Chandrakumara

Insee Ecocycle Lanka (Private) Limited/Level 23, Access Tower II, 278/4, Union Place, Colombo 02, Sri Lanka
themiya.sandaru003@gmail.com

Abstract. The hard plastic waste management in Sri Lanka experiences environmental, economic, and social challenges due to poorly controlled waste generation which reaches 1.5 million tons each year. The focus of recycling programs operates in cities because rural areas do not have formal waste collection systems in place. Ineffective recycling from unrecognized informal waste collectors occurs because they receive no training and no access to technology or recognition despite collecting 60% of waste. This review examines hard plastic recycling practices in Sri Lanka by analyzing recent developments through system-based methods that discover gaps before proposing solutions supported by peer-reviewed studies and official government documentation as well as international industry information. The adoption of Extended Producer Responsibility frameworks combined with essential recommendations should focus on uniting informal waste management with formal operations while enabling recycling advancements coupled with regional collection facilities. Sri Lanka needs both the enforcement of waste management rules and educational initiatives to enhance community engagement and decrease pollution from improper waste elimination. This review uses international methods from Japan and the European Union to define a circular economy transition route that Sri Lanka can use to achieve both environmental goals and worldwide sustainability objectives and establish foundations for future research on scalable innovative technologies.

Keywords: Hard plastics · recycling · sustainability · circular economy · waste management

1 Introduction

Plastic pollution has emerged as a critical environmental crisis in the 21st century, impacting ecosystems, economies, and public health globally. The production of plastic material exceeds 353 million metric tons annually; therefore, poor disposal and management result in significant environmental degradation (Ferronato et al., 2024). Plastic as the hard plastics like polypropylene (PP), High-Density Polyethylene (HDPE), Polyvinyl Chloride (PVC), and polycarbonate (PC) is used in packaging, construction, and other industries because of their strength, versatility, and economic advantages. However, they imply long-term negative impacts when disposed of improperly, including resource depletion,

R. Dissanayake et al. (Eds.): ICRES 2025, PTT, pp. 57–66, 2025.
https://doi.org/10.1007/978-981-96-9551-5_5

pollution, and contributions to global warming through greenhouse gas emissions during incineration (Borrelle et al., 2020)

Globally, recycling is still insufficient, and today only 9% of all the plastic waste produced is recycled. About 22% of plastics are discarded through open dumping and burning though, nearly 44% are buried in landfills further leading to problems such as water pollution and soil depletion amongst others due to the emission of toxic gasses (Law et al., 2020). Developed countries such as Japan, Germany, and most of the members of the European Union have stepped up their efforts to recycle and the results have been high rates of recycling of over 40% because of strengthened legislation on the issue of environmental conservation, technological advancement, and high public participation. In contrast, developing nations struggle with inadequate infrastructure and policy frameworks, leading to significantly lower recycling rates.

Sri Lanka faces similar challenges, compounded by its unique geographical, economic, and social factors. As an island nation with a population exceeding 22 million, Sri Lanka generates approximately 1.5 million tons of plastic waste annually. Hard plastics, which are predominantly used in packaging, agriculture, and household goods, form a substantial portion of this waste. However, data from the Central Environmental Authority (CEA) reveals that over 60% of this waste is not adequately managed. Improper disposal practices, including open dumping, incineration, and unregulated landfilling, are widespread, contributing significantly to marine plastic pollution. As highlighted in the papers Sri Lanka is among the five countries most contributing to the mismanaged plastic waste flow into the ocean illustrating the imperative need to enhance waste disposal management systems (Jambeck et al., 2015).

The informal sector dominates waste collection and recycling in Sri Lanka, yet its operations are constrained by limited access to technology, resources, and government support (Fernando et al., 2022). Recycling infrastructure is concentrated in urban areas, leaving rural regions underserved. Furthermore, the absence of advanced recycling technologies, such as chemical recycling, limits the country's ability to process certain types of hard plastics. Public awareness about waste segregation and recycling remains low, with studies indicating that less than 15% of households practice source segregation, a fundamental step in effective recycling (Jayasinghe et al., 2022).

Existing policies, such as the ban on single-use plastics and the establishment of extended producer responsibility (EPR) frameworks, represent initial steps toward addressing plastic waste. However, their enforcement is rarely effective because the funding and institutional capabilities for their implementation are indigent. Sri Lanka does not have a coherent national policy approach for addressing the problem that entails policies, technologies, and public participation. Other challenges include a lack of one central strategy and weak partnerships among the relevant stakeholders, such as government, private sector organizations, and informal waste collectors.

Addressing these challenges requires a multifaceted approach. Drawing lessons from global best practices, such as Japan's zero-waste policies and the European Union's circular economy model, Sri Lanka can develop localized solutions that are both scalable and sustainable. The objective of this review is to analyze the current state of hard plastics recycling in Sri Lanka, to outline the primary gaps in the environment in terms of infrastructure, policy, and public involvement, and to present specific recommendations for the improvement of the resource efficiency and the environmental outlook of the country. By aligning national efforts with global sustainability goals, Sri Lanka has the potential to reduce plastic pollution and transition toward a circular economy significantly.

2 Methodology

2.1 Search Strategy

A systematic search strategy was employed to identify and retrieve relevant literature on hard plastics recycling, focusing on both global and Sri Lankan contexts. Google Scholar, Science Direct, Springer Link, and Research Gate were used sources to include only quality work and peer-reviewed articles. Moreover, governmental reports from the Central Environmental Authority (CEA) and international organizations such as the European Commission and the United Nations Environment Programme (UNEP) were reviewed to provide insights into policy frameworks and local practices.

The search was conducted using targeted keywords, including hard plastics recycling, plastic waste management Sri Lanka, informal recycling sector, circular economy and plastic waste, and advanced recycling technologies.

To capture both historical trends and recent developments, the search timeframe was restricted to publications between 2010 and 2024. The chosen range enabled the use of key studies in waste management while focusing on the recent developments within this field. Supplementary searches included grey literature such as conference proceedings, case studies, and reports from global organizations like IPEN, which provided context-specific data on Sri Lanka.

2.2 Selection Criteria

The identification and selection of studies followed a screening process to ensure high-quality, relevant literature.

- Studies addressing hard plastics recycling, plastic waste management, or related environmental impacts were considered.
- Priority was given to research directly related to Sri Lanka, while global studies were included if they provided comparative insights or best practices applicable to the Sri Lankan context.
- Only studies offering quantitative or qualitative data on recycling rates, technological adoption, stakeholder roles, or policy frameworks were considered.
- Peer-reviewed journal articles, government reports, institutional publications, and high-quality reviews were selected for analysis.

3 Overview of Hard Plastics Recycling

Recycling of hard plastics such as high-density polyethylene, polypropylene, and polycarbonate is highly important in reducing the impacts of plastic waste. Globally, hard plastics constitute a significant portion of the waste stream, yet their recycling rates remain disappointingly low (Borrelle et al., 2020). Research indicates that only 9% of global plastic waste is recycled effectively, while the majority is landfilled or mismanaged (Law et al., 2020). The environmental impacts of this problem include soil and water pollution, emission of toxic gases from open burning, and the appearance of microplastics in marine ecosystems (Kafa et al., 2020).

In Sri Lanka, the challenges associated with hard plastics recycling are distinct. The country's reliance on informal waste collection systems results in inconsistent recycling rates and high levels of contamination in recyclable materials (Fernando et al., 2022). The infrastructure required to facilitate extensive recycling is insufficient, especially in rural regions with limited access to formal waste management systems, despite increased awareness of plastic pollution. Additionally, the ability to treat specific kinds of hard plastics is limited by the lack of sophisticated recycling methods like chemical recycling (Rahimi & García, 2017). A comparative analysis with developed nations like Germany and Japan demonstrates the possible advantages of implementing a circular economy strategy, which reduces waste and efficiently reuses or recycles resources.

Research highlights the misuse of circular economy principles in developing nations, where open dumping and burning continue to be the most common disposal methods. A critical review of previous studies demonstrates the complexity of the plastic waste crisis and the variety of strategies to address it (Ferronato et al., 2024). The study recommends scaling up successful pilot projects and fostering international collaboration to enhance recycling practices. Similarly, it highlights the need for robust collection systems and public engagement initiatives to address the underlying challenges in global plastic waste management (Kibria et al., 2023).

(Kafa et al., 2020) examine the barriers to implementing circular economy principles, identifying economic constraints and the lack of advanced recycling technologies as critical challenges. The role of informal waste pickers, who are often marginalized despite their significant contributions to waste collection and recycling. The integration of informal sectors into formal waste management systems is proposed as a viable strategy for improving recycling rates and social equity (Hopewell et al., 2010).

In the Sri Lankan context, reports by the Central Environmental Authority and IPEN provide insights into the composition and management of plastic waste. These studies demonstrate the urgent need for community involvement, policy changes, and technology investments to address the nation's particular problems (Table 1).

Table 1. Key findings from global and Sri Lankan studies, highlighting the challenges, solutions, and their applicability to the local context

Study	Region	Challenges Identified	Proposed Solutions	Applicability to Sri Lanka
(Fernando et al., 2022)	Developing countries	Lack of circular economy implementation	Scale up pilot projects Foster international collaboration Increase public awareness	Pilot projects can be initiated in urban and rural areas International partnerships can aid technology transfer
(Ferronato et al., 2024)	Global	Low recycling rates Inefficient collection systems Limited public engagement	Improve waste collection infrastructure Conduct public education campaigns Integrate education into policy initiatives	Address poor public awareness and weak collection systems Integrate education into policy initiatives
(Kafa et al., 2020)	Europe & Asia	Economic and technological barriers to circular economy adoption	Encourage investment in recycling technologies Develop local markets for recycled products	Focus on introducing affordable recycling technologies Build demand for recycled products locally
(Law et al., 2020)	Global	High levels of mismanaged plastic waste (landfilling, open dumping, burning)	Invest in environmentally sound disposal and recycling systems Strengthen regulatory frameworks	Replace open dumping and burning practices with formal waste treatment facilities Enforce existing regulations

(*continued*)

Table 1. (*continued*)

Study	Region	Challenges Identified	Proposed Solutions	Applicability to Sri Lanka
(National Plastic Waste Inventory for Sri Lanka, 2024)	Sri Lanka	Lack of accurate data on plastic waste streams Insufficient infrastructure	Conduct regular waste audits Invest in collection and recycling infrastructure	Improve data collection and reporting Expand recycling infrastructure, especially in rural areas
(A European Strategy for Plastics in a Circular Economy, 2019)	Europe	Over-reliance on single-use plastics	Introduce strict single-use plastic bans Strengthen recycling incentives for businesses	Enforce strict bans on single-use plastics Develop financial incentives for recycling businesses

4 Current State of Recycling in Sri Lanka

Addressing Sri Lanka's expanding plastic waste issue is highly challenging due to the country's inadequate and dispersed recycling infrastructure. Hard plastics account for a significant amount of the nation's 1.5 million tons of plastic waste generated every year, which is why recycling activities are concentrated in cities like Galle, Kandy, and Colombo. However, large amounts of plastic trash go unrecorded or are not effectively managed in rural areas, which make up almost 80% of the country's landmass and lack access to official collection and recycling facilities (Plastic Waste Management Country Situation Report-Sri Lanka, 2020; National Plastic Waste Inventory for Sri Lanka, 2024).

Informal waste pickers, who make up nearly 60% of waste recovery operations in Sri Lanka, are primarily responsible for the collection and processing of plastic material. Because these individuals lack access to modern sorting and recycling technologies, recovery rates are weak, and recyclable materials are highly contaminated. For instance, research shows that hard plastic contamination rates frequently surpass 25%, which drastically lowers the quality and marketability of recovered goods (Fernando et al., 2022).

The limited adoption of advanced technologies, such as chemical recycling, exacerbates these challenges. Most recycling in Sri Lanka relies on basic mechanical processes, including shredding and pelletizing, which are often incapable of handling contaminated or mixed plastic streams effectively. In contrast, countries like Japan and Germany have incorporated advanced recycling techniques, achieving recycling rates of over 80% for specific types of plastics (A European Strategy for Plastics in a Circular Economy, 2019).

Despite these limitations, the government has made some progress in promoting recycling. The CEA has implemented policies such as the ban on single-use plastics and the introduction of waste segregation initiatives at the household level. However, enforcement remains inconsistent due to insufficient funding, lack of trained personnel, and weak political will. For example, while the single-use plastics ban has been in effect since 2021, reports indicate that compliance rates are below 40%, with illegal production and distribution of banned items persisting in rural markets (National Plastic Waste Inventory for Sri Lanka, 2024).

Private recycling companies play an important role in the sector but face significant economic challenges. Fluctuations in global oil prices, which impact the cost of virgin plastics, directly affect the profitability of recycled materials. High operational costs, limited access to financial incentives, and inadequate demand for recycled products further deter investment in advanced recycling technologies. In 2022, it was reported that less than 15% of recycled plastic products in Sri Lanka met export-grade quality standards, highlighting the urgent need for improvements in processing efficiency and quality control (Fernando et al., 2022).

The informal sector, which is critical to the country's recycling ecosystem, is often overlooked in formal policy frameworks. Informal waste pickers, numbering around 15,000 to 20,000 individuals, operate under precarious conditions, earning minimal incomes and lacking access to healthcare, safety equipment, or training. Their contributions, however, are significant; informal collectors recover an estimated 300,000 tons of plastic waste annually, accounting for nearly 50% of the country's total recyclable plastic waste. Integrating this sector into formal waste management systems could substantially improve collection and recycling rates while addressing social equity concerns ((Plastic Waste Management Country Situation Report-Sri Lanka, 2020; National Plastic Waste Inventory for Sri Lanka, 2024).

5 Results

The findings of this review reveal critical insights into the state of hard plastics recycling in Sri Lanka:

5.1 Recycling Rates and Infrastructure

Sri Lanka's hard plastic recycling rate is still at 10%, much below world leaders like the European Union (42%), and Japan (84%). The majority of recycling facilities are located in metropolitan areas; more than half of operating centres are in Colombo alone. But because rural regions are still underserved, hard plastics, especially high-density polyethylene and polypropylene are frequently dumped in the open and burned there ((Plastic Waste Management Country Situation Report-Sri Lanka, 2020);(National Plastic Waste Inventory for Sri Lanka, 2024); Jayasinghe et al., 2023).

5.2 Stakeholder Contributions

With about 60% of all recyclable plastic collected each year coming from the informal waste sector, this sector is essential to the recovery of plastic waste. Informal collectors' efficiency and economic potential are nonetheless constrained by their lack of official registration, safety precautions, and technological access. Policies including home segregation initiatives and prohibitions on single-use plastics have been implemented by government organizations, such as the Central Environmental Authority. However, awareness programs have not successfully reached rural people, and implementation of these regulations is still variable (Ranatunga et al., 2023).

5.3 Technology Adoption

The review draws attention to the dependence on simple mechanical recycling techniques like pelletizing and shredding. Sri Lanka rarely has any advanced technology, such as automated sorting systems and chemical recycling, which are essential for processing contaminated or mixed plastics. The lack of these technologies lowers the quality of recycled items and their market value by contributing to contamination rates that exceed 25% (Samarasinghe et al., 2021).

6 Discussion

The results highlight the pressing need for an integrated approach to enhance Sri Lanka's recycling of hard plastics. The study reveals several connected themes:

6.1 Infrastructure Development

Lessons from Japan's decentralized waste management model suggest that localized solutions tailored to regional needs are key to improving efficiency and accessibility. One important challenge is the geographical disparity in recycling infrastructure, with rural areas relying on informal collectors who lack the resources to manage waste effectively and urban areas having relatively better access to formal recycling facilities. This disparity can be addressed by investing in regional recycling hubs equipped with advanced technologies (A European Strategy for Plastics in a Circular Economy, 2019).

6.2 Strengthening Stakeholder Integration

The establishment of a structured recycling ecosystem is complicated by the scattered nature of stakeholder contributions. It is possible to greatly increase collection rates and lower pollution by formalizing the work of informal garbage collectors through collaborations with private sector stakeholders, financial incentives, and training programs. The potential advantages of such tactics are demonstrated by successful integration models from nations like India, where over 60% of plastic recycling is handled by the unorganized sector.

6.3 Policy Enforcement and Innovation

Although the government has implemented several laws aimed at managing plastic waste, their efficacy is compromised by inadequate enforcement. Extended Producer Responsibility programs that strengthen regulatory frameworks might encourage producers to embrace sustainable practices, such as creating more recyclable items. Furthermore, financial tools like tax exemptions for businesses that use recycled materials and subsidies for recycling technology can promote private sector involvement.

6.4 Public Engagement

One major obstacle to efficient recycling is the lack of public participation and awareness. Increased community engagement may be achieved through nationwide awareness efforts that highlight the financial and environmental advantages of recycling, along with incentives like reduced municipal garbage costs for families who practice segregation. Educational initiatives targeting schools and community organizations can also play a vital role in changing cultural attitudes toward waste management (Fernando et al., 2022).

6.5 Adopting Advanced Technologies

Beyond simple mechanical procedures, Sri Lanka's recycling ecosystem has to incorporate cutting-edge technology like chemical recycling, which can manage complicated and polluted plastic waste. Scalable solutions may be made possible by pilot projects that evaluate the viability of such technologies in regional settings with the help of foreign capital and experience. Successful chemical recycling system implementations in nations like Germany provide important lessons for incorporating these technologies into national waste management plans (Rahimi & García, 2017; Tsuchimoto & Kajikawa, 2022).

7 Conclusion

Despite producing about 1.5 million tons of plastic waste yearly, Sri Lanka's hard plastics recycling system confronts numerous obstacles, with recycling rates staying below 10%. The sector's potential is highlighted by the dependence on informal garbage collectors, who account for almost 60% of waste recovery; nevertheless, their efforts are hampered by a lack of official recognition, training, and access to cutting-edge technologies. The lack of sophisticated recycling techniques, including chemical recycling, restricts the processing of contaminated and mixed plastics, and recycling equipment is still mostly located in metropolitan areas, underserving rural communities. High rates of open dumping and marine plastic pollution are also caused by insufficient implementation of laws like the prohibition on single-use plastics and a lack of public knowledge regarding waste segregation.

By implementing focused interventions, Sri Lanka may greatly enhance its recycling environment, taking inspiration from world leaders such as the European Union and Japan. These include investing in cutting-edge technologies to improve recycling rates and lower pollution, integrating informal waste handlers into formal systems, and putting EPR frameworks into practice. Important infrastructural and participation shortages can be filled by growing regional recycling hubs and launching national public awareness initiatives. Sri Lanka may move toward a more sustainable waste management system, lessening its effects on the environment and advancing global sustainability goals, by embracing a circular economy strategy and encouraging cooperation between public and private sectors as well as communities.

References

A european strategy for plastics in a circular economy (2019). https://ec.europa.eu/environment/circular-economy/

Borrelle, S.B., et al.: Predicted growth in plastic waste exceeds efforts to mitigate plastic pollution (2020). https://www

Fernando, S.P., Kuruppu, I.V., Aiome, G.V.N., Dissanayaka, D.M.A.C., Wijesinghe, W. A.A.S.: Practices followed to manage plastic waste including shopping bags and lunch sheets in sri lanka: a preliminary study in western province. Vidyodaya J. Hum. Soc. Sci. **07**(01), 24–43 (2022). https://doi.org/10.31357/fhss/vjhss.v07i01.02

Ferronato, N., et al.: A review of plastic waste circular actions in seven developing countries to achieve sustainable development goals. In Waste Manage. Res. **42**(6), 436–458. SAGE Publications Ltd (2024). https://doi.org/10.1177/0734242X231188664

Hopewell, J., Dvorak, R., Kosior, E.: Plastics recycling: Challenges and opportunities. Philosophic. Transa. Royal Soc. B: Biolog. Sci. **364**(1526), 2115–2126. Royal Society (2010). https://doi.org/10.1098/rstb.2008.0311

Jambeck, J.R., et al.: Plastic waste inputs from land into the ocean. Science **347**(6223), 768–771 (2015). https://doi.org/10.1126/science.1260352

Jayasinghe, R.R., et al.: Exploring the plastic collection and recycling trends in Sri Lanka. Recycling **8**(6) (2023). https://doi.org/10.3390/recycling8060094

Jayasinghe, R.R., et al.: Analysis of the community behavioural patterns in management of household plastic waste due to the COVID-19 pandemic in Sri Lanka. Case Studies in Chem. Environ. Eng. **6** (2022). https://doi.org/10.1016/j.cscee.2022.100246

Kafa, N., Jaegler, A., Sarkis, J.: Harnessing corporate sustainability decision-making complexity: a field study of complementary approaches. Sustainability (Switzerland) **12**(24), 1–23 (2020). https://doi.org/10.3390/su122410584

Kibria, M.G., Masuk, N.I., Safayet, R., Nguyen, H.Q., Mourshed, M.: Plastic waste: challenges and opportunities to mitigate pollution and effective management. Int. J. Environ. Res. **17**(1). Springer Science and Business Media Deutschland GmbH (2023). https://doi.org/10.1007/s41742-023-00507-z

Law, K.L., Starr, N., Siegler, T.R., Jambeck, J.R., Mallos, N.J., Leonard, G.H.: The United States contribution of plastic waste to land and ocean. In Sci. Adv. **6** (2020). http://advances.sciencemag.org/

National Plastic Waste Inventory for Sri Lanka (2024)

Plastic waste management country situation report-Sri Lanka. (2020). https://www.unep.org/interactive/beat-plastic-pollution/

Rahimi, A., García, J.M.: Chemical recycling of waste plastics for new materials production. Nat. Rev. Chem. **1**(6), 0046 (2017). https://doi.org/10.1038/s41570-017-0046

Ranatunga, R. R. M. K. P., et al.: ranking sri lanka among the world's top mismanaged waste polluters: does model data change the story? Sustainability (Switzerland), **15**(3) (2023).. https://doi.org/10.3390/su15032687

Samarasinghe, K., Pawan Kumar, S., Visvanathan, C.: Evaluation of circular economy potential of plastic waste in Sri Lanka. Environ. Qual. Manage. **31**(1), 99–107 (2021). https://doi.org/10.1002/tqem.21732

Tsuchimoto, I., Kajikawa, Y.: Recycling of plastic waste: a systematic review using bibliometric analysis. Sustainability (Switzerland) **14**(24) (2022). MDPI. https://doi.org/10.3390/su142416340

Biosorption of Methylene Blue in Activated Sludge

S. M. M. Siyam, G. H. T. C. Madumal, W. W. W. T. M. Fernando, and R. P. T. P. Keerthisinghe(✉)

Department of Chemical and Process Engineering, University of Moratuwa, Katubedda, Sri Lanka
tharushik@uom.lk

Abstract. Methylene Blue (MB) is one of the most extensively used synthetic dyes, which poses significant environmental hazards and health concerns due to its persistent nature and toxicity. Thus, removing MB from industrial effluents is crucial. This study focused on the fate of MB in activated sludge using laboratory-developed activated sludge reactors. The activated sludge bioreactors could remove >90 and >75% of MB at 10 and 20 ppm of MB concentrations, respectively. Further experiments were conducted to elucidate the mechanism of MB removal in activated sludge, which showed that biosorption plays a vital role in removing MB in activated sludge, with no clear evidence for biodegradation. Langmuir, Freundlich's, and Redlich-Peterson adsorption models were applied to predict the biosorption behaviour, and the results showed a strong correlation between the observed and predicted values ($R^2 = 0.99$). Fourier transform infrared (FTIR) spectroscopy analysis further confirmed the biosorption of MB in activated sludge. This research provides important insights into the optimization of the activated sludge process for dye removal and, thus, forms a framework for enhancing the efficiency of wastewater treatment systems. This investigation allows us to broaden our current knowledge regarding the interaction between synthetic dyes and biological processes for wastewater treatment and supports the development of better and more eco-friendly practices in managing wastewater.

Keyword: Methylene blue · activated sludge · adsorption · wastewater

1 Introduction

For the preservation of the quality of natural water resources and to maintain the safeguard of environmental and human health, wastewater treatment is essential. The persistence and hazardous nature of the contaminants, particularly those involving synthetic dyes in industrial activities, pose a serious threat to aquatic environments. Furthermore, such contaminants create significant challenges to the wastewater treatment process. Among the various synthetic dyes, methylene blue (MB) is of particular concern. MB is extensively used in the textile, printing, and pharmaceutical industries due to its vivid color and chemical stability (Fito, Abrham and Angassa, 2020a). The concentration of methylene blue found in textile industry effluent can vary widely. Typically, untreated

R. Dissanayake et al. (Eds.): ICRES 2025, PTT, pp. 67–78, 2025.
https://doi.org/10.1007/978-981-96-9551-5_6

effluent from textile industries can have methylene blue concentrations ranging from 10 to 100 mg/L (Fito, Abrham and Angassa, 2020b).

However, the release of MB into aquatic ecosystems can have severe environmental repercussions, including toxicity to marine life, disruption of photosynthetic processes, and potential bioaccumulation in the food chain. For instance, the presence of MB in water bodies has been shown to disrupt metabolic functions in aquatic organisms, leading to oxidative stress, impaired growth, and a reduction in biodiversity (Aghalari et al., 2020; Fito, Abrham and Angassa, 2020a). Therefore, the efficient removal of MB from industrial effluents is paramount to prevent its harmful environmental impacts. One promising approach for the removal of MB from wastewater is the activated sludge Process, a widely employed biological treatment method in wastewater treatment plants (Fito, Abrham and Angassa, 2020b) The specific efficacy of activated sludge in removing synthetic dyes like MB remains less certain, as these dyes are often resistant to conventional biological degradation due to their complex aromatic structures and synthetic origins (Andreas, 2019).

This research aims to comprehensively investigate the removal of MB in the activated sludge, focusing on elucidating the removal kinetics and the underlying mechanisms. Understanding kinetics will enable the assessment of the treatment efficiency and the identification of factors that influence MB removal rates. By comparing the removal performance under varying conditions, such as the presence or absence of microbial activity, this study aims to determine the dominant pathway responsible for MB removal. Fits the data into a series describing the removal kinetics of MB in the activated sludge process based on the experimental data. The findings from this research will provide valuable insights into the suitability of the activated sludge process for the removal of MB from wastewater. Ultimately, this research aims to advance our understanding of the interactions between synthetic dyes and biological treatment processes, thereby contributing to the broader goal of enhancing the sustainability and efficacy of wastewater management practices.

2 Materials and Methods

2.1 Bioreactor Setup and MB Exposure

Three sets of bioreactors were established for the control and MB-fed (10 and 20 ppm) bioreactors. Activated sludge obtained from an aeration tank of a municipal wastewater treatment facility in Western Province, Sri Lanka, was used as the inoculum for the bioreactors. Synthetic wastewater, composed of 1.8 g/L of glucose (equivalent to 2 g chemical oxygen demand [COD]/*l*), a mineral medium solution contained (g/L) K_2HPO_4 (0.6), KH_2PO_4 (0.34), $CaCl_2$ (0.053), $MgSO_4.7H_2O$ (0.27), $FeSO_4 .7H_2O$ (0.001), NH_4Cl (0.03), and trace minerals were used as the growth media of the activated sludge microbiome (Keerthisinghe et al., 2019)). MB-fed bioreactors were fed with 10 and 20 ppm of MB while the control reactor was maintained with no MB addition. The pressure and the temperature of the reactors were maintained at atmospheric conditions. Air was supplied through a 0.22 μm filter continuously for aeration and proper circulation to avoid the dead zone at the bottom of the bioreactors. 15 ml mixed liquor samples were withdrawn from the reactors every 30 min, centrifuged and the supernatant was used to determine the concentration of MB.

2.2 Batch Experiment to Identify MB Removal Pathways

Batch experiments were conducted to determine the removal routes of MB in activated sludge. Four batch test conditions were established with different inoculums, as detailed in Table 1. Active sludge obtained directly from the wastewater treatment facility was washed and used as active sludge for reactors 1 and 4. For reactor 2, sludge was inactivated by autoclaving at 121 C for 20 min (Nguyen, Nghiem and Oh, 2018). The exposure concentrations of MB were selected as 20 mg/L for reactors 1, 2, and 3. 15 ml mixed liquor samples were withdrawn from the reactors every 2 h, centrifuged and the supernatant was used to determine the concentration of MB.

Table 1. Batch test conditions for MB removal experiments.

Reactor number	Synthetic Wastewater	Active sludge	Inactive sludge	MB
1	+	+	–	+
2	+	–	+	+
3	+	–	–	+
4	+	+	–	–

2.3 Analytical Methods

The surface functional groups of the waste-activated sludge were detected by Fourier transform infrared (FT-IR) spectroscopy (SHIMADZU FT-IR 8400). The spectra were recorded from 4,000 to 400 cm^{-1}. Activated sludge samples exposed to 10 and 20 ppm and a control sample were analyzed using FTIR. The KBr method was used in the FTIR since it is the most suitable method for analyzing the functional groups of activated sludge (Kowalski et al., 2018). The treated sludge samples were thoroughly washed using deionized water before the FTIR analysis. Multiple washing cycles were included in this method to remove loosely bound or unabsorbed MB molecules. This method primarily ensures that the FTIR spectra only represent the chemically absorbed MB. The FTIR spectra of washed sludge samples were compared to the pure MB solution spectrum found in the literature to further confirm whether the observed peaks were due to the adsorption or residual MB. A UV–UV-visible spectrophotometer (UV 1800, SHIMADZU) was used to analyze the concentration of MB in the reactors. A suitable wavelength of 664 nm was determined through a spectrum run using the UV-visible spectrophotometer to detect the absorbance of MB.

2.4 Modeling the Adsorption Kinetics of MB in Activated Sludge

Activated sludge was exposed to MB concentrations of 10, 20, 50, 70, and 100 ppm for the data collection for the modeling of adsorption kinetics. The experiment utilized the same equipment setup as previously described in Section 2.1. To quantify MB uptake, the wet weight of the biomass was measured, allowing calculation of the MB concentration per gram of biomass. The experimental data obtained were fitted into three common adsorption kinetic models, which include Langmuir, Freundlich, and Redlich-Peterson adsorption models. (Montanher, Oliveira, and Rollemberg 2005).

3 Results and Discussion

3.1 Removal of MB in activated sludge

Three sets of reactors; control with no MB exposure and with 10 and 20 ppm of MB exposure, were maintained to investigate whether activated sludge can remove the MB *in vitro*. The results indicated that the activated sludge could efficiently remove MB from the wastewater (Fig. 1).

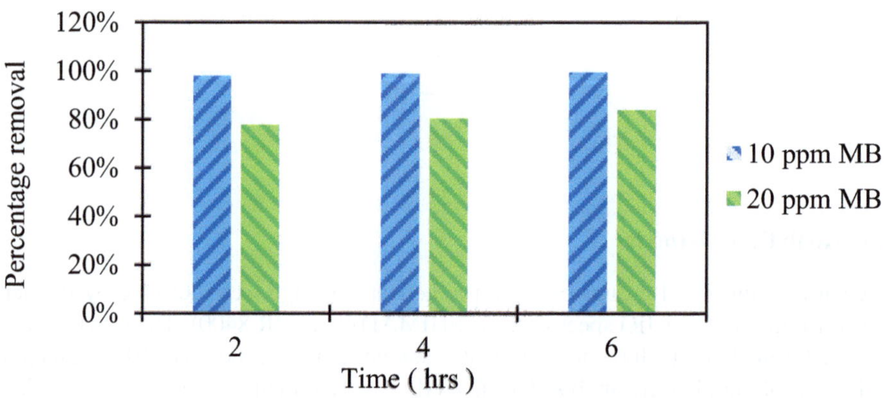

Fig. 1. Percentage removal of MB in activated sludge

The bioreactor with 20 ppm MB could remove 78%, 83%, and 84% of MB from wastewater in 3, 4, and 6 h, respectively, based on the single measurement obtained. Furthermore, the bioreactor with 10 ppm MB could remove 99% of MB from wastewater in 6 h. While indicating efficient removal, these results demonstrated an increase in removal percentage with time, based on the results obtained from a single measurement. The removal of MB can be due to different reasons, including biodegradation by activated sludge microbiome, biosorption, or abiotic removal. These results suggest that the activated sludge system effectively removes MB and extending the MB-activated sludge contact time enhances the overall removal efficiency.

3.2 Mechanism of MB Removal in Activated Sludge

After confirming the significant removal of MB in activated sludge, the possible removal routes were investigated with bioreactors under different conditions. One set was inoculated with active sludge, and the MB removal is due to possible biodegradation, biosorption, and abiotic removals, including vaporization. Another set was inoculated with inactive sludge, where the removal through biodegradation was eliminated. Abiotic control with no sludge will show the removals due to abiotic mechanisms. The measurement of MB concentration in the liquid phase of the reactors over a specific period indicated significant MB removal in the reactors with both active and inactive sludge, according to Fig. 2.

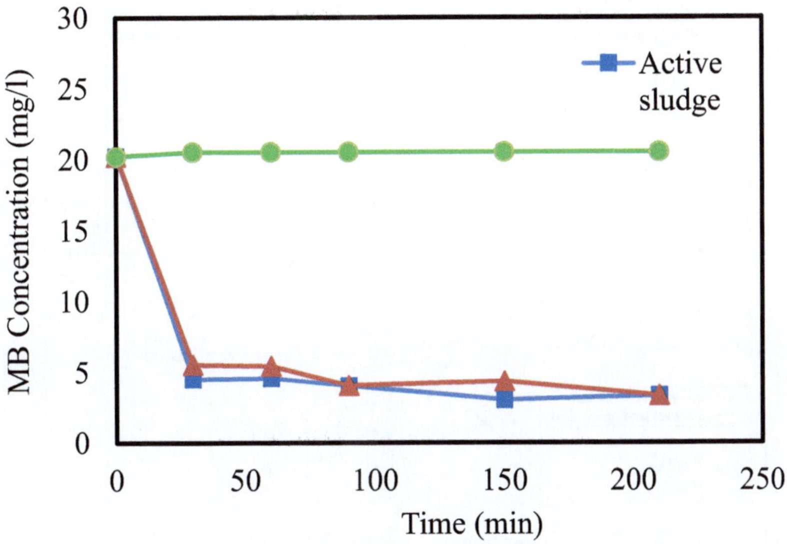

Fig. 2. Variation of MB concentration with time in three sets of reactors with active sludge, inactive sludge and abiotic control.

In the active sludge reactor, a rapid reduction in MB concentration was observed during the initial 30 min, with levels decreasing from ~20 mg/L to 4.5 mg/L. Following the initial 30 min, the rate of MB reduction markedly slowed, with only a further decrease of approximately 1.2 mg/L observed over the next 180 min, bringing the total duration to 210 minutes. In the inactive sludge reactor, the MB concentration similarly exhibited a rapid initial decrease from ~20 mg/L to 5.5 mg/L within the first 30 min. These results are based on a single measurement at each time point. However, there was no significant difference between the MB concentration in both reactors with active and inactive sludge after around 210 min. This observation suggests that biosorption of MB is the principal mechanism in play, and there is no evident biodegradation. In contrast, the third reactor, which contained only wastewater and MB without sludge, showed no observable reduction in MB concentration, confirming the absence of significant abiotic removal pathways. These findings suggest that both active and inactive sludge effectively contributes to the removal of MB from wastewater, highlighting the critical role of sludge in facilitating dye reduction through biosorption.

3.3 Adsorption Kinetic Modeling

Adsorption isotherms provide insights into the sludge's biosorption capacity. Fitting experimental data into adsorption isotherms helps evaluate the efficiency and interactions between the activated sludge and MB. The experimental results were fitted to three major adsorption isotherms: Langmuir, Freundlich, and RedLich-Peterson models (Fig. 3).

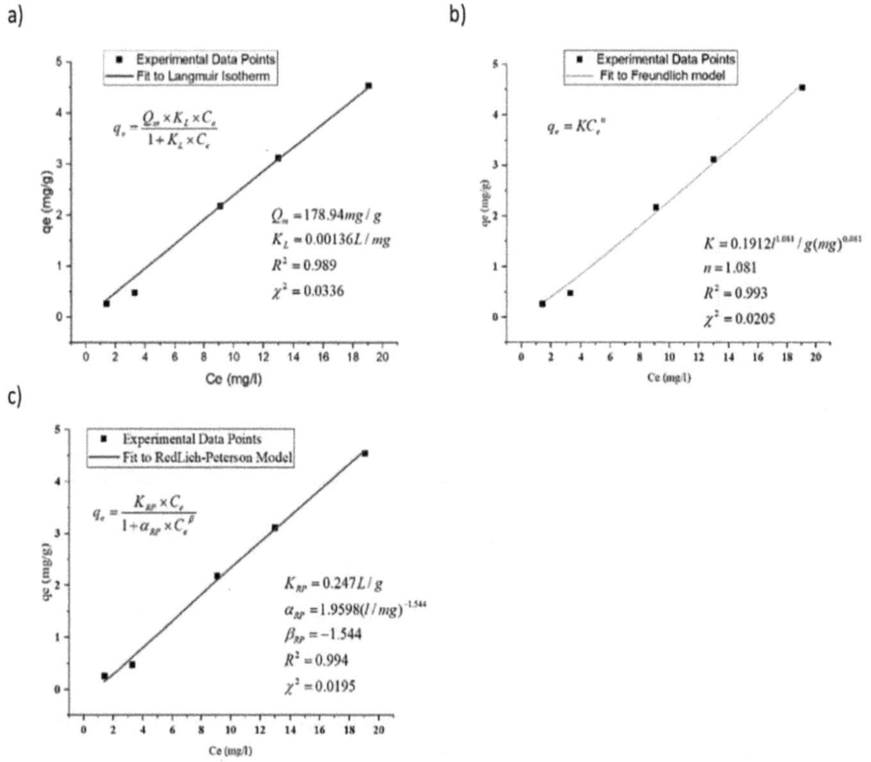

Fig. 3. Experimental data points fit into the a) Langmuir isotherm, b) Freundlich model, and c) Redlich-Peterson model.

3.3.1 Langmuir Isotherm

The Langmuir Isotherm models are used to describe the adsorption behavior of MB onto the sludge surface. As per this model, the adsorption involves the formation of a single layer of adsorbate (MB) on the adsorbent surface (sludge surface). It assumes that the adsorbent surface consists of identical adsorption sites. Equation 1 describes the

Langmuir isotherm (Auta and Hameed, 2014).

$$q_e = \frac{Q_m K_L C_e}{1 + K_L C_e} \quad (1)$$

q_e – Amount of MB bound per gram of biomass at equilibrium (mg/g)

C_e – Residual MB concentration in solution after binding (mg/l).

Q_m – Maximum amount of MB per unit weight of biomass to form a complete monolayer on the surface-bound at high Ce K_L – Constant related to the affinity of the binding sites.

According to Fig. 3a, the maximum adsorption capacity, Q_m was determined to be 178.94 mg/g, indicating a high potential for MB uptake by the active sludge. The Langmuir constant, K_L, which relates to the affinity of the biosorbent for the adsorbate, was found to be 0.00136 L/mg. The coefficient of determination ($R^2 = 0.989$) signifies an excellent fit of the data to the Langmuir model, suggesting that the adsorption of MB onto active sludge follows a highly predictable pattern as described by the Langmuir isotherm.

3.3.2 Freundlich Model

The Freundlich model is an empirical equation based on adsorption on a heterogeneous surface relating to the fact that binding sites are not equivalent or independent. Equation 2 describes the Freundlich model for the adsorption of MB in sludge (Dutta et al., 2021).

$$q_e = KC_n \quad (2)$$

K – Freundlich parameter related to adsorption capacity.

n – Freundlich parameter related to adsorption intensity.

Unlike the Langmuir model, the Freundlich isotherm is based on multilayer adsorption over heterogeneous surfaces, suggesting that the adsorption sites have varying affinities for MB. From the model, the Freundlich constant $K = 0.1912$ represents the adsorption capacity, while the value of $n = 1.081$ indicates the intensity of the adsorption process (Fig. 3b). Since n is greater than 1, this implies that the adsorption is favorable and becomes more efficient as the concentration of MB increases. The fit of the experimental data to the Freundlich isotherm is confirmed by the high correlation coefficient $R^2 = 0.993$, which indicates excellent alignment between the observed and predicted values.

3.3.3 Redlich-Peterson Model

The Redlich-Peterson model indicates a linear dependence on concentration in the numerator and an exponential function in the denominator which has been proposed to improve the fit by the Langmuir and Freundlich models (Auta and Hameed, 2014; Peterson, 1945) shown in Eq. 3.

$$q_e = \frac{K_{RP} C_e}{1 + \alpha_{RP} C_e^{\beta}} \quad (3)$$

K_{RP}, α_{RP}, β – Redlich-Peterson parameters.

β lies between 0 and 1. Further, when β becomes 1, the model is converted to Langmuir isotherm.

In this case, the constants obtained from the model fitting are as follows: the adsorption capacity constant $K_{RP} = 0.247$ $\beta_{RP} = -1.544$ which reflects the extent of non-linearity in the adsorption process (Fig. 3c). The Redlich-Peterson model is flexible, and when β_{RP} approaches 1, the model behaves like the Langmuir isotherm, but in this case, the value suggests significant deviation from ideal monolayer adsorption. The high correlation coefficient $R^2 = 0.994$ indicates an almost perfect fit between the experimental data and the model, demonstrating that the Redlich-Peterson isotherm is highly effective in capturing the behavior of MB biosorption onto active sludge.

3.4 FTIR Analysis of Active Sludge

The FTIR analysis and spectra comparison of control (no MB) activated sludge exposed to MB revealed the active groups in activated sludge, MB, and further confirmed the adsorption of MB into activated sludge (Fig. 4). The peak at 3411 cm^{-1} is related to the OH bond and the peak near the wavelength of 2900 cm^{-1} is for CH and CH_2 aliphatic stretching groups. The C = O stretching group is also visible in this spectroscopy due to the peak at 1645 cm^{-1} wavelength. C = N or N-H also can be found due to the presence of the peak at the wavelength of 1541 cm^{-1}. The peak near 1386 cm^{-1} is due to the CH and CH_2 aliphatic bending groups. The peak at 1041 cm^{-1} is due to the C-O bond stretching. All these mentioned wavelengths corresponding to the bonds are mostly common in a sample of activated sludge due to the presence of microorganisms, however, the value of absorbance can vary due to the structures of each microorganism (Kowalski et al., 2018; Mecozzi and Sturchio, 2017).

According to Fig. 4b, there is a shift between the two graphs in the absorbance axis. All the bands showed a higher absorbance in the 10 ppm exposed activated sludge than the activated sludge sample that had no MB. The peak at 1049 cm^{-1} is higher in the activated sludge sample with MB than in the sample without MB. This is due to interactions between MB's functional groups (such as sulfonic or aromatic groups) and the sludge matrix, indicating the adsorption of MB. The 1385 cm^{-1} peak can be seen in both graphs, although the activated sludge sample with MB has a higher peak than the other sample that does not contain MB. This difference is due to the aromatic methyl groups of methylene blue interacting with the sludge. The 1541 cm^{-1} peak corresponds to N-H bending or interactions between MB's amine groups and the active sludge. In conclusion, differences in the absorbance are due to the binding of MB with the active sites of activated sludge. This aligns with expectations of MB adsorption driven by chemical interactions like π-π stacking, electrostatic forces, and hydrogen bonding (El-Sharkawy, Soliman and Al-Amer, 2007; Kowalski et al., 2018; Mecozzi and Sturchio, 2017).

A comparison of the 20 ppm MB exposed active sludge sample with the control without MB in Fig. 4c revealed the same pattern as the 10 ppm exposed sample. Therefore, it is evident that MB is absorbed into the active sites of the active sludge. Furthermore, Fig. 4d indicated that the absorbance of every band corresponding to the bonds of the 20-ppm activated sludge sample is higher than the 10-ppm activated sludge sample. The peak at 1049 cm^{-1} is higher in the activated sludge sample with 20 ppm MB than the sample with 10 ppm MB due to the higher interactions with activated sludge at higher MB concentrations. The 1385 cm^{-1} peak can be seen in both graphs, but the activated sludge sample with 20 ppm MB has a higher peak than the 10 ppm MB. This difference is due to the more and more aromatic methyl groups of MB interacting with the sludge. The 1541 cm^{-1} peak corresponds to N-H bending or interactions between MB's amine groups and the active sludge also has a higher absorption value in the 20 ppm sample than the 10 ppm sample. Therefore, there is a higher absorption of MB into the active sites of activated sludge when we are increasing the concentration of MB (El-Sharkawy, Soliman and Al-Amer, 2007; Kowalski et al., 2018; Mecozzi and Sturchio, 2017).

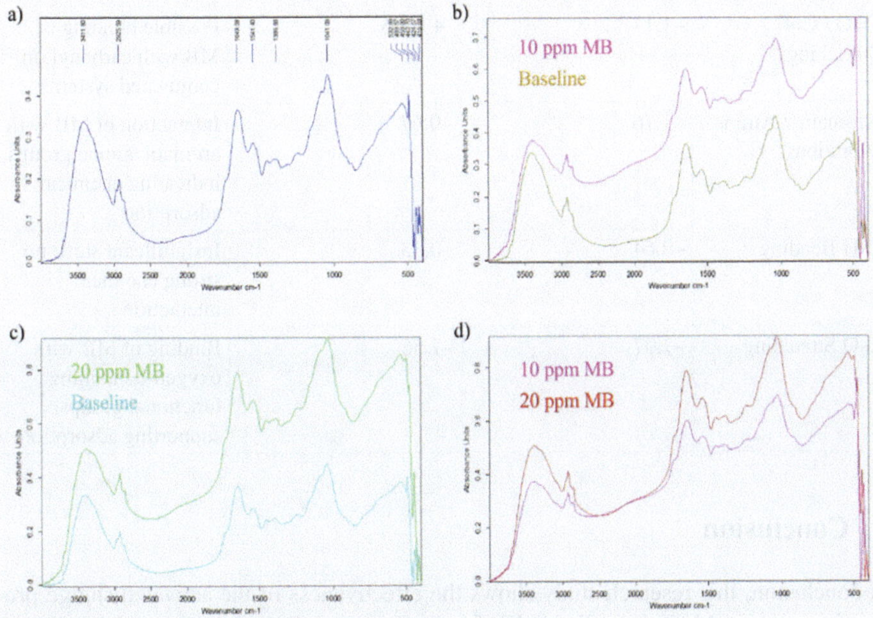

Fig. 4. FTIR analysis of activated sludge with a) no MB (base line), b)10 ppm MB with baseline, c) 20 ppm MB with baseline and d) 10 and 20 ppm MB

Comparison of the x-axis of FTIR spectra of MB exposed and control with no MB further revealed the adsorption sites of MB in activated sludge, which are summarized in Table 2. In the 20 ppm MB exposed active sludge sample, the magnitude of peak shift is more noticeable than the 10 ppm MB with active sludge. O-H /N-H stretching peak shifted by 17.24 cm^{-1} at 20 ppm indicating a stronger interaction at higher MB concentrations. This analysis shows that absorption capacity increases with the MB concentration. The FTIR analysis shows that MB Absorption happens via hydrogen bonding and electrostatic interaction between MB functional groups and the hydroxyl, amine, and aromatic groups present in the sludge matrix.

Table 2. Summary of the shift of wavelengths corresponding to bonds of 10 ppm and 20 ppm activated sludge sample compared to the base.

Functional Group	Shift of 10 ppm from the base (cm^{-1})	Shift of 20 ppm from the base (cm^{-1})	Indication
Hydroxyl (-OH) / Amine (-NH)	–11.87	–17.24	Interaction between MB and hydroxyl or amine groups on the sludge surface
Aliphatic C-H Stretching	–0.69	–0.29	Minimal interaction; suggests weak or non-specific binding
C=O / C=C Stretching	–1.17	+ 0.06	Possible bonding of MB with carbonyl or conjugated systems
Aromatic / Amine Vibrations	–1.76	–0.67	Interaction of MB with aromatic/amine groups, indicating chemical adsorption
C-H Bending	–0.64	–0.33	Insignificant shift; no strong chemical interaction
C-O Stretching	–7.07	–7.56	Binding of MB with oxygen-containing functional groups, supporting adsorption

4 Conclusion

In conclusion, this research study shows the effectiveness of the activated sludge process in removing Methylene Blue (MB) from wastewater. Research successfully showed that the MB removal process is mainly through biosorption, and there was no sufficient evidence to show the biodegradation. In the first 30 min, a rapid reduction of MB concentration was demonstrated in the reactors with active sludge and inactive sludge, while no significant difference in the MB concentration was observed between the two sets of reactors. This scenario highlights that biosorption was the primary mechanism during this phase.

FTIR analysis validated MB adsorption onto activated sludge by examining functional group interactions. Peaks corresponding to hydroxyl (-OH), amine (-NH), carbonyl (C=O), and aromatic groups showed shifts in wavenumber and absorbance, confirming chemical binding. Increased MB concentration (20 ppm vs. 10 ppm) resulted in higher absorbance and more pronounced shifts, indicating stronger interactions at higher concentrations. Key mechanisms include hydrogen bonding, electrostatic forces, and π-π interactions. Minimal shifts in aliphatic C-H stretching suggest weak non-specific binding.

These results confirm MB adsorption is concentration-dependent, highlighting the role of activated sludge's functional groups in effectively removing MB from aqueous solutions. The findings of this research highlight biosorption is important in the MB removal process in activated sludge. These results are important for enhancing the wastewater treatment processes. Finally, this research provides a strong foundation for creating better sludge-based treatment methods, particularly the anaerobic digestion of MB-activated sludge complex. These advanced methods are important to keep water clean and they ensure that public health is protected.

References

Aghalari, Z., et al.: Effectiveness of wastewater treatment systems in removing microbial agents: a systematic review. Globalization Health. BioMed Central Ltd. (2020). https://doi.org/10.1186/s12992-020-0546-y

Andreas, S.: Wastewater treatment processes—techniques, technologies, challenges faced, and alternative solutions. In: Soft Computing Techniques in Solid Waste and Wastewater Management, pp.35–53 (2020). https://doi.org/10.1016/B978-0-12-824463-0.00004-5

Auta, M., Hameed, B.H.: Chitosan-clay composite as highly effective and low-cost adsorbent for batch and fixed-bed adsorption of methylene blue. Chem. Eng. J. **237**, 352–361 (2014). https://doi.org/10.1016/j.cej.2013.09.066

Das, R., et al.: Remediation of antiseptic components in wastewater by photocatalysis using TiO2 nanoparticles. Ind. Eng. Chem. Res. **53**(8), 3012–3020 (2014). https://doi.org/10.1021/ie403817z

Dutta, S., et al.: Recent advances on the removal of dyes from wastewater using various adsorbents: a critical review. Mater. Adv. Royal Soc. Chem. **2**(14), 4497–4531 (2021). https://doi.org/10.1039/d1ma00354b

El-Sharkawy, E.A., Soliman, A.Y., Al-Amer, K.M.: Comparative study for the removal of methylene blue via adsorption and photocatalytic degradation. J. Colloid Interface Sci. **310**(2), 498–508 (2007). https://doi.org/10.1016/j.jcis.2007.02.013

Fito, J., Abrham, S., Angassa, K.: Adsorption of methylene blue from textile industrial wastewater onto activated carbon of parthenium hysterophorus. Int. J. Environ. Res. **14**(5), 501–511 (2020). https://doi.org/10.1007/s41742-020-00273-2

Gobi, K., Mashitah, M.D., Vadivelu, V.M.: Adsorptive removal of methylene blue using novel adsorbent from palm oil mill effluent waste activated sludge: equilibrium, thermodynamics and kinetic studies. Chem. Eng. J. **171**(3), 1246–1252 (2011). https://doi.org/10.1016/j.cej.2011.05.036

Keerthisinghe, T.P., et al.: Antiseptic chlorhexidine in activated sludge: biosorption, antimicrobial susceptibility, and alteration of community structure. J. Environ. Manage. **237**, 629–635 (2019). https://doi.org/10.1016/j.jenvman.2019.02.043

Kowalski, M., et al.: Qualitative analysis of activated sludge using FT-IR technique. Chem. Pap. **72**(11), 2699–2706 (2018). https://doi.org/10.1007/s11696-018-0514-7

Mecozzi, M., Sturchio, E.: Computer assisted examination of infrared and near infrared spectra to assess structural and molecular changes in biological samples exposed to pollutants: a case of study. J. Imaging **3**(1) (2017). https://doi.org/10.3390/jimaging3010011

Montanher, S.F., Oliveira, E.A., Rollemberg, M.C.: Removal of metal ions from aqueous solutions by sorption onto rice bran. J. Hazard. Mater. **117**(2–3), 207–211 (2005). https://doi.org/10.1016/J.JHAZMAT.2004.09.015

Nguyen, L.N., Nghiem, L.D., Oh, S.: Aerobic biotransformation of the antibiotic ciprofloxacin by Bradyrhizobium sp. isolated from activated sludge. Chemosphere **211**, 600–607 (2018). https://doi.org/10.1016/j.chessmosphere.2018.08.004
Oladoye, P.O., et al.: Methylene blue dye: toxicity and potential elimination technology from wastewater'. Results Eng. **16**, 100678 (2022). https://doi.org/10.1016/J.RINENG.2022.100678

Partial Replacement of Cement by Ceramic and Porcelain Waste in Concrete Interlocking Paving Block Production for Sustainability

H. M. B. Krishantha and A. M. L. N. Gunathilaka(✉)

Department of Civil Engineering, The Open University of Sri Lanka, Nawala, Sri Lanka
amlgu@ou.ac.lk

Abstract. Management of construction waste, reduction of CO_2 emissions and reduction of energy consumption in cement production are some sustainability goals relevant to the construction industry in the present context. Various studies have highlighted the possibility of using ceramic and porcelain waste as a Supplementary Cementous Material (SCM) for concrete. Therefore, studying the possibility of adopting it for concrete interlocking paving blocks would be another initiative for achieving sustainability goals. Due to the desired characteristics of concrete interlocking paving blocks, it is widely used as a paving material in Sri Lanka. Therefore, this study is focused on studying the possibility of using ceramic and porcelain waste as a SCM for their production and identifying the optimum replacement percentages in this regard. According to the experimental study carried out in this regard relevant to strength class 3 of paving blocks, it was found that replacing cement with 15% of ceramic waste in powder form (by weight) as the optimum replacement percentage since further increment of replacement percentage would lead to dissatisfaction of strength of characterises as per local standard of interlocking paving blocks. It was further noted that especially the early age compressive strength of interlocking paving blocks reduced when it is used ceramic waste as a SCM even though it does not matter for the compliance of interlocking paving blocks for local standards.

Keywords: Concrete interlocking paving blocks · Supplementary Cementitious Material · Sustainability · Ceramic and porcelain waste · Compressive strength

1 Introduction

In Sri Lanka, the management of ceramic and porcelain waste has become an issue due to the growing production and consumption of these materials. In the ceramic tile production industry, during the after-kilning process, tile pieces and unusable tile squares are discarded as waste, leading to the generation of about 3 tons of garbage per day at certain factories in Sri Lanka (Liyanage and Ranasinghe, 2019). Additionally, ceramic and porcelain waste are generated from the removal/demolition of floors and walls with tiles during renovations/demolition for new construction and accidental breakages of tiles and porcelain items during construction and other usages. Presently, there is no

R. Dissanayake et al. (Eds.): ICRES 2025, PTT, pp. 79–87, 2025.
https://doi.org/10.1007/978-981-96-9551-5_7

defined process to recycle or reuse this waste and common usage is the utilization of these materials for landfilling with other construction waste. However, various studies (Raval et al, 2013, Bhargav, 2020, Rani, 2016, Ghonaim et al., 2023, Reig et al, 2015, Liyanage & Ranasinghe, 2019) have highlighted that ceramic and porcelain waste have the potential to be used as a partial replacement ingredient for cement in concrete and concrete related products. Figure 1 shows a stock of ceramic waste pilled at a site.

Fig. 1. A stock of ceramic waste pilled at a site

Concrete interlocking paving blocks have become a popular paving material in the country due to their considerably high rainwater infiltration, aesthetically pleasing appearance, competitive market price, and relatively easier replacement/repairing ability, etc. However, with the escalation of prices of raw materials such as cement and aggregates especially with the financial crises of the country, studying possibilities of using alternative low-cost raw materials for paving block production has become a timely need. Further, the partial replacement of cement by other alternative materials has become essential in sustainability perspective, since cement production is widely recognized as a high energy consuming and significant CO_2 emitting industry, which leads to global warming. Studies done by Raval et al, 2013, it has indicated that it is possible to partially replace cement by the ceramic waste in powder form up to 30% of Grade 25 concrete. Further, a chemical composition analysis performed of ceramic waste in powder form (see Table 1) has indicated the possibility of considering ceramic waste as a Supplementary Cementitious Material (SCM).

As per ASTM C618–22, 2022, if the percentage of ($SiO_2 + Al_2O_3 + Fe_2O_3$) is greater than or equal to 70% in a material, it can be considered as a Class N SCM. Therefore, there is a potential to use ceramic waste to partially replace cement in concrete. As already mentioned, some previous research studies have highlighted usage of ceramic waste as a SCM for concrete. However, the possibility of using these findings of these studies directly for Concrete interlocking paving blocks is questionable as certain specific requirements and conditions such as maintaining water absorption percentage below 6%,

Table 1. Chemical Composition of Ceramic Watse in powder form (Raval et al, 2013)

Materials	Tile(%)	Cement(%)
SiO2	63.29	20.52
Al2O3	18.29	5.46
Fe2O3	4.32	3.46
CaO	4.46	65.04
MgO	0.72	1.35
P2O5	0.16	0.04
K_2O	2.18	0.68
Na_2O	0.75	0.17
SO3	0.10	2.18
Cl−	0.005	0.002
TiO2	0.61	0.27
SrO2	0.02	0.03
Mn2O3	0.05	0.07
L.O.I	1.61	1.16

usage of course aggregates in the range of 10–12 mm, etc. are applicable for concrete paving blocks.

Accordingly, this study focused on identification of optimum partial replacement percentage of cement by ceramic waste in powder form for Concrete interlocking paving blocks through an experimental study.

2 Methodology

According to the Sri Lankan standard for Concrete interlocking paving blocks (SLS 1425: Part 1, 2011), four (04) strength classes have been specified. The minimum strength class recommended for Concrete interlocking paving blocks subjected to vehicular traffic is strength class 3 and the minimum average compressive strength required for interlocking blocks in that class is 30 N/mm2. Therefore, most of the Concrete interlocking paving blocks produced in the country belong to this strength class other than Concrete interlocking paving blocks produced for the usage of residential requirements. Accordingly, it was decided to focus this study on interlocking blocks belonging to strength class 3 of SLS 1425: Part 1, 2011. Also, SLS 1425: Part 1, 2011 specifies two main types of paving blocks based on the shape of the product. Those two types are Rectangular (R) and Non-Rectangular (NR). Since the rectangular type is a very common and consistent type, it was decided to focus on this study of that type. The standard size for this block is 200 mm (L) x 100 mm (W) x 80 mm (T).

The Design of Experiments (DoE) method was used for the development of mix design for the production of concrete interlocking blocks as a local previous study

(Baskaren and Gopinath, 2013) had indicated that the DoE method would facilitate the achievement of target strength in a more precise manner. Initially, the mix design was developed for the control sample with an assumed target minimum average strength of 30 N/mm2. As per literature (Alengaram, 2022), availability of sufficient water should be ensured when a SCM is used for the concrete production. Also, a study carried out by Sampath and Gunathilaka, 2024 has highlighted that water/cement ratio of 0.50 would be the optimum water/cement ratio when Rice Husk Ash (RHA) is used as a SCM for concrete interlocking block production. Accordingly, 0.5 was selected as the water/cement ratio for this study. In addition to control samples, another six (06) mix proportions were selected by changing partial replacement percentages of cement by ceramic waste powder upto 35% by weight. A summary of those mix proportions per 1m3 of concrete is presented in Table 2.

It used locally available Ordinary Portland Cement (OPC) complied with SLS 107: Specification for Ordinary Portland Cement for this study. Well graded Crushed coarse aggregates and fine aggregates (river sand) complied with BS 812: Part 2: 1995 and BS 882:1992 were used for this experimental study. However, the maximum particle size of coarse aggregates was maintained at about 10–12 mm as it was the standard practice of the industry even though SLS 1425: Part 1, 2011 did not specify any such requirements.

Table 2. Summary of mix proportions

Sample type	Ceramic Waste %	Cement (kg)	Ceramic Waste in powder form(kg)	Fine aggregate (kg)	Coarse aggregate (kg)
Control	0	320	0	630	1340
Alt.1	10	288	32	630	1340
Alt.2	15	272	48	630	1340
Alt.3	20	256	64	630	1340
Alt.4	25	240	80	630	1340
Alt.5	30	224	96	630	1340
Alt.5	35	208	112	630	1340

The ceramic waste powder for the study was produced using mix of ceramic tile waste, discarded ceramic sanitaryware, and discarded porcelain products for domestic uses. First, the collected waste was crashed using an industrial compactor and subsequently, those were crushed using a conventional metal crusher. Finally, the outcome was sent through a 75-micron sieve and particles passing through that sieve were taken as the ceramic waste powder for the study. All blocks for the study were produced manually using steel moulds. After the standard curing process (immersing in a water bath immediately after initial setting time) for the required number of days, mechanical and physical testing were performed as necessary.

3 Result and Discussion

The workability of all mixes (control sample and five alternative mixes) was checked through a slump cone test even though local standard does not mention any such requirement. Results of the considered mix proportions are presented in Table 3. As per results of Table 3, it is evident that partial replacement of cement by ceramic waste powder up to 35% is marking very minimal effect on workability.

Table 3. Results of workability test

Sample type	Cement(%)	Ceramic waste Replacement (%)	Slump (mm)
Control	100	0	Zero slump
Alt.1	90	10	3
Alt.2	85	15	2
Alt.3	80	20	Zero slump
Alt.4	75	25	Zero slump
Alt.5	70	30	Zero slump
Alt.5	65	35	2

As per SLS 1425: Part 2, 2011, it is required to check paving blocks through a visual inspection to identify any defects, discolouration, etc. Accordingly, a close visual inspection was carried out, but any defect or variation of colour with changes in mix proportions was not observed.

Normally, when a SCM is introduced to concrete, properties of concrete depend on factors like filler effects, dilution and the reactivity. Hence, strength properties of concrete interlocking paving blocks would depend on the combined effect of these factors. Therefore, results of mechanical and other characteristics of this study should be analysed in this context.

The main mechanical test for the compliance of paving blocks specified in SLS 1425: Part 1, 2011 is the compressive strength test of blocks. Accordingly, this test was performed on 7th day and 28th day of blocks. Results of these tests are presented in Fig. 2. As per Fig. 2, the highest average compressive strength has been reported in samples where cement was replaced by ceramic waste up to 10%. Beyond 10% replacement, the 28-days compressive strength has gradually reduced. However, samples with 15% of ceramic waste have also achieved a strength above 30 N/mm^2. In the case of 7 days strength, almost gradual reduction of strength can be observed with the increase of replacement percentage of cement by ceramic waste. According to literature (FHWA-HIF- 16–001, 2016), the production of concrete (when using OPC) would be taken place by formation of Calcium Silicate Hydrate (SCH) and Calcium Hydroxide (CH). Even though CSH is desirable product, CH is not desirable. However, when a SCM is available the CH produced from OPC would react with pozzolanic materials and produce desired CSH. However, this secondary reaction would take place at slow pace in concrete in early stages. Hence, when cement was partially replaced by ceramic waste powder as a

SCM, amount of SCH production would reduce at early stages, even though it provided pozzolanic materials to react with CH to improve final strength. This would be the probable reason for reduction of early age strength in all samples where ceramic waste powder was introduced. A similar observation had been reported in a study carried out on usage of Rice Husk Ash as a SCM for concrete interlocking blocks by Sampath and Gunathilaka, 2024.However, improvement of the CSH formation from the secondary reaction of CH and pozzolanic materials (received from ceramic waste), would be the probable reasons for improvement of 28 day strength until partial replacement recaches to the 10% as it can observed in Fig. 2.

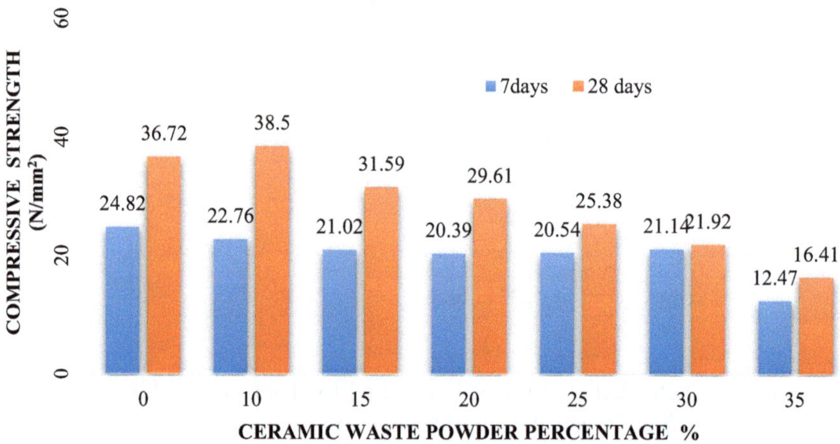

Fig. 2. 7 days and 28 days Compressive strength of concrete interlocking paving blocks

Even though SLS 1425: Part 1, 2011 does not specify a flexural strength test for paving blocks, some other international codes such as IS 15658: 2006 (Indian Standard for paving blocks) specify flexural strength test too. In the real usage of paving blocks, there is a considerably high probability of developing flexural stresses in them due to unevenness of the bedding layer, the presence of pocket of lose soils below blocks, etc. at sites. Hence, samples of control mix and mixes with ceramic waste up to 25% were tested as per the testing procedure specified in IS 15658: 2006. A graphical representation of the specified test setup and photograph of actual test setup is shown in Fig. 3.

The results of the test are presented in Fig. 4. The highest flexural strength (5.41 N/mm^2) was reported in the control sample. All samples having ceramic waste reported a flexural strength below 5 N/mm^2, but those values are still satisfactory even for heavy duty/industrial roads as per Indian Standard. Also, a minimal variation of flexural strength is visible with the variation of partial replacement percentages as it can observed in Fig. 4.

Figure 5 shows the water absorption test results conducted for paving block samples. As SLS 1425: Part 1, 2011 the water absorption percentage of blocks shall be less than 6% for the acceptance. According to test results, all considered samples have shown water absorption percentages below the above value. However, the lowest water absorption percentage was noted with samples having 10% partial replacement percentage.

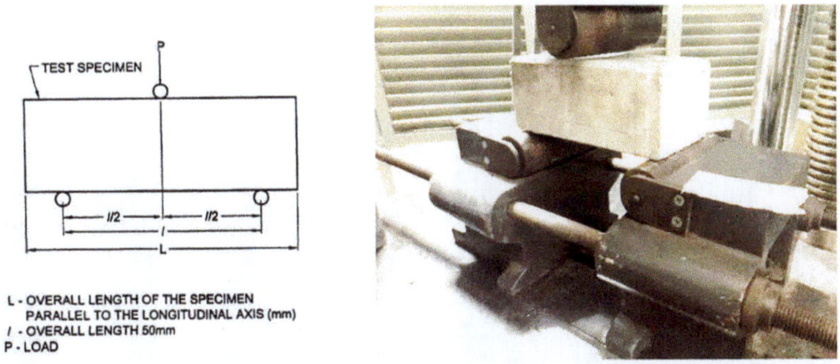

Fig. 3. Test setup for testing flexural strength of concrete interlocking paving blocks

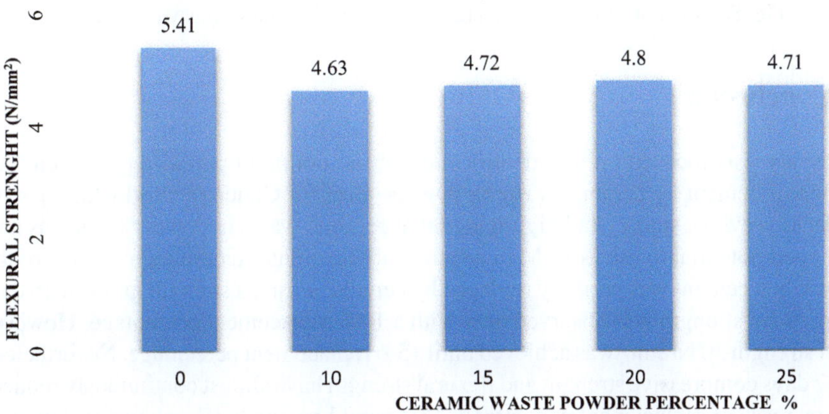

Fig. 4. 28 days flexural strength of concrete interlocking paving blocks

Incidentally, highest 28 days compressive strength was also reported in the same set of samples. However, any consistent pattern in the variation of water absorption with partial replacement percentages cannot be identified.

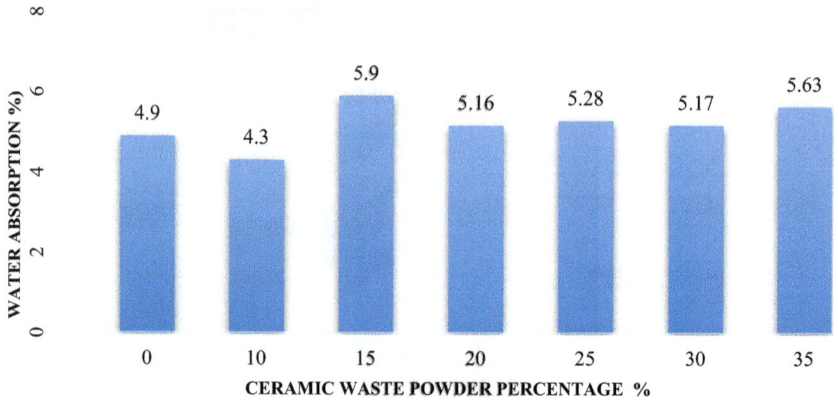

Fig. 5. Water absorption percentages of concrete interlocking paving blocks.

4 Conclusion

This study was focused on the identification of the optimum partial replacement percentage of cement by ceramic waste in powder form for Concrete interlocking paving blocks as previous studies had highlighted that ceramic waste in powder form was having a good potential to use as SCM in concrete applications. According to results of this study, when cement was partially replaced by ceramic waste, a strength gains in 28 days compressive strength was observed only with a 10% replacement percentage. However, target strength 30 N/mm^2 was achieved until 15% replacement percentage. Nevertheless, both 7 days compressive strength and flexural strength have almost continuously reduced with increase of replacement percentage. This would be due to the relatively slow pace of pozzolanic reaction with ingredients of ceramic waste powder during the early stages. Since 7 days strength is not a governing matter unless it is needed to dispatch blocks in a short curing period, and all reported flexural strengths are within satisfactory range for heavy duty/industrial roads as per Indian Standard, usage of ceramic waste as a partial replacement of cement in interlocking paving block production noted to be an acceptable approach in desired strength perspectives of concrete interlocking paving block. In the case of the water absorption requirement, all replacement percentages have satisfied the specified limit of the local standard.

In terms of sustainability and financial perspectives, maximizing the replacement percentage of cement would be the best-case scenario as it helps to reduce both environmental concerns related to cement production and production cost of blocks. Accordingly, replacing cement by 15% of ceramic waste in powder form (by weight) can be considered as the optimum cases, since it was the maximum replacement percentage that has satisfied all specified requirements considered in this study. This proposed partial replacement of cement in concrete interlocking block production would be ultimately beneficial for the society at large in many ways as reduction of usage of natural resources, reduction of emission of greenhouse gases and the reduction of production cost of the product.

References

Liyanage, S., Ranasinghe, D.: Identification of the possibility to reuse the ceramic lazed tile was as a ecofriendly raw materials for manufacturing industry in Sri Lanka. In: 12th International Conference of Faculty of Architecture Research Unit (FARU), December 2019 Conference: University of Moratuwa, Sri Lanka (2019)

Raval, A.D., Patel, I.N., Pitroda, J.: Eco-efficient concretes: use of ceramic powder as a partial replacement of cement. Int. J. Innovative Technol. Exploring Eng. (IJITEE) 3(2) (2013). ISSN: 2278–3075

Bhargav, M.: Experimental investigation to substitute of cement with ceramic tiles powder in concrete. Int. J. Res. Appl. Sci. Eng. Technol. **8**(9), 302–307 (2020). https://doi.org/10.22214/ijraset.2020.31363

Rani, M.S.: A study on ceramic waste powder. Int. J. Civ. Eng. **3**(7), 1–6 (2016). https://doi.org/10.14445/23488352/IJCE-V3I7P101

Ghonaim, S., Morsy, R.: Utilization of ceramic waste material as cement substitution in concrete. Buildings **13**(8), 2067 (2023). https://doi.org/10.3390/buildings13082067

Reig, L., Mas, M.A., Monzó, J., Borrachero, M.V., Payá, J.: Ceramic tiles waste as replacement material in Portland cement. Adv. Cem. Res. **28**(4), 221–232 (2015). https://doi.org/10.1680/adcr.15.00021

ASTM International, Standard specification. For Coal Fly Ash and Raw or Calcined Natural Pozzolana for use in concrete, ASTM C618–22. West Conshohocken, PA

Baskaran, K., Gopinath, K.: Study on applicability of ACI and DoE mix design methods for paving blocks. Annu. Trans. IESL, 127–134 (2013)

Alengaram, U.J.: Effect of non traditional supplementary cementitious materials in concrete. Chapter Metric Overview (2022). https://doi.org/10.5772/intechopen.111900

Sampath, J.A.K., Gunathilaka, A.M.L.N.: Identification of the optimum partial replacement percentage and w/c ratio for the partial replacement of cement by RHA in concrete interlocking paving block production. Modulus **34**(02), 13–17 (2024)

SLS 1425-Part 1:2011 Sri Lanka Standard -Specifications for concrete paving blocks Prat 1: Requirements

SLS 1425-Part 2:2011 Sri Lanka Standard -Specifications for concrete paving blocks Prat 1: Testing

IS 15658:2006 Indian Standards – Pre cast concrete blocks for paving –Specifications

SLS 107: 2015 Sri Lanka Standard – Specification for Ordinary Portland Cement (Fifth revision)

BS 812–1: 1991,– British Standard Testing Aggregates Part 1: Methods for Determination of Particle Size and Shape , 75th Edition (1991)

BS 882:1992 Specification for aggregates from natural sources for concrete (1992)

FHWA-HIF- 16–001, Supplementary Cementous Materials , Tech Brief – US Department of Transport, February (2016)

The Influence of Polypropylene Plastic Aggregates on the Mechanical Properties of Normal and High Strength Concrete

K. G. D. I. H. Thilakarathna(✉), T. Abirami, C. S. Bandara, and P. B. R. Dissanayake

Department of Civil Engineering, Faculty of Engineering, University of Peradeniya, Peradeniya 20400, Sri Lanka
hashanthilakarathna24@gmail.com

Abstract. The global accumulation of plastic waste presents significant environmental challenges, contributing to both land and marine pollution. Polypropylene (PP) has been identified to be a plastic type with higher production rates with the lowest recycling percentages. Since PP is relatively tough, researchers are exploring the use of recycled PP waste plastic as an alternative to concrete aggregate, where it may be a potential solution to both plastic waste mitigation and address the growing scarcity of natural aggregates. This study investigates the effect of 20% of volumetric replacement of natural coarse aggregates (NCA) with recycled Polypropylene plastic aggregates (PPA) in both normal strength concrete (NSC) and high strength concrete (HSC). The experimental studies focused on evaluating workability, compressive strength and splitting tensile strength. The target strength classes for NSC and HSC were C25/30 and C55/60. The results revealed that for both NSC and HSC the workability reduced with the incorporation of PPA, mainly caused by the irregular shape and sharp edges of PPA. Thus, the use of a superplasticizer was essential in achieving the specified standards. However, it was essential to control its dosage and use appropriate compaction techniques to maintain the homogeneity and performance of concrete without any loss in strength due to segregation. With 20% replacement of NCA with PPA, 21% and 30% reductions were observed in compressive strength and tensile strength respectively for NSC. Similarly, for HSC, 25% and 36% respective reductions were present.

Keywords: Polypropylene plastic waste · Natural coarse aggregate · Normal strength concrete · High strength concrete · Workability

1 Introduction

The idea of using recycled plastic as an aggregate in concrete has been explored in several studies, with the aim of reducing both plastic waste and the demand for natural resource extraction. As reported by Abu-Saleem et al. (2021) and several other authors, high-density polyethylene (HDPE), polyethylene terephthalate (PET) and polypropylene (PP) are plastic types with high potential to be incorporated in concrete. Though it has been reported that HDPE and PP are the most produced plastic materials, a very low

R. Dissanayake et al. (Eds.): ICRES 2025, PTT, pp. 88–100, 2025.
https://doi.org/10.1007/978-981-96-9551-5_8

percentage (around 0.6%) of PP waste has been recycled annually. Thus, several authors have evaluated the feasibility of PP waste in concrete in different forms. Yin et al. (2016) as a recycled fiber, Hossain et al. (2019) and Shahjalal et al. (2020) as a virgin fiber, Yang et al. (2015) as a fractional substitution of natural fine aggregate (NFA) are some of the available literatures. However, only a few authors have reported on the effects of partial replacement of natural coarse aggregate (NCA) with recycled Polypropylene plastic aggregates (PPA).

Islam et al. (2022) and Abu-Saleem et al. (2021) have studied the effect of PP on concrete properties of normal strength concrete (NSC) as a partial replacement of NCA and reported that the workability increased with PPA addition. Islam et al. (2022) has observed an improvement in workability with PPA. They have attributed this, to the smooth surfaces of processed PPA and improved rheological properties of concrete by the hydrophobic nature of PP. However, Abu-Saleem (2021) has incorporated PP waste directly into the NSC, as a replacement of NCA and observed a reduction in workability. They have concluded that this reduction had caused by irregular flaky shapes of PP waste as well as the angularity in PPAs, since they were used directly. These PPAs had been produced by just crushing PP plastic waste originated from different sources (ex: plastic chairs and vehicle parts). Further, Kou et al. (2009) had reported that plastic aggregates had no significant influence on workability, where slump values of concrete with plastic aggregates were similar to that of conventional concrete.

Similar to concrete with HDPE and PET, the concrete incorporated with PPA, showed a reduction in both compressive strength and splitting tensile strength with increasing PPA as reported by many including Ozbakkaloglu et al. (2017). This strength reductions are caused by the poor interfacial transition zones (ITZs) between the cement paste and PPA as reported by Abu-Saleem et al. (2021) and Abirami et al. (2024). These weak ITZs have been resulted due to the restrained hydrated reaction near PPA surface due to hydrophobic nature of PP. As observed by Islam et al. (2022), the cube compressive strength has been reduced by 29% and 30% at 7 days, 20% and 21.8% at 28 days for 10% and 20% replacement levels respectively of NCA with PPA. Further, splitting tensile strength (STS) has decreased with PPA by 14.6% and 24.4% compared to concrete without plastic. Similarly, several other authors have also reported that the strength reductions in concrete can be limited to up to 20% replacement of NCA with PPA and desired strength classes can be achieved in concrete incorporated with plastic waste.

With the increasing interest in utilizing plastic waste in concrete, some authors have studied the effects on mechanical properties of high strength concrete (HSC) incorporated with PPA. Faraj et al. (2019) has reported on the mechanical and durability properties of HSC, incorporated with PP plastic waste. They have concluded that the negative impacts of PPA incorporation on the workability of HSC are not significantly present as much as in NSC, due to the effect of a high dosage of superplasticizer on the concrete mix. However, reductions in compressive strength and splitting tensile strength are observed in HSC with PPA. Thus, they have recommended further study on the behavior of HSC with PPA.

Therefore, this study investigates the effects on mechanical properties of NSC and HSC, incorporating PPA as a 20% partial replacement of NCA. The study focuses on evaluating key physical and mechanical properties such as workability, compressive strength and splitting tensile strength (STS).

In total, six concrete batches were mixed, where four batches for NSC and two batches for HSC. For both NSC and HSC, 0% and 20% replacement levels of NCA with PPA, were evaluated. For NSC, two batches with and without the addition of superplasticizer were casted to investigate the effects of admixtures on the properties of concrete with PP. For HSC, both two batches were cast with the addition of a superplasticizer.

1.1 Concrete Material

The study consumed Ordinary Portland Cement (OPC), CEM type-I, as per BS EN 197–1:2003 as the binding material in concrete. The specific gravity of cement was evaluated as 3.15 adhering to the ASTM C188. For NSC, a water-reducing admixture (Plastobuild ES, Finex Engineering, Sri Lanka) meeting the specifications of EN 934–2 and ASTM C494 Types A and F was utilized. The specific gravity of the admixture was 1.20. For HSC, high performance superplasticizer (MasterGlenium SKY 8233, Finex Engineering, Sri Lanka) was incorporated which met the requirements of EN 934–2 and ASTM C494 Type F.

For the natural fine aggregate (NFA), well-graded river sand was used. The maximum size of NFA was 4.75 mm. The specific gravity of NFA was evaluated according to ASTM C125, and it was 2.65. As per ASTM C136, the fineness modulus was 3.1. For the experimental program, two types of coarse aggregates were used, (i) NCA and (ii) PPA. Crushed rock with a specific gravity of 2.85, was used as the NCA. Maximum nominal size of NCA was 12.5 mm and its water absorption was measured as 0.34%.

Fig. 1. (a) Mechanically shredded plastic flakes, (b) Polypropylene plastic aggregates (PPA).

PPAs were the shredded PP plastic wastes (as shown in Fig. 1(a)) extracted from different sources at the Plastic and Polythene Waste Recycling Centre in Gohagoda, Kandy, Sri Lanka. The main sources of PP waste plastic were plastic chairs, electronic wastes and vehicle parts as per the data provided by the waste collectors. The PP plastic waste has been crushed by using a mechanically operated shredder, resulting irregularly shaped, flaky PPAs with sharp edges as shown in Fig. 1b. The maximum nominal size of the PPA was 8 mm where the specific gravity was 0.89 and the water absorption was 0.001%.

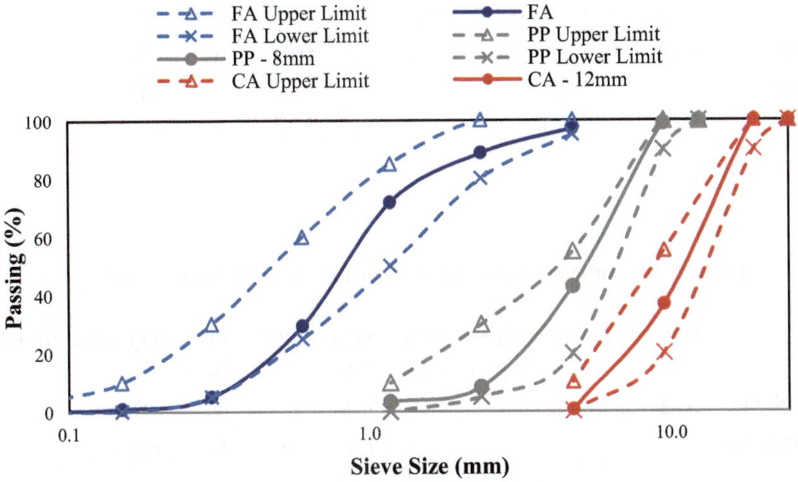

Fig. 2. Particle size distribution of natural fine aggregate (NFA), polypropylene plastic aggregate (PPA) and natural coarse aggregate (NCA) along with aggregate limits specified in ASTM C33.

The sieve analysis was performed as per ASTM C136/C136M-19 and the particle size distribution of NFA, NCA and PPA is given in Fig. 2. Each aggregate stratified the respective limits specified in ASTM C33/C33M – 18. However, there was significant difference in the maximum nominal sizes as well the particle distribution of NCA and PPA. Thus, the incorporated PPA would have act as an intermediate sized aggregate rather than an identical replacement for NCA.

1.2 Mix Proportions

The experimental program aimed at producing both NSC and HSC. For NSC, the target strength class was C25/30. The water-cement ratio was evaluated as 0.55. The target slump was ranged from 75 to 100 mm and mix proportioning was based on ACI 211.1. Four batches of concrete were prepared, including (i) no replacement of NCA without superplasticizer, (ii) no replacement of NCA with superplasticizer, (iii) 20% replacement of NCA without superplasticizer and (iv) 20% replacement of NCA with superplasticizer. A summary of the mix proportions for NSC is given in Table 1. NS-PP20-SP indicates 20% of NCA volume has been replaced with PPA, with an addition of superplasticizer for normal strength concrete mix. The NS-PP0-SP mix was used only for the XRD analysis and no other tests were performed.

For HSC, two batches were prepared, including (i) no replacement of NCA, (ii) 20% replacement of NCA. A summary of the mix proportions for HSC is given in Table 2. Both mixes were added with high range superplasticizer, to achieve the desired workability with less water-cement (W/C) ratio. The W/C ratio was assumed to be 0.32 and the target strength class was C55/60.

Table 1. Mix proportions of concrete per 1 m^3 for normal strength concrete.

Batch	W/C Ratio	Cement (kg)	Water (kg)	NFA (kg)	NCA (kg)	PP %	PP (kg)	Superplasticizer (ml)
NS-PP0	0.55	393	216	886	832	0	0	0.00
NS-PP0-SP								393
NS-PP20					780	20	52	0
NS-PP20-SP								1180

Table 2. Mix proportions of concrete per 1 m^3 for high strength concrete.

Batch	W/C Ratio	Cement (kg)	Water (kg)	NFA (kg)	NCA (kg)	PP %	PP (kg)	Superplasticizer (ml)
HS-PP0-SP	0.32	550	176	784	784	0	0	3025
HS-PP20-SP					731	20	52	3025

1.3 Mixing, Casting and Testing

Before mixing the coarse aggregates were brought to their saturated surface dry conditions. The sand was used in its air-dry conditions. A pan mixture was used for the mixing process and all concrete specimens were cast in the lab. Per batch, five standard-size (100 × 100 × 100 mm) cubes were casted to evaluate the cube compressive strength (f_{cu}) at 7 and 28 days respectively. The cube specimens were prepared by following BS 1881–125:1986. Three cylinders (100 × 200 mm) per batch were cast to determine concrete splitting tensile strength ($f_{ct,sp}$), by following ASTM C192/C192M. With the effects of admixtures all the mixtures achieved their desired workability expect NS-PP-20. The specimens were cast in steel molds and immediately covered with a polythene sheet after casting. The specimens were stripped from the molds after 24 h and placed in a water tank for curing. All the specimens were taken out from the curing tanks before testing. The cube compressive strength was determined by following BS 1881–116 and splitting tensile strength was evaluated according to ASTM C496–17. A centric compressive load at a rate of 0.25 kN/s was applied for compressive and split tensile tests. The strength parameters were calculated based on the actual dimensions of the specimens and measured ultimate loads.

1.4 XRD Analysis

XRD analysis was employed to identify different phases in the materials, especially hydration products in concrete. In this study, XRD analysis was done for NS-PP0 and NS-PP0-SP batches, to investigate the causes for strength variation due to the addition of Plastobuild ES type superplasticizer. XRD analysis was carried out using a diffractometer (Instrument Centre, Faculty of Applied Science, University of Sri Jayewardenepura, Sri

Lanka) equipped with Kβ X-ray source (operating at 40 kV and 30 mA). The testing was performed at room temperature on powered samples at angles from 5^0 to 85^0 (2θ) at a scan rate of 2^0/ min by using the Rigaku Ultima IV instrument model.

2 Results and Discussion

The summary of the compressive and tensile strength results for NSC and HSC is given in Table 3 and Table 4 respectively.

Table 3. Summary of test results for normal strength concrete.

Batch	Slump (mm)	Slump Variation (%)	f_{cu} (MPa) 7 days	$C_{P/N,\ 7\ day}$	f_{cu} (MPa) 28 days	$C_{28/7}$	$C_{P/N,\ 28\ day}$	$f_{ct,sp}$ (MPa)	$T_{P/N,\ 28\ day}$
NS-PP0	85	–	20.15	–	40.25	2.00	–	2.64	–
NS-PP20	55	–35	16.64	0.83	33.63	2.02	0.84	2.06	0.88
NS-PP20-SP	170	+ 100	15.98	0.79	31.96	2.00	0.79	1.85	0.70

Table 4. Summary of test results for high strength concrete.

Batch	Slump (mm)	Slump Variation (%)	f_{cu} (MPa) 7 days	$C_{P/N,\ 7\ day}$	f_{cu} (MPa) 28 days	$C_{28/7}$	$C_{P/N,\ 28\ day}$	$f_{ct,sp}$ (MPa)	$T_{P/N,\ 28\ day}$
HS-PP0-SP	230	–	58.5	–	73.3	1.25	–	3.70	–
HS-PP20-SP	220	–4	45.5	0.78	55.1	1.21	0.75	2.36	0.64

2.1 Normal Strength Concrete (NSC)

2.1.1 Concrete Workability

The slump value for different batches of concrete is shown in Fig. 3. With a 20% replacement of NCA with PPA the slump value has been reduced by 35%. The addition of a superplasticizer has increased its slump beyond the desired limits.

There are contradictory observations regarding the workability of concrete incorporated with PPA. Some authors (ex:Islam and Shahjalal, (2021)) reported that workability increased with increasing PPA and others (ex: Abu-Saleem et al. (2021)) observed a reduction in workability with increasing PP content in concrete, which is the observation in this study also. Islam and Shahjalal, (2021) had replaced NCA with PPA which is produced in same size of NCA after post processing the PP waste. These post processing techniques had included crushing of PP waste, heating, cooling and again crushing PP waste to produce PPA with the required sizes. Thus, the PPAs were smooth, round and regular in shape as shown in Fig. 4a. On the other hand, Abu-Saleem et al. (2021) had used the crushed PP plastic waste as a direct substitution for NCA without any post processing (same as in this study). Thus, the PPAs were rough, flaky and irregular in shapes as shown in Fig. 4b. Therefore, it can be concluded that the effect on application of post

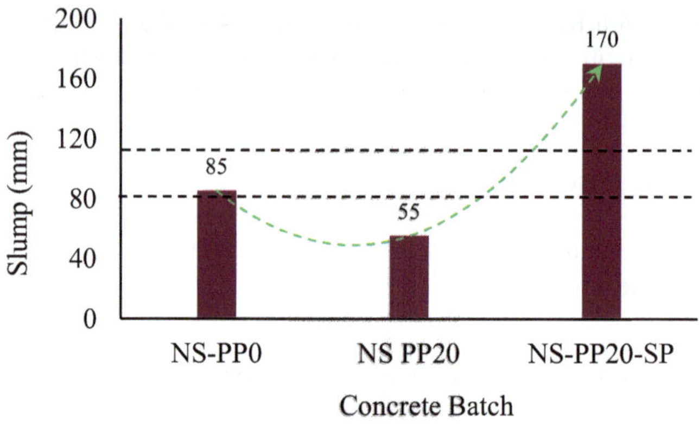

Fig. 3. Variation in the slump of different mixtures for NSC.

processing methods would effect on the workability of concrete with PPA. However, the cost and energy associated with post processing techniques, would make those process to be less sustainable. Thus, the studies on direct application of crushed plastic as PPA should be encouraged to make it as a feasible industrial application.

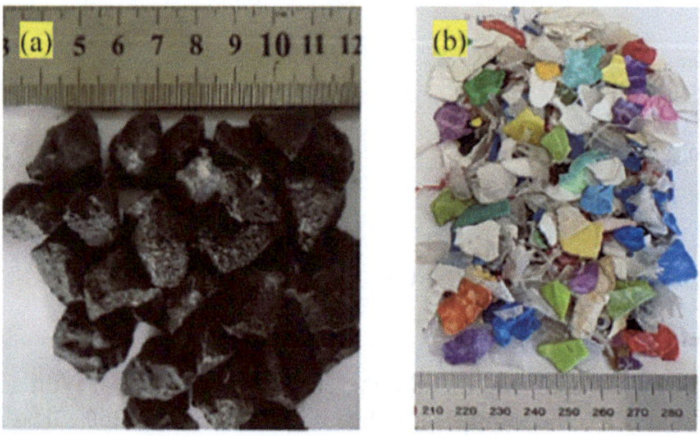

Fig. 4. (a) Post processed PPAs by Islam and Shahjalal, (2021), (b) Crushed PP waste as PPA by Abu-Saleem et al. (2021).

Figure 5 illustrates the deviation of the particle size distribution of total aggregate mix, which indicates that, PPA has altered the relative fineness of the aggregate mix in different boundaries. The inclusion of PPA have resulted a finer mix in coarse regions and, coarser mix in fine regions. These alternations could have resulted in densifying the concrete mix, by limiting the spaces for relative moves of the aggregates, by reducing the workability of concrete.

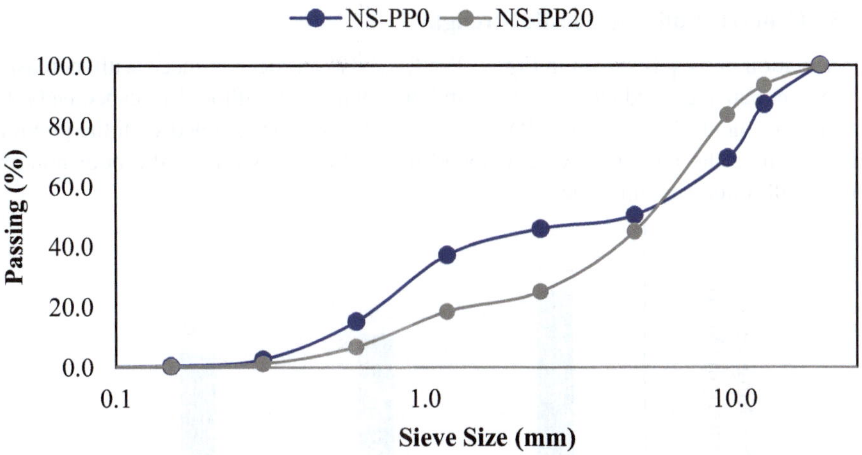

Fig. 5. Particle size distribution in PP0 and PP20 mixes for NSC.

2.1.2 Concrete Cube Compressive Strength

As shown in Fig. 6, f_{cu} (at 7 days and 28 days) decreases with increasing PPA percentages. The compressive strength of concrete with PPA has been reduced by 17% and 21% at 7 days and 15% and 21% at 28 days, with and without SP respectively. Similar strength reduction has been observed in previous studies, including Islam et al. (2022), Abu-Saleem et al. (2021) and many other authors. This reduction resulted from the weak ITZ between cement paste and PPA. However, NS-PP20-SP showed a slight reduction of compressive strength relative to NS-PP20 which have been resulted by reduced hydration reaction at PPAs due to the hydrophobic nature of PPA causing more weaker ITZs in concrete with PPA. Thus, further studies are required to improve PPA surface texture to strengthen the bond between PPA and cement paste.

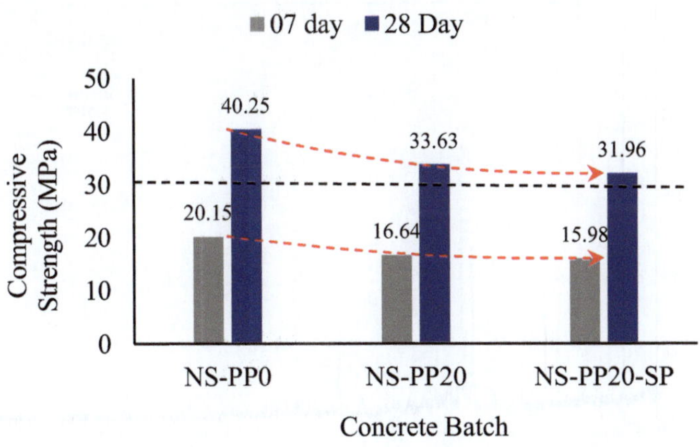

Fig. 6. Variation in cube compressive strength (7 days and 28 days) of different NSC mixtures.

2.1.3 Concrete Splitting Tensile Strength

The variation of $f_{ct,sp}$ is shown in Fig. 7. The $f_{ct,sp}$ also has been reduced with increasing PP content causing a reduction of 12% and 30% with and without SP respectively, for 20% replacement of NCA with PPA. These observations are aligned with the previous studies, where the authors have concluded this reduction is due to the poor bonding between PPA and cement paste.

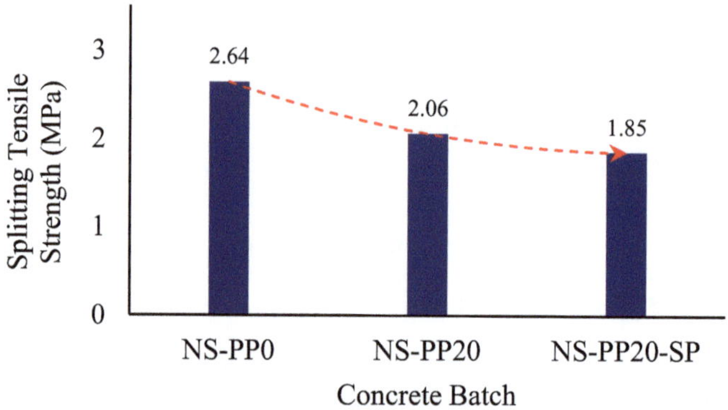

Fig. 7. Variation in splitting tensile strength of different NSC mixtures.

2.1.4 XRD Analysis

Figures 8 and 9 illustrate the XRD patterns of powdered samples corresponding to NS-PP0 and NS-PP0-SP concrete, respectively. The X-ray diffractograms were analyzed using Profex 5.2.5 and it confirmed the presence of different hydrated and un-hydrated phases in concrete, such as calcium silicate hydrate (CSH), portlandite (CH), ettringite, quarts, calcite, C3S, C2S and so on. The presence of CSH and ettringite in NS-PP0 was higher than that in concrete with superplasticizer (NS-PP0-SP). Thus, further studies are required to make conclusions about the effect of superplasticizers on concrete strength.

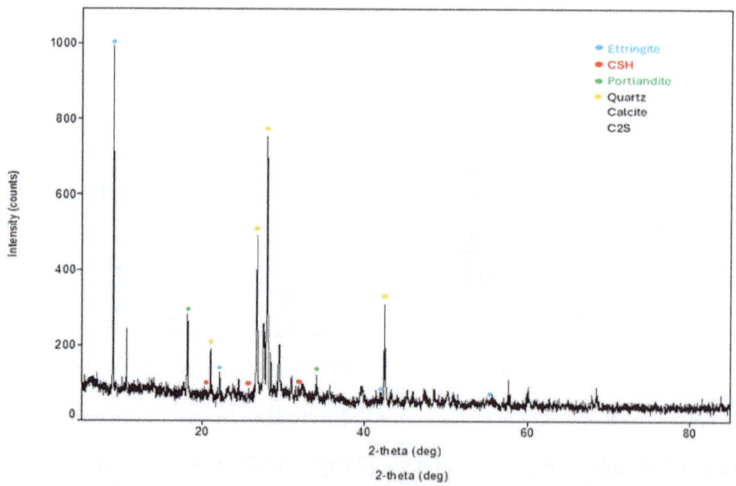

Fig. 8. Hydrated and un-hydrated phases in NS-PP0 concrete.

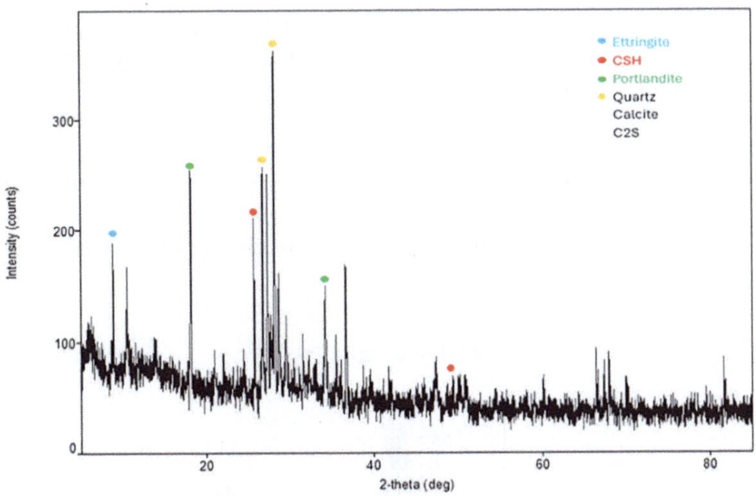

Fig. 9. Hydrated and un-hydrated phases in NS-PP0-SP concrete.

2.2 High Strength Concrete (HSC)

2.2.1 Concrete Workability

As shown in Fig. 10, the workability has been reduced with the addition of PPA in concrete. However, this reduction is about 4%, which can be neglected by comparing the workability reduction in NSC due to the incorporation of PPA. However, it should be noted that, the high-range superplasticizer had been added for both the mixes (with and without PP for HSC), to achieve high strength with a desired workability.

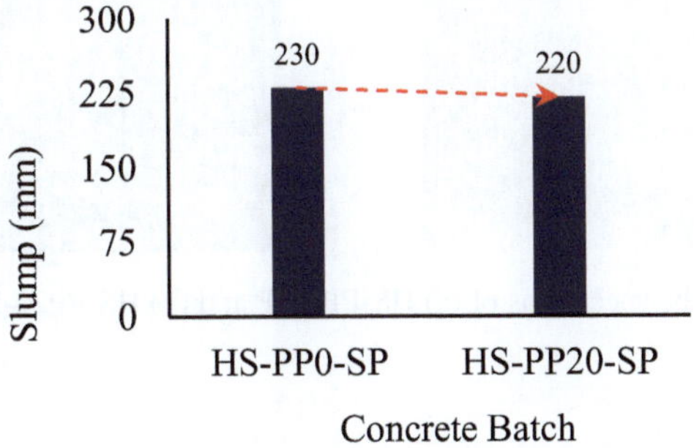

Fig. 10. Variation in the slump of different mixtures for HSC.

2.2.2 Concrete Cube Compressive Strength

Figure 11 presents f_{cu} values at 7 days and 28 days in HSC. The reduction of f_{cu} in concrete with PPA can be attributed to the weak ITZ. Though it was able to achieve the desired strength limit (30 MPa) in NSC with 20% replacement of NCA, the required strength was not achieved in HSC (60 MPa). Thus, further studies are required to determine the optimum addition levels of PPA in HSC.

Further, it was observed that there is a possibility of segregation of PPA in concrete mix, compared to conventional concrete (as shown in Figs. 12a and 12b). This could allow the less dense PPAs to float on the mixture during the compaction with mechanical vibration. This segregation could negatively impact on the appearance as well as the strength of concrete with PPA, , due to ununiform distribution of the materials within the concrete.

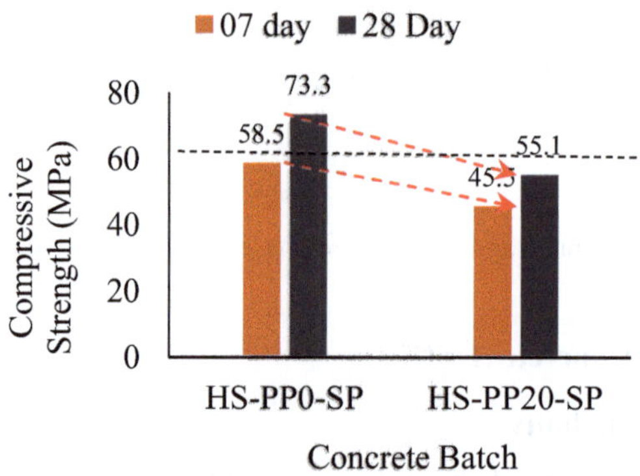

Fig. 11. Variation in cube compressive strength (7 days and 28 days) of different HSC mixtures.

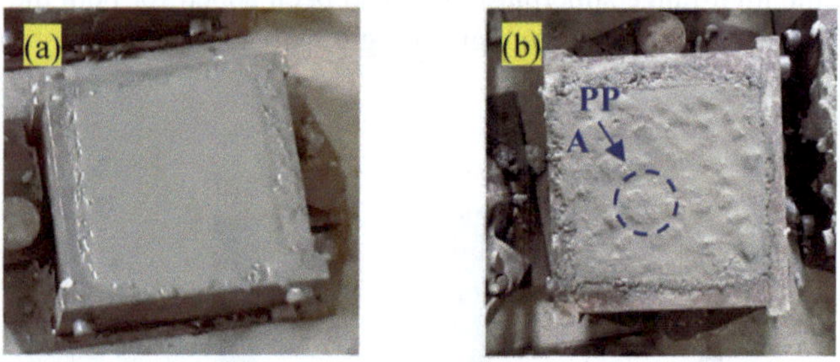

Fig. 12. Surface of cube specimens of (a) HS-PP0-SP and (b) HS-PP20-SP after compaction.

2.2.3 Concrete Splitting Tensile Strength

As given in Fig. 13, $f_{ct,sp}$ decreases with the inclusion of PPA, where the strength reduction is about 36%. Thus, PPAs have more negative effects in terms of the tensile properties of HSC, than the compressive strength. Further, it was observed that, the effects of PPA incorporation on $f_{ct,sp}$ of HSC, are more detrimental than, its influence on $f_{ct,sp}$ of NSC.

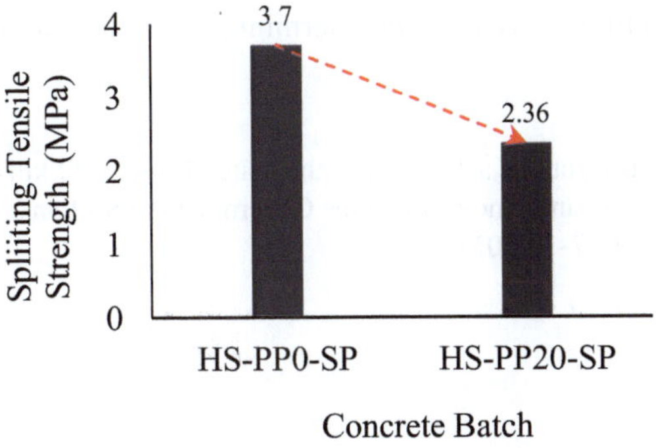

Fig. 13. Variation in splitting tensile strength of different HSC mixtures.

3 Conclusion

This study investigated the effect of the replacement of natural coarse aggregate (NCA) with recycled Polypropene plastic aggregates (PPA) in normal-strength concrete (NSC) and high strength concrete (HSC). An optimum 20% volumetric replacement of NCA was considered by referring to past literature. Additionally, the effect of superplasticizers (SP) on the performance of NSC and HSC with PPA was investigated. The key findings of this study are as follows.

1. Workability in NSC with PPA was drastically reduced due to the irregular shape and sharp edges of the PCA, which negatively affected the flowability of the mix. However, the dosage of 300 ml of superplasticizer (Plastobuild ES, Finex Engineering, Sri Lanka) per 100 kg of cement, significantly improved the slump of the concrete from 55 mm to 170 mm without segregation, which helped compensate for the reduced workability due to the PPA.
2. The impact of PPA inclusion on workability was minimal in HSC. The addition of a high range superplasticizer (Masterglenium SKY 8233, Finex Engineering, Sri Lanka) enhanced the workability, increasing the slump to around 230 mm. However, as the superplasticizer content was high (500 ml per 100 kg of cement), some of the PPA was segregated and floated to the top concrete mix during compaction using mechanical vibration. This should be further studied, to assess the feasibility of concrete with PPA for industrial applications.
3. This study highlights that superplasticizer can significantly improve the workability of concrete with PPA, particularly in NSC. However, their dosage should be carefully controlled to avoid segregation, especially in HSC. Additionally, achieving uniform distribution of PPA in the concrete requires appropriate compaction methods, as PPA can float in the mix under certain conditions.
4. Both compressive strength and splitting tensile strength decreased with the addition of PPA into concrete. The tensile properties of concrete had more negative impacts than compressive properties and these detrimental effects were more prominent in HSC than NSC. Further, 20% replacements of NCA with PPA still could achieve the desired strength in NSC but not for HSC. That implies, further studies on HSC

incorporated with PPA is essential for determining optimum NCA replacement levels for HSC.

Acknowledgements. The authors acknowledge the State Ministry of Skills Development, Vocational Education, Research, and Innovation of the Government of Sri Lanka for granting financial support (MSVRI/RES/03/07–12/2021).

References

Abirami, T., Thilakarathna, K.G.D.I.H., Weragama, W.A.R.U., Bandara, C.S., Yapa, H.D., Dissanayake, P.B.R.: Mechanical properties of green concrete containing polypropylene plastic waste, pp. 117–130 (2024). https://doi.org/10.1007/978-981-97-3737-6_10

Abu-Saleem, M., Zhuge, Y., Hassanli, R., Ellis, M., Rahman, M., Levett, P.: Evaluation of concrete performance with different types of recycled plastic waste for kerb application. Constr. Build. Mater. **293**, 123477 (2021). https://doi.org/10.1016/J.CONBUILDMAT.2021.123477

Faraj, R.H., Sherwani, A.F.H., Daraei, A.: Mechanical, fracture and durability properties of self-compacting high strength concrete containing recycled polypropylene plastic particles. J. Building Eng. **25**, 100808 (2019). https://doi.org/10.1016/J.JOBE.2019.100808

Hossain, F.M.Z., Shahjalal, M., Islam, K., Tiznobaik, M., Alam, M.S.: Mechanical properties of recycled aggregate concrete containing crumb rubber and polypropylene fiber. Constr. Build. Mater. **225**, 983–996 (2019)

Islam, M.J., Shahjalal, M.: Effect of polypropylene plastic on concrete properties as a partial replacement of stone and brick aggregate. Case Stud. Constr. Mater. **15**, e00627 (2021). https://doi.org/10.1016/J.CSCM.2021.E00627

Islam, M.J., Shahjalal, M., Haque, N.M.A.: Mechanical and durability properties of concrete with recycled polypropylene waste plastic as a partial replacement of coarse aggregate. J. Build. Eng. **54**, 104597 (2022). https://doi.org/10.1016/J.JOBE.2022.104597

Kou, S.C., Lee, G., Poon, C.S., Lai, W.L.: Properties of lightweight aggregate concrete prepared with PVC granules derived from scraped PVC pipes. Waste Manage. **29**, 621–628 (2009)

Ozbakkaloglu, T., Gu, L., Gholampour, A.: Short-term mechanical properties of concrete containing recycled polypropylene coarse aggregates under ambient and elevated temperature. J. Mater. Civ. Eng. **29**, 04017191 (2017)

Shahjalal, M., Islam, K., Ahmed, K.S., Tamanna, K., Alam, M.S.: Mechanical characterization of rubberized fiber reinforced recycled aggregate concrete for bridge barriers. In: IABSE-JSCE Joint Conference on Advances in Bridge Engineering-IV. Dhaka. Bangladesh (2020)

Yang, S., Yue, X., Liu, X., Tong, Y.: Properties of self-compacting lightweight concrete containing recycled plastic particles. Constr. Build. Mater. **84**, 444–453 (2015). https://doi.org/10.1016/j.conbuildmat.2015.03.038

Yin, S., Tuladhar, R., Sheehan, M., Combe, M., Collister, T.: A life cycle assessment of recycled polypropylene fibre in concrete footpaths. J. Clean. Prod. **112**, 2231–2242 (2016)

Processing of Activated Carbon from Pine-Needle and Pine-Cone to Remove Fluoride from Water

R. S. M. Perera, R. A. T. Isuranaga(✉), and C. S. Kalpage

Faculty of Engineering, University of Peradeniya, Peradeniya, Sri Lanka
ratisuranga@gmail.com

Abstract. Fluoride is an essential ion for human health, but both its deficiency and excess intake can lead to health problems. The primary source of fluoride for humans is drinking water, and in many regions, excessive fluoride in water poses significant health risks, such as dental and skeletal fluorosis. Effective removal of fluoride from water is therefore critical, particularly in areas where naturally occurring fluoride levels exceed safe limits. Aim of this study was to remove fluoride ions from aqueous solutions using activated carbon (AC). Pine needles and pine cones were selected as the carbon precursors due to their availability and low cost. These materials were chemically activated using zinc chloride ($ZnCl_2$) at various impregnation ratios (1:1, 1:2, and 1:3) and carbonization temperatures (500 °C, 600 °C, and 700 °C).The resulting activated carbon was characterized using different analytical techniques, ensuring its suitability for fluoride adsorption. A synthesized fluoride solution, prepared using sodium fluoride (NaF), was treated with the AC, and fluoride concentrations were measured using the SPADNS method. The adsorption data were analysed with various isotherm models, and it was found that the Freundlich model provided the best fit for the equilibrium data. The maximum surface areas obtained for the activated carbons were 750 m^2/g for the pine needle-derived carbon and 498 m^2/g for the pine cone-derived carbon, both produced at a 1:3 impregnation ratio and carbonized at 600 °C. Ultimately, the study demonstrated that pine needles and pine cones are viable low-cost precursors for producing effective activated carbon, offering an eco-friendly and efficient solution for fluoride removal from water.

Keyword: Activated Carbon · Pine Cone · Pine Needle · Fluoride Removal

1 Introduction

Water, covering about three-fourths of the Earth's surface, is crucial for all life forms. Its chemical properties determine its suitability for various uses, with fluoride excess in drinking water posing significant health risks. Fluoride contamination can arise from natural and human-made sources. (Rao, 2003) There is a potential link between drinking water sources and chronic kidney disease in Sri Lanka. (Balasubramanya et al., 2020) While fluoride can be beneficial in small amounts, excessive fluoride in drinking water

R. Dissanayake et al. (Eds.): ICRES 2025, PTT, pp. 101–114, 2025.
https://doi.org/10.1007/978-981-96-9551-5_9

poses significant health risks, such as dental and skeletal fluorosis, which can cause severe tooth and bone damage. Fluoride contamination can stem from both natural sources, such as leaching from the earth's crust, and human activities. This issue is particularly prevalent in many countries, where high fluoride levels in groundwater are widespread. (Piriya et al., 2021).

Several methods have been developed to mitigate the harmful effects of fluorosis. These include coagulation-precipitation, membrane separation, ion exchange, adsorption, and advanced techniques like electro-dialysis and electrochemical processes. Among these, approaches that leverage the fluoride exchange properties of appetites such as those incorporating bone constituents, ion exchange mechanisms, and the adsorption capabilities of Aluminium compounds have shown significant potential for efficiently reducing excess fluoride levels. (Arfin et al., 2016; MAIER, 1947).

Among these, adsorption has been demonstrated as a practical method due to its cost-effectiveness, ease of operation, high removal capacity, and the ability to reuse the adsorbent. Researchers are increasingly exploring adsorption techniques, including the use of high-valency metals to functionalize sorbents, for their potential to produce high-quality water efficiently. These attributes motivate researchers to further investigate and develop adsorbents to explore their full potential. (Habuda-Stanić et al., 2014; He et al., 2020).

Activated carbon is traditionally produced from coal, a non-renewable resource. However, recent efforts are focused on substituting coal-based activated carbon with alternatives derived from renewable biomass waste (Jiunn Boon Yong et al., 2022). Many researchers have studied the use of activated carbons for adsorbing and removing inorganic substances from water. (Chakrapani et al., 2010) Materials like corncobs(Song et al., 2013), date seeds (Ogungbenro et al., 2020), banana trunks (Danish et al., 2018), peanut shells (Wilson et al., 2006), and mango seeds have been used to prepare activated carbon.

The primary objective of this study was to develop activated carbon from pine cones and needles through a controlled carbonization and activation process to achieve a high surface area and porosity. Specific activation methods were employed to enhance the adsorption properties of the synthesized carbon, ensuring optimal fluoride removal efficiency. A comprehensive characterization of the physical and chemical properties, including surface area and relative activeness, was conducted to assess the quality of the activated carbon. Additionally, systematic adsorption studies were performed to evaluate its effectiveness in removing fluoride from wastewater, demonstrating its potential as a cost-effective and sustainable solution for water purification.

Activated carbon can be created through physical, chemical, physiochemical, or microwave-assisted activation of biochar and biomass. Physical activation involves the use of heat and gases like steam, CO_2, N_2, or their mixtures. Chemical activation utilizes agents such as acids, bases, or metal oxides, while physiochemical activation combines heat with chemical treatments. Microwave-assisted activation employs microwave radiation. The effectiveness of activation is influenced by factors such as particle size, retention time, impregnation ratio, process setup, activation duration, precursor characteristics, and the chemicals used. To enhance the properties of activated carbon, researchers are exploring methods like loading nanoparticles onto its surface, which could improve its efficiency in removing organic and inorganic substances from water. (Reza et al., 2020).

2 Materials and Method

The pine needles and pinecones used in this experiment were collected from the Hanthana mountain range around the University of Peradeniya. The pine needles were first washed with distilled water and then dried in an oven at 80 °C until they reached a stable weight. After drying, they were ground to a particle size less than 2 mm. The pine needles and cones were impregnated with $ZnCl_2$ in a 1:1, 1:2 and 1:3 mass ratios. This mixture was then soaked in a specific quantity of deionized water and stirred for 4 h to ensure thorough blending. After the material was fully impregnated, it was dried again at 80 °C until a constant mass was achieved. The nitrogen (N_2) flow was used for the carbonization process. In the carbonization experiments, pre-treated pine needles and cones with $ZnCl_2$ were heated from 25 °C to target temperatures of 500 °C, 600 °C, 700 °C at the same heating rate of 10 °C per minute and held at the chosen temperature for 1 h. After the carbonization process, the sample was cooled under a flow of nitrogen (N_2). The solid products were then treated by stirring with a 0.1 mol/L HCl solution for 2 h to remove inorganic components. After the acid treatment, the sample was neutralized with deionized water and dried until it reached a constant weight. The yields of the products were determined using Eq. (1), by the mass difference between the activated carbon.

$$Y_{AC} = \frac{m_{AC}}{m_{in}} \times 100\% \tag{1}$$

m_{AC}: mass of prepared activated carbon; m_{in}: initial mass of pine needles and cones.

3 Results and Discussion

3.1 Yield of the Activated Carbon

The yield (Y_{AC}) of activated carbon from pine needles is influenced by both temperature and the ratio of $ZnCl_2$ to biomass as shown in Fig. 1. At a 1:1 ratio, the yield starts relatively high at 500 °C and gradually decreases as the temperature increases to 700 °C, indicating that higher temperatures reduce yield in this ratio. For the 1:2 ratio, shows a more consistent decline as the temperature increases from 500 °C to 700 °C. At the 1:3 ratio, the yield behaves, with a peak yield at 500 °C before dropping again at 700 °C. This suggests that increasing the $ZnCl_2$ ratio affects the yield differently across temperatures.

According to Fig. 2 pinecones, the trend is somewhat similar but shows less variation across temperature ranges compared to pine needles. At the 1:1 ratio, the yield starts high at 500 °C and gradually decreases as the temperature increases to 700 °C. In the 1:2 ratio, the yield reaches its maximum at 500 °C, showing a reduction of yield when decreased to 700 °C. For the 1:3 ratio, the yield peaks at 500 °C, which contrasts with pine needles.

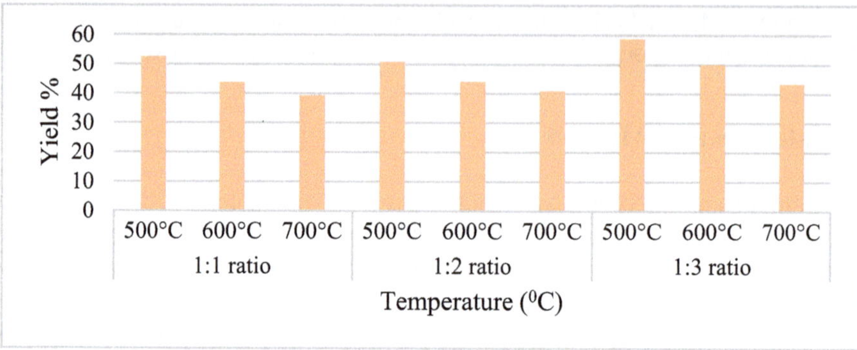

Fig. 1. Yield of activated carbon derived from Pine Needles

In conclusion, increasing the temperature from 500 °C to 700 °C during biomass processing significantly enhances carbon burning-off, tar volatilization, and devolatilization. This temperature rise promotes the transformation of biomass into volatile products (both liquid and gas), thereby reducing the char yield. However, excessively high activation temperatures can have detrimental effects, including a decrease in carbon yield.

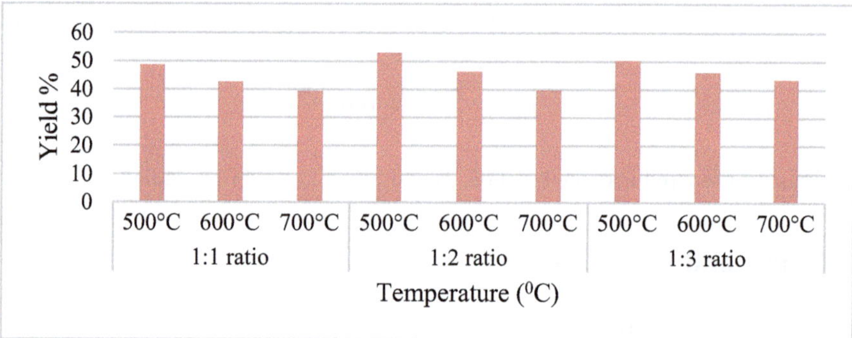

Fig. 2. Yield of activated carbon derived from Pine Cones

This occurs because high temperatures accelerate the conversion of solid-phase cross-linked substances into volatile gaseous products. Therefore, optimizing the activation temperature is crucial to balance the char yield and specific surface area while maximizing the yield of volatile components. Proper temperature control ensures efficient biomass conversion while mitigating the loss of valuable solid-phase material. (Açikyildiz et al., 2014; Gao et al., 2020; Li et al., 2023).

3.2 Iodine Number

The adsorption capacity of activated carbons can be assessed using their iodine numbers. This metric indicates how much iodine (in milligrams) can be adsorbed by 1 g of activated

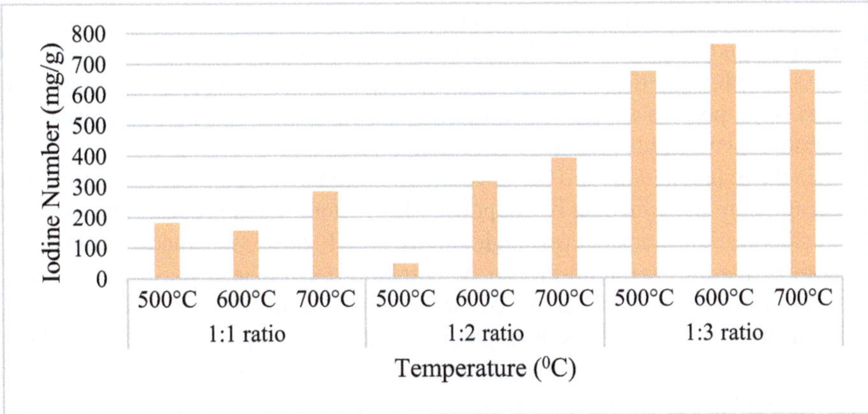

Fig. 3. Iodine number values for Pine needle derived activated carbon

carbon under specific conditions. The method applied in this study follows the ASTM standard D4607–94 (ASTM D4607–94). The iodine number provides an estimate of the activated carbon's adsorption ability, particularly in relation to its microporosity, when interacting with a condensed (liquid) phase like water. (Siragi et al., 2021) Figs. 3 and 4 depicts the iodine numbers of activated carbon derived from pine needles and cones.

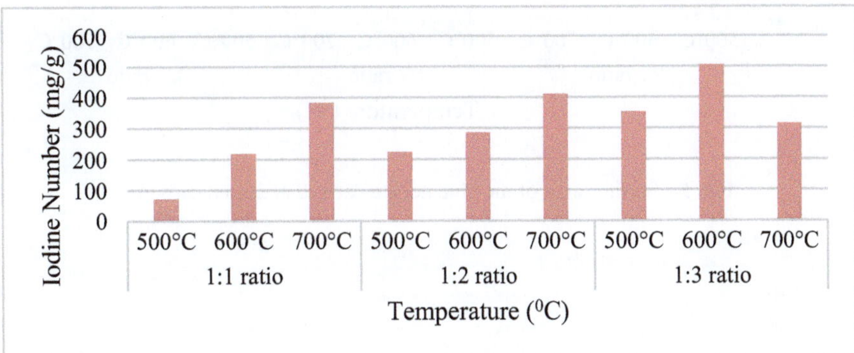

Fig. 4. Iodine number values for Pine cone derived activated carbon

Figure 4 shows Iodine number values for Pinecone derived activated carbon. The importance of micro- and mesoporous volumes, surface area, and the careful selection of both the activating agent and raw material has been emphasized. (Din et al., 2017)

$$S_{BET} = 0.9946 \times IN - 4.91 \quad (2)$$

The Eq. (2) (Mianowski et al., 2007) converts the iodine number into surface area, which is determined using BET isotherms. It estimates the total surface area of activated carbon based on its iodine adsorption capacity. Figures 5 and 6 shows the obtained results for the pine needles and cones after converting them into surface areas.

At lower impregnation ratios, the total pore volume is predominantly composed of micropores, indicating that smaller, microporous structures are more common at these ratios. As the impregnation ratio increases, mesopores start to play a more significant role in the total pore volume, across all carbonization temperatures. This suggests that higher ratios encourage the formation of mesopores.

When the impregnation ratio becomes larger, mesopore development becomes more prominent due to the widening of the pores. This pattern remains consistent across all tested carbonization temperatures, highlighting how higher impregnation ratios drive mesopore formation. (Moreno-Piraján & Giraldo, 2010) The highest surface areas obtained were 750 m^2/g for the activated carbon derived from pine needles and 498 m^2/g for that derived from pinecones, both prepared with a 1:3 impregnation ratio and carbonized at 600 °C. For further analysis and adsorption capacity isotherms, these two activated carbons were chosen for evaluation.

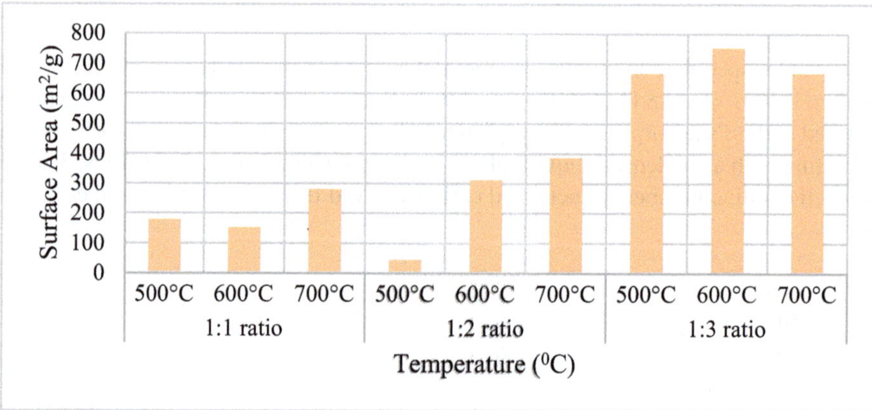

Fig. 5. Surface area of the pine needle derived activated carbon

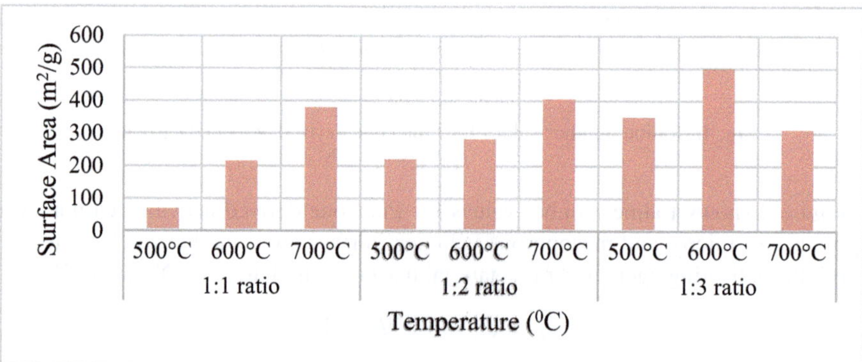

Fig. 6. Surface area of the pine cone derived activated carbon

3.3 TG and DTG Analysis

Thermogravimetric analysis (TGA) measures the mass loss of biomass during heating, providing insights into thermal degradation. Pyrolysis, a key process for energy recovery, can yield valuable pyrolytic liquids. This study examines the pyrolysis of biomass and its components, exploring the influence of molecular structure on thermal degradation. (Zainal *et al.*, 2021; Williams and Besler, 1993) The TGA and DTG curves for pine needles and cones are shown in the Fig. 7 and 8. The Thermogravimetric Analysis (TGA) and Derivative Thermogravimetric (DTG) curves for pine needles and pine cones show similar patterns. There is a noticeable weight loss starting around 100 °C, followed by another significant weight loss between 200 °C and 300 °C. A third weight loss occurs at approximately 600 °C. The initial weight loss around 100 °C is likely due to the removal of moisture and very light volatile components from the biomass.

The subsequent weight loss between 200 °C and 300 °C is probably due to the breakdown of hemicellulose and cellulose, as well as reactions involving $ZnCl_2$ that create aromatic compounds. The weight loss around 600 °C is due to degradation that produces char residue, which could be linked to the breakdown of lignin or the decomposition of $ZnCl_2$ derivatives, including the evaporation of metallic zinc.

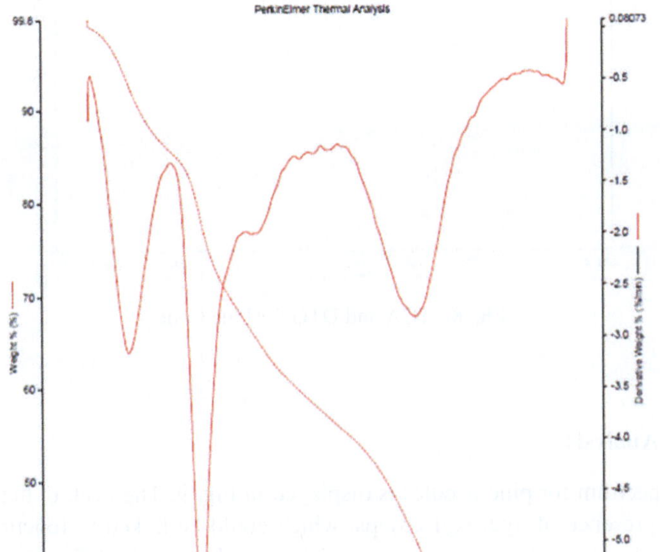

Fig. 7. TGA and DTG for Pine Needles

This peak in weight loss at higher temperatures may also indicate the collapse and shrinkage of the pore structure of activated carbon, leading to a reduced specific surface area. This pattern has been observed in similar studies of other types of biomass. (Li et al., 2023; Jaroniec et al., 1996; Varma and Mondal, 2016) Further research on the zinc mass balance and analysis of the gaseous products in the experimental processes would provide valuable insights.

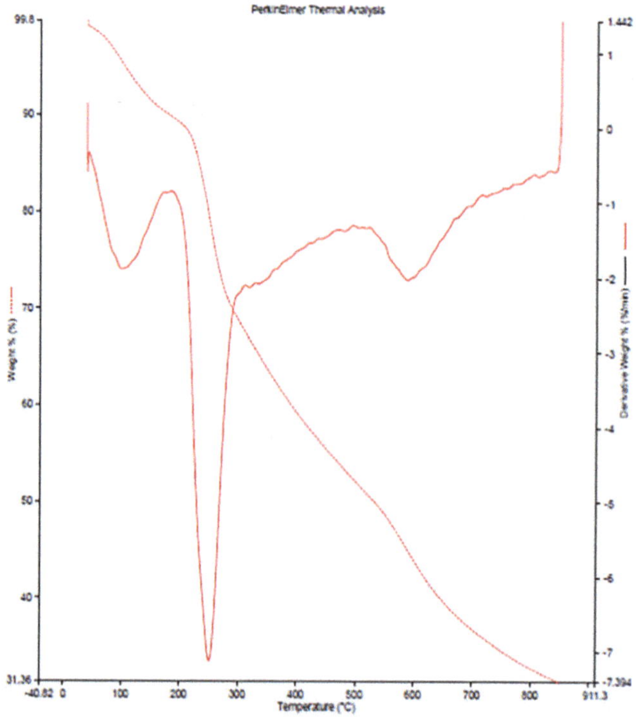

Fig. 8. TGA and DTG for Pine Cone

3.4 FTIR Analysis

The FTIR spectrum for pine needles is displayed in Fig. 9. The peak in the 3750 cm^{-1} suggest the presence of hydroxyl groups, which could be linked to functional groups such as organic acids, alcohols, or phenols. The wavelength of 2335 cm^{-1} range, there are long, sharp peaks that correspond to the stretching vibrations of C-H bonds from different groups. Additionally, small peaks appearing 1558 cm^{-1} are associated with $C = O$ stretching vibrations of aromatic compounds. Lastly, a peak in the 1134 cm^{-1} further supports the presence of specific functional groups related to the C-O structure of the adsorbent. This combination of peaks and vibrations helps to identify the chemical composition and changes in the pine needles, which likely occurred during the activation process. (Senewirathna et al., 2022; Shen et al., 2018).

The FTIR spectrum for pine cones is shown in Fig. 10. A moderate broad peak is observed around 3489 cm^{-1}, which suggests the presence of hydroxyl groups, potentially linked to functional groups such as organic acids, alcohols, or phenols. The wavelength 2670 cm^{-1} long sharp peaks appear, indicating the stretching vibrations of C-H bonds from different groups. These broad and distinct peaks suggest that the adsorbent may have undergone a chemical activation process. Additionally, peak is in the 1555 cm^{-1} range are associated with C = O stretching vibrations of aromatic compounds. Lastly, a peak in the 1139 cm^{-1} provides further evidence of functional groups present within the adsorbent material. These spectral features give insight into the chemical structure and functional changes that the pine cones underwent during activation.(Momčilović et al., 2011; Shen et al., 2018).

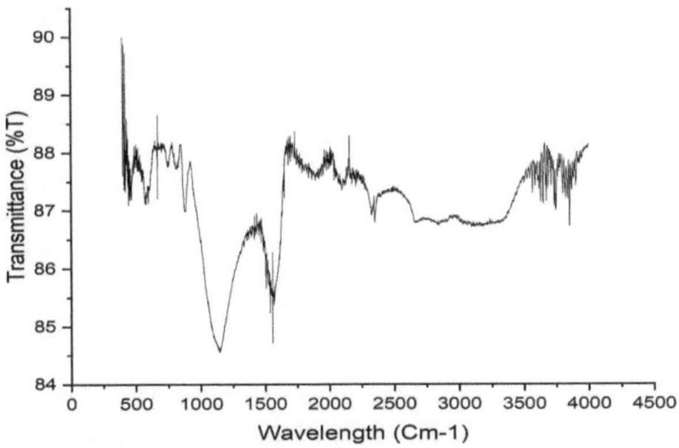

Fig. 9. FTIR results of pine needles derived activated carbon

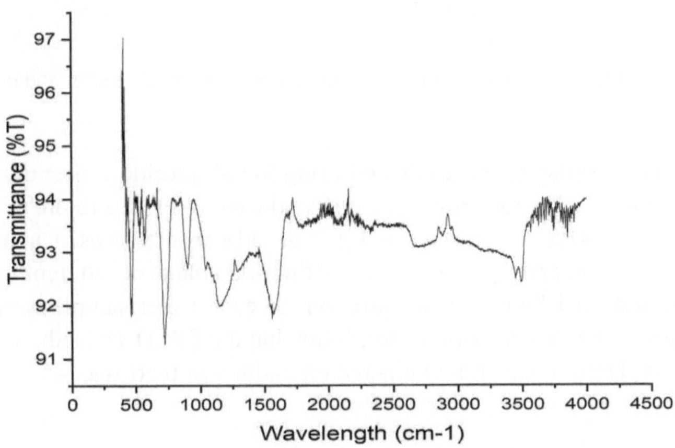

Fig. 10. FTIR results of pinecones derived activated carbon

Although most of the functional groups present in the activated carbons derived from both pine needles and pine cones are similar, there are noticeable differences in the O-H and C-H bond wavelengths. These variations indicate that different compounds are responsible for these bonds, suggesting distinct chemical structures in the activated carbons produced from the two materials. This difference in structure implies that, while both materials yield effective activated carbon, the specific composition and arrangement of the compounds in each type of carbon are unique to the respective precursor.

3.5 Isotherm Study

Modelling adsorption isotherm data is crucial for predicting and assessing adsorption performance, which helps optimize adsorption mechanisms, determine adsorbent capacities, and design efficient adsorption systems. Commonly used two-parameter isotherm models for this purpose include Langmuir, Freundlich, Temkin, and Dubinin-Radushkevich.(Chen, 2015) In this study the Langmuir and Freundlich isotherms were determined.

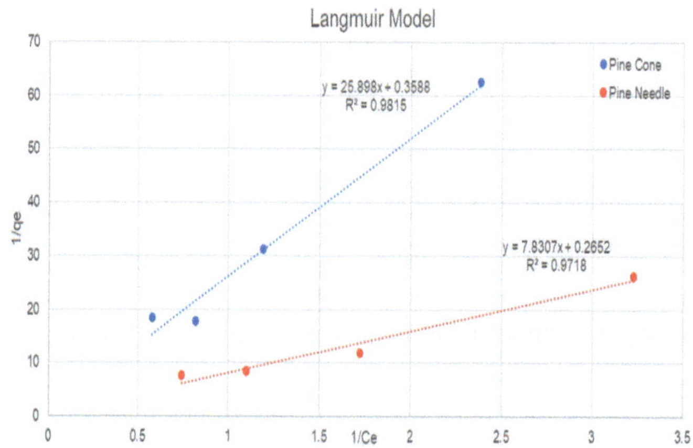

Fig. 11. Langmuir isotherm for the activated carbon of pine needles and cones

The adsorption isotherm was evaluated using initial fluoride concentrations of 0.5, 1.0, 1.5, and 2.0 mg/L. The experiments were conducted at a temperature of 30 °C, with a shaking speed of 150 rpm. A dosage of 1 g of the adsorbent was used, and the contact times were set at 1 min and 30 min to observe fluoride uptake at different intervals.

The initial and final fluoride concentrations in each water sample were measured using a multiparameter bench photometer, following the SPADNS method for accurate fluoride analysis. The obtained Langmuir and Freundlich isotherms are shown in Figs. 11 and 12.

The adsorption isotherm analysis reveals important insights into the fluoride removal capacities of activated carbon derived from pine needles and pine cones as shown in Table 1. The Langmuir isotherm model indicates that the maximum adsorption capacity (Q_max) for the pine needle-derived activated carbon is 3.772 mg/L, significantly higher than the 2.787 mg/L observed for the pine cone-derived carbon. This suggests that the activated carbon from pine needles offers a greater efficiency in capturing fluoride ions, potentially due to its enhanced surface area and unique structural characteristics. The Freundlich isotherm values further support this finding, highlighting the superior adsorption properties of the pine needle-derived carbon in comparison to its counterpart.

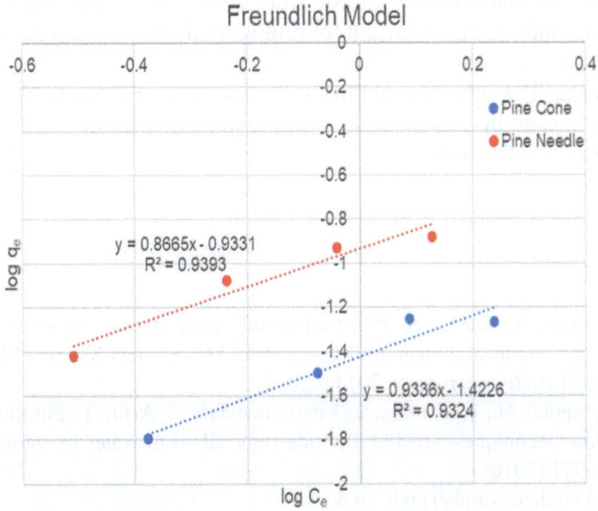

Fig. 12. Freundlich isotherm for the activated carbon of pine needles and cones

Table 1. The Langmuir and Freundlich isotherm values for the activated carbon derived from pine needles and cones

Isotherm		Q_{max}	K_L	R^2
Langmuir Isotherm	Pine Cone	2.787	0.115	0.982
	Pine Needle	3.772	0.547	0.972
		K_F	n	R^2
Freundlich Isotherm	Pine Cone	0.038	1.071	0.932
	Pine Needle	0.117	1.154	0.939

4 Conclusion

The analysis indicates that activated carbon derived from pine needles and cones is an effective adsorbent for fluoride ions in water. The maximum surface areas achieved were 750 m^2/g for the pine needle-derived carbon and 498 m^2/g for the pine cone-derived carbon, both produced at a 1:3 impregnation ratio and carbonized at 600 °C. Optimal conditions for achieving the highest fluoride removal efficiency were identified as a contact time of 1 h and 30 min, an adsorbent dosage of 1 g, and a temperature of 30 °C. The maximum adsorption capacities (Q_max) were found to be 3.772 mg/L for the pine needle-derived activated carbon and 2.787 mg/L for the pine cone-derived carbon. These findings demonstrate that activated carbon from both pine needles and cones can serve as a cost-effective and efficient solution for removing fluoride ions from drinking water.

Acknowledgements. The authors would like to express their gratitude to Mr. Janith Bandara for his invaluable support in conducting the iodine number test, which was instrumental in the successful completion of this work.

References

Açikyildiz, M., Gürses, A., Karaca, S.: Preparation and characterization of activated carbon from plant wastes with chemical activation. Microporous Mesoporous Mater. **198**, 45–49 (2014). https://doi.org/10.1016/j.micromeso.2014.07.018

Arfin, T., Environmental, N., Waghmare, S., Environmental, N., Arfin, T.: Fluoride removal from water by various techniques : review fluoride removal from water by various techniques : review **2**, 560–571 (2016)

ASTM D4607–94 (Iodine Number).pdf. (n.d.)

Balasubramanya, S., Stifel, D., Horbulyk, T., Kafle, K.: Chronic kidney disease and household behaviors in Sri Lanka: historical choices of drinking water and agrochemical use. Econ. Hum. Biol. **37**, 100862 (2020). https://doi.org/10.1016/j.ehb.2020.100862

Chakrapani, C., Babu, C.S., Vani, K.N.K., Rao, K.S.: Adsorption kinetics for the removal of fluoride from aqueous solution by activated carbon adsorbents derived from the peels of selected citrus fruits. E-J. Chem. **7**, S419–S427 (2010). https://doi.org/10.1155/2010/582150

Chen, X.: Modeling of experimental adsorption isotherm data. Information (Switzerland) **6**(1), 14–22 (2015). https://doi.org/10.3390/info6010014

Danish, M., et al.: Optimization of banana trunk-activated carbon production for methylene blue-contaminated water treatment. Appl Water Sci **8**(1), 1–11 (2018). https://doi.org/10.1007/s13201-018-0644-7

Din, M.I., Ashraf, S., Intisar, A.: Comparative study of different activation treatments for the preparation of activated carbon: a mini-review. Sci. Prog. **100**(3), 299–312 (2017). https://doi.org/10.3184/003685017X14967570531606

Gao, Y., Yue, Q., Gao, B., Li, A.: Insight into activated carbon from different kinds of chemical activating agents: a review. Sci. Total. Environ. **746**, 141094 (2020). https://doi.org/10.1016/j.scitotenv.2020.141094

Habuda-Stanić, M., Ravančić, M., Flanagan, A.: A review on adsorption of fluoride from aqueous solution. Materials **7**(9), 6317–6366 (2014). https://doi.org/10.3390/ma7096317

He, J., et al.: Review of fluoride removal from water environment by adsorption. J. Environ. Chem. Eng. **8**(6), 104512 (2020). https://doi.org/10.1016/j.jece.2020.104516

Jaroniec, M., Gilpin, R.K., Ramler, J., Choma, J.: Characterization of microporous carbons by using TGA curves measured under controlled conditions. Thermochim. Acta **272**(1–2), 65–73 (1996). https://doi.org/10.1016/0040-6031(95)02460-3

Yong, J.B., Tan, L.S., Tan, J.: Comparative life cycle assessment of biomass-based and coal-based activated carbon production. Prog. Energy Environ. **20**(1), 1–15 (2022). https://doi.org/10.37934/progee.20.1.115

Li, B., et al.: Activation of pine needles with zinc chloride: evolution of functionalities and structures of activated carbon versus increasing temperature. Fuel Process. Technol. **252**, 107987 (2023). https://doi.org/10.1016/j.fuproc.2023.107987

MAIER, F.J.: Methods of removing fluorides from water. Am. J. Publ. Health Nation's Health **37**(12), 1559–1566 (1947). https://doi.org/10.2105/AJPH.37.12.1559

Mianowski, A., Owczarek, M., Marecka, A.: Surface area of activated carbon determined by the iodine adsorption number. Energy Sour. Part A: Recovery Utilization Environ. Effects **29**(9), 839–850 (2007). https://doi.org/10.1080/00908310500430901

Momčilović, M., Purenović, M., Bojić, A., Zarubica, A., Randelovid, M.: Removal of lead(II) ions from aqueous solutions by adsorption onto pine cone activated carbon. Desalination **276**(1–3), 53–59 (2011). https://doi.org/10.1016/j.desal.2011.03.013

Moreno-Piraján, J.C., Giraldo, L.: Adsorption of copper from aqueous solution by activated carbons obtained by pyrolysis of cassava peel. J. Anal. Appl. Pyrol. **87**(2), 188–193 (2010). https://doi.org/10.1016/j.jaap.2009.12.004

Ogungbenro, A.E., Quang, D.V., Al-Ali, K.A., Vega, L.F., Abu-Zahra, M.R.M.: Synthesis and characterization of activated carbon from biomass date seeds for carbon dioxide adsorption. J. Environ. Chem. Eng. **8**(5), 104257 (2020). https://doi.org/10.1016/j.jece.2020.104257

Piriya, R.S., Jayabalakrishnan, R.M., Maheswari, M., Boomiraj, K., Oumabady, S.: Coconut shell derived ZnCl2activated carbon for malachite green dye removal. Water Sci. Technol. **83**(5), 1167–1182 (2021). https://doi.org/10.2166/wst.2021.050

Rao, N.: Fluoride and environment-a review. In: Proceedings of the Third International Conference on Environment and Health, Chennai, pp. 15–17 (2003)

Reza, M.S., et al.: Preparation of activated carbon from biomass and its' applications in water and gas purification, a review. Arab J. Basic Appl. Sci. **27**(1), 208–238 (2020). https://doi.org/10.1080/25765299.2020.1766799

Senewirathna, D.S.G.D., Thuraisingam, S., Prabagar, S., Prabagar, J.: Fluoride removal in drinking water using activated carbon prepared from palmyrah (Borassus flabellifer) nut shells. Curr. Res. Green Sustain. Chem. **5**, 100304 (2022). https://doi.org/10.1016/j.crgsc.2022.100304

Shen, J., Huang, G., An, C., Xin, X., Huang, C., Rosendahl, S.: Removal of tetrabromobisphenol a by adsorption on pinecone-derived activated charcoals: synchrotron FTIR, kinetics and surface functionality analyses. Biores. Technol. **247**, 812–820 (2018). https://doi.org/10.1016/j.biortech.2017.09.177

Siragi, D.B.M., Desmecht, D., Hima, H.I., Mamane, O.S., Natatou, I.: Optimization of activated carbons prepared from <i>Parinari macrophylla</i>. Shells. Mater. Sci. Appl. **12**(05), 207–222 (2021). https://doi.org/10.4236/msa.2021.125014

Song, M., et al.: The comparison of two activation techniques to prepare activated carbon from corn cob. Biomass Bioenerg. **48**, 250–256 (2013). https://doi.org/10.1016/j.biombioe.2012.11.007

Varma, A.K., Mondal, P.: Physicochemical characterization and kinetic study of pine needle for pyrolysis process. J. Therm. Anal. Calorim. **124**(1), 487–497 (2016). https://doi.org/10.1007/s10973-015-5126-7

Williams, P.T., Besler, S.: Thermogravimetric analysis of the components of biomass. Adv. Thermochemical Biomass Conversion. **2**, 771–783 (1993). https://doi.org/10.1007/978-94-011-1336-6_60

Wilson, K., Yang, H., Seo, C.W., Marshall, W.E.: Select metal adsorption by activated carbon made from peanut shells. Biores. Technol. **97**(18), 2266–2270 (2006). https://doi.org/10.1016/j.biortech.2005.10.043

Circularity in Urban Planning and Infrastructure

Sustainable Pervious Concrete: Taguchi Optimization Meets Recycled Aggregates

Asoharasa Janarth[1(✉)], Daniel Niruban Subramaniam[2], and Navaratnarajah Sathiparan[2]

[1] Faculty of Technology, Eastern University, Vantharumoolai, Chenkalady, Sri Lanka
Janartha@esn.ac.lk

[2] Department of Civil Engineering, Faculty of Engineering, University of Jaffna, Ariviyal Nager, Kilinochchi, Sri Lanka

Abstract. Pervious concrete is critical for reducing urban flooding and in increasing groundwater recharge as it allows water to permeate through its structure. It also helps reduce surface runoff, which addresses urban waterlogging and pollution issues. This study investigates the effects of aggregate proportioning, compaction effort, and aggregate-to-cement (A/C) ratios on the porosity and compressive strength of pervious concrete. The study uses Natural Aggregates (NA) and Recycled Concrete Aggregates (RCA) from Sri Lanka, optimizing the mix design using Taguchi method to balance structural strength and porosity. Pervious concrete samples were prepared with varying proportions of smaller aggregates (10%, 30%, and 50%), A/C ratios (3.0, 4.0, and 5.0), and compaction levels (0, 224.4 Nm, and 561.0 Nm). Samples were cured for 28 days and tested for porosity and compressive strength. Taguchi method, along with analysis of variance, employed to identify optimal mix parameters. Results indicate that higher compaction levels and lower A/C ratios (3.0) enhance compressive strength, while higher proportions of smaller aggregates and lower compaction improve porosity. Optimized mix (30% smaller aggregates, A/C ratio of 5.0, and compaction at 224.4 Nm) balances both properties. For optimized mix concrete made with RCA exhibited 29% less compressive strength compared to NA concrete but showed higher porosity due to adhered mortar. This study highlights trade-offs between strength and permeability, offering insights for sustainable applications of pervious concrete in drainage and pavement systems.

Keywords: Compressive strength · Pervious concrete · Porosity · Recycled concrete aggregate · Taguchi method

1 Introduction

Permeable pavement is an emerging concept in Low Impact Developments (LIDs) to manage stormwater runoff, and flash floods and recharge in-situ groundwater during rain events. It is a concrete made up of aggregate and binder only, sometimes a small percentage of fine aggregates. Due to its distinctive porous structure and existence of vertically and horizontally interconnected pores, which allow water to move through,

R. Dissanayake et al. (Eds.): ICRES 2025, PTT, pp. 117–130, 2025.
https://doi.org/10.1007/978-981-96-9551-5_10

less surface runoff and natural infiltration are generated. In addition, their porous nature, they traps pollutants and particulates in the voids, aiding removal of pollutants. This mobilizes heavy metals, hydrocarbons, and suspended/microparticles into groundwater, also minimizing the direct accumulation of contaminants in the surface water bodies and improving water quality through natural treatment processes (Zhang, Li et al. 2021, Rasmussen, Lykkemark et al. 2023, Mitchell and Jayakaran 2024).

Permeability and compressive strength are major performance indicators of pervious concrete. Several variables, such as aggregate properties, binder characteristics, aggregate-to-cement ratio, water-to-cement ratio, and compaction energy contribute in deciding pervious concrete performance. Despite the existence of standards including BS EN 206, ASTM C1688 (Density and Void Content of Freshly Mixed Pervious Concrete), and ACI 522R-10 (Guide to Pervious Concrete), a widely recognized mix design procedure for pervious concrete is lacking. Although these standards offer recommendations, they very rarely address individual circumstances, material variability, or performance needs. Variability in void content that affects permeability, compressive strength, and decreased durability are some of the limitations (Liu, Chi et al. 2020). Furthermore, it is difficult to optimize the mix for a variety of applications and climates due to the lack of clear correlations between the mix proportions and desirable qualities like strength and porosity.

1.1 Aggregate Size

As aggregates contribute more than 80% volume of pervious concrete (Zaetang, Wongsa et al. 2013), larger aggregates improve permeability with wider voids, however, decrease strength due to fewer contact area, while smaller aggregates increase strength through better packing but decrease permeability. The best performance is achieved by balancing aggregate size, compaction, and mix proportion; smaller aggregates improve packing density, stress distribution, and the quality of the interfacial transition zone (ITZ), which increases strength; larger aggregates may weaken load transfer and ITZ due to reduced contact areas and binder coverage, which compromises overall performance (Yu, Sun et al. 2019, Shen, Zheng et al. 2020, Yavuz and Yazıcı 2023).

1.2 Aggregate to Cement Ratio

Strength, permeability, and durability of pervious concrete are all significantly impacted by the aggregate-to-cement (A/C) ratio. Increased void content from a greater A/C ratio improves permeability but lowers compressive strength because of weakened aggregate bonds and less cement paste coverage as shown in Fig. 1.

On the other hand, a lower A/C ratio results in a denser matrix and stronger aggregate bonding, which increases strength and durability. However, as voids are constrained, permeability decreases. Strength and permeability must be balanced for particular applications to maximize the A/C ratio. While too much cement paste can undermine structural integrity, too much paste can clog voids and reduce infiltration capacity (Yin, Chen et al. 2018, Sajeevan and Subramaniam 2023).

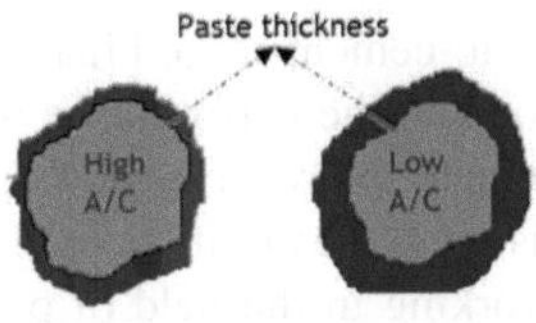

Fig. 1. Variation of paste thickness with A/C

1.3 Compaction Energy

Compaction energy has a major impact on pervious concrete's permeability and strength. Increased compaction energy increases compressive strength while decreasing permeability by improving aggregate interlock and reducing void content. On the other hand, inadequate compaction weakens structural integrity by leaving too many voids (Anburuvel and Niruban Subramaniam 2022, Subramaniam, Hareindirasarma et al. 2022, Mikami, Pieralisi et al. 2024). For the intended uses, achieving the ideal compaction energy guarantees a balance between structural integrity and efficient permeability.

1.4 Alternatives Utilized for Aggregates

Alternatives to conventional aggregates in pervious concrete, such as recycled materials, industrial by-products, and organic or synthetic aggregates, enhance sustainability but are inherent with challenges. Although recycled materials like crushed concrete and ceramic waste encourage the reuse of resources, their irregular forms and weaker bonding may cause them lose significant fraction of compressive strength (Thomas, Setién et al. 2013, Zhang, Feng et al. 2020, Zhang, Wang et al. 2023). Steel slag and other industrial by-products increase strength and durability, but they also need to be processed carefully and may change permeability. High permeability and sustainability are provided by organic materials, such as coconut shells, but mechanical strength is sacrificed. Although they are more expensive, novel materials like expanded clay or polymer aggregates improve durability. Optimizing mix design and careful material selection are necessary to strike a balance between sustainability and performance.

Construction and demolition waste has been considered a crucial environmental impact and a financial defect. Waste generated is more challenging to manage due to difficult engineering structures and with the increase of demand in growing and developing communities. Production of recycled concrete aggregates is a sustainable approach from an environmental and economic point of view. However, poor performance of concrete with recycled aggregates is a significant drawback to replacing natural aggregates with recycled aggregates. The structural applications of recycled concrete are currently quite limited and widely used for pipe bedding, backfilling, landscaping, and non-load-bearing applications. Application of RCA in low-strength concrete is the best way to initiate this in the industry as the cement motor adheres to crates a weak bond in concrete (Meddah, Al-Harthy et al. 2020, Nedeljković, Visser et al. 2021).

With all consideration adequacy of low compressive strength in the permeable pavement applications signifies the possibility of incorporating RCA in producing pervious concrete. This study investigates compressive strength and porosity of pervious concrete, emphasizing the role of compaction energy along with mix design variables such as aggregate size and aggregate-to-cement ratio. Finally, the experimental design was optimized using Taguchi method and the performance was compared for the pervious concrete cast using natural aggregate and recycled concrete aggregate for the optimized mix design. Significance of this study lies in its potential to offer valuable insights to researchers and practitioners working in the field of porous concrete technology and design by providing a pathway to produce permeable pavements with optimized performances, aligning with the specific requirements of various applications and ensuring that the influence of aggregate gradation is systematically considered in the design process.

2 Methodology

2.1 Experimental Program

This section describes the materials, experimental schedule, test methods, and optimization techniques used in this study. Specifically, aggregate selection, effort for compaction, specimen preparation, and determination of porosity and compressive strength are described.

2.1.1 Materials Used

Rock crushed Natural Aggregates (NA) with a specific gravity of 2.81 from the western province of Sri Lanka, Recycled Concrete Aggregate (RCA) with a specific gravity of 2.23 from the COWAM plant, Galle, Sri Lanka, and Type 1 Ordinary Portland Cement (OPC) as a binder with a specific gravity of 3.15, and water from the institution's water supply was used to make the pervious concrete samples. To make sure there was no free chlorine in the water, tests were conducted. Table 1 provides an overview of the OPC's chemical components.

Table 1. Chemical composition of cement (Binder)

Chemical composition	CaO	SiO_2	Al_2O_3	Fe_2O_3	SO_3	MgO	Na_2O	K_2O
By Weight (%)	66.45	20.9	4.50	3.53	2.52	1.270	0.40	0.38

The aggregates were examined according to BS standards for the strength and associated properties. The assessment values for aggregate properties and the standard procedures for both aggregate types are compiled in Table 2.

2.1.2 Mix Design

To cast pervious concrete mixes, natural aggregates were sieved to separate particles with sizes ranging from 12 – 18, and 18 – 25 mm. Smaller aggregates (12–18) were then proportionally mixed with larger aggregates (18–25) in 10, 30 and 50%. Aggregate-to-cement ratios (A/C) of 3.0, 4.0, and 5.0 were then used to prepare the mix. A standard 20-min machine concrete mixing method was used throughout this study. To ensure zero

Table 2. Properties of natural and recycled concrete aggregates

Properties/ Standard	NA	RCA
Bulk density/ BS-EN-1097(kg/m^3)	1587.69	1413.35
Specific gravity/ BS-EN-1097	2.81	2.23
Absorption/ BS-EN-1097	0.19%	2.21%
Moisture content/ BS-EN-1097	0.34%	1.24%
Aggregate Impact value-AIV/ BS-EN-1097	26.54%	31.84%
Aggregate Crushing Value-ACV/BS-EN-1097	30.81%	37.11%
Flakiness Index/ BS-EN-933	14.92	13.43
Elongation Index/ BS -EN-933	23.92%	20.72%

slumps in the mix, a water-to-cement ratio of 0.30 was used to avoid segregation after a few trial mixes. Wet mix from the machine was then cast into 150 mm x 150 mm x 150 mm cubes. Each above-mentioned mix design was further separated into three groups, and each group was given compaction from a 2.5 kg standard proctor. It was dropped freely from a height of 305 mm for each blow, and each group was given 0, 30, and 75 blows of compaction in total. Three samples were cast to each mix design derived from Fig. 2.

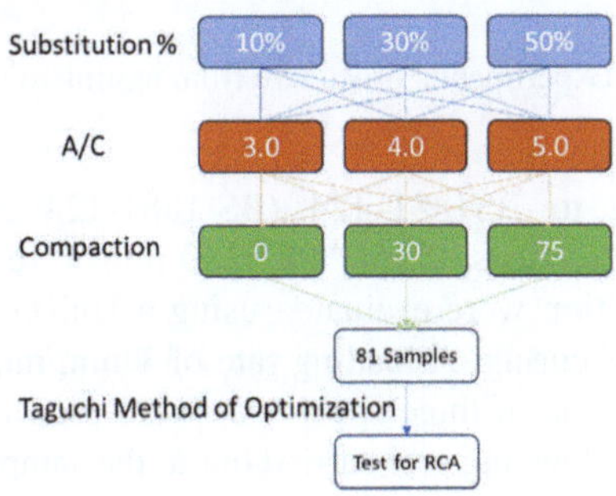

Fig. 2. Chart of mix-design

After compaction, excess concrete was removed with a flat rod, the casting procedure was shown in Fig. 3. Then the pervious concrete cubes were cured at room temperature for 24 h and subsequently, the cubes were de-molded and immersed in clean water for 28 days. After curing, the specimens were removed from the curing tank, resulting in 81 cube samples.

2.1.3 Testing

The air-dried samples were undergone for porosity measurement, a simple water displacement method was used to determine the porosity using Eq. (1).

$$Porosity = \frac{V_c - (L_i - L_f)A}{V_c} \quad (1)$$

where: L_i – Initial water height, L_f – Final water height, A – Cross-sectional area of the sample, V_c – Volume of the Cube sample

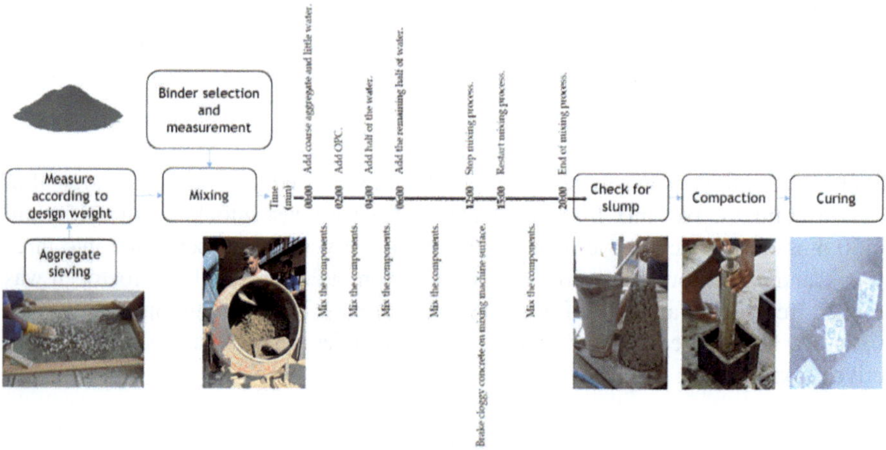

Fig. 3. Experimental procedure from casting to testing

Thereafter, according to BS1881–124 (BS-1881-124 2015), the compressive strengths of cured concrete cubes of 150*150*150 mm were tested. Three cubes for each mortar mix composition were evaluated using a Universal Testing Machine for compressive strength after curing. A loading rate of 1 mm/min was used to apply the axial load. The average value of three specimens were used for further analysis. The same method of casting and testing was carried out to the samples produced from RCA for the optimized mix design.

2.2 Taguchi Method of Optimization

It is method to improve quality of experimental design by systematically studying the influence of multiple independent variables and their interactions. Developed by Genichi Taguchi, it employs orthogonal arrays to design experiments, reducing the number of trials needed while ensuring comprehensive factor exploration. This method focuses on identifying robust settings that minimize variability and optimize performance under real-world conditions. Key elements include the selection of control factors, noise factors, and performance metrics (signal-to-noise ratio). The signal-to-noise ratio quantifies the stability of results, prioritizing conditions that maximize performance consistency. Analysis of variance (ANOVA) is often used to determine the significance of factors and their contributions to the output. The Taguchi method is widely applied in engineering, manufacturing, and process optimization, offering an efficient and cost-effective way to achieve quality improvements while accounting for variability in uncontrollable external factors.

3 Results and Discussion

3.1 Experimental Results

Porosity and compressive strength are two performance indicators, the distributions and summary statistics of which are shown in Fig. 4. For porosity (measured as a volume fraction), the distribution appears moderately normal with a mean of 0.2337 and a standard deviation of 0.1046, indicating relatively low variability. With a mean of 10.51 MPa and a standard deviation of 5.33 MPa, compressive strength is approximately normal, showing greater variability. Both variables' density plots and box plots visually illustrate the data's central tendency and spread. Relationships between porosity (a) and compressive strength (b) in pervious concrete under varying conditions of aggregate proportions, compaction levels, and aggregate-to-cement (A/C) ratios are illustrated in Fig. 5. Each row represents a different proportion of smaller aggregates, 10% (0.1), 30% (0.3), and 50% (0.5) with the remainder comprising larger aggregates. The x-axis represents A/C ratios (3, 4, and 5), while the y-axis represents compaction levels ranging from no compaction (0 Nm) to higher levels (224.4 and 561.0 Nm).

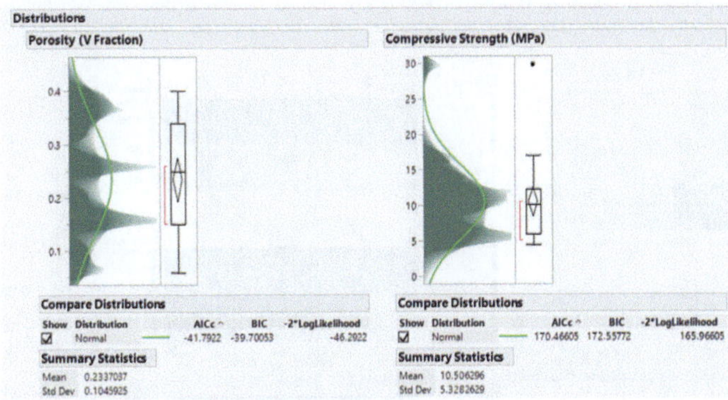

Fig. 4. Distribution of results for porosity(left) and compressive strength (right)

These maps highlight the influence of these variables on porosity and compressive strength. Proportioning of the mix with smaller aggregates influenced porosity, as the proportion of smaller aggregates increases from 10% to 50%, porosity increases. This outcome might be the result of smaller aggregates' tendency to obstruct appropriate packing when utilized in larger quantities because of their larger surface area and less interlocking capacity. This finding is consistent with research in the literature showing that a mix that is dominated by smaller aggregates may become less compacted, which increases the number of void spaces. Higher A/C ratios (moving from 3 to 5) generally result in higher porosity levels. This is due to the increased aggregate content that prevents optimal packing of aggregates. Higher compaction levels (224.4 Nm and 561.0 Nm) significantly reduce porosity across all aggregate proportions and A/C ratios. This agrees with studies showing that compaction decreases void spaces, enhancing density and reducing porosity. Compressive strength tends to drop slightly when the percentage of smaller aggregates rises from 30% to 50%. This is probably due to the fact that smaller aggregates increase porosity, which lowers the mix's overall density and strength(Yavuz and Yazıcı 2023). Across all compaction and aggregate proportion levels, highest compressive strength is achieved with an A/C ratio of 3.0. This is explained by the fact that mixes with A/C 3.0 have high cement paste, which enables improved aggregate packing and less porosity. Cement paste content drops with a greater A/C ratio (4.0 or 5.0), which can result in weaker bonding and lower compressive strength because of less aggregate interlocking supported by (Wijekoon, Janarth et al. 2025). Compressive strength is always higher at higher compaction levels. This is to be expected since compaction enhances the mechanical interlock between aggregates, which is essential for strength development, decreases voids, and increases density.

Percentage of smaller aggregates showcased a significance as other independent parameters AC ratio and compaction, have on the performance of pervious concrete. While lower A/C ratios and higher compaction levels maximize strength, increased smaller aggregate proportions improve porosity but decrease compressive strength. By eliminating repetitive tasks, a methodical approach to optimizing the mix design can save time and money while guaranteeing optimal performance for the intended applications from the available resources.

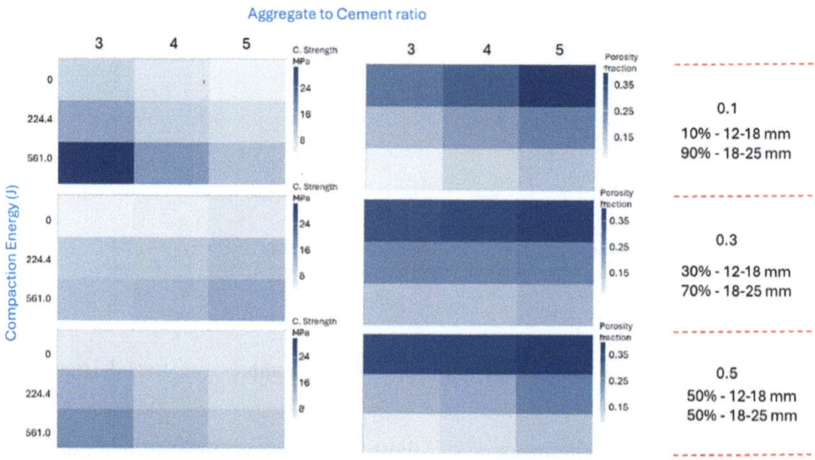

Fig. 5. Variation of compressive strength (C Strength) and porosity with A/C and compaction for each proportion of aggregate

3.2 Mix-Design Optimization

The provided graphs depict the Taguchi optimization process applied to maximize both compressive strength and porosity in a pervious concrete mix. The method uses signal-to-noise (SN) ratios under the "Larger is Better" criterion to identify optimal levels of mix-design parameters such as compaction, aggregate-to-cement ratio, and proportion of smaller aggregates (aggregate size). The plots include the main effects (left) and interaction effects (right) for each response. Each factor's contribution to the performance characteristic (e.g., compressive strength) is assessed individually.

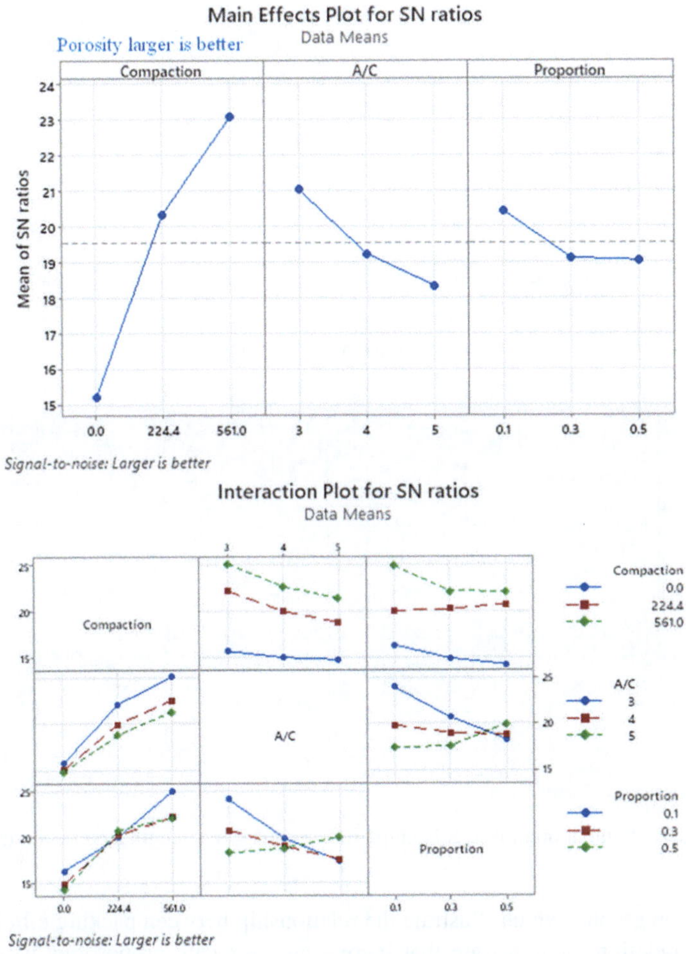

Fig. 6. Taguchi optimized S/N graph for porosity larger is better

To identify the best performing mix-designs the SN ratio for each factor is considered from the main plot and then from the interaction plot, it is essential to confirm whether combining the highest-performing levels is consistent with trends. From Figs. 6 and

7, highest compressive strength is achieved with the mix design consisting of 10% (0.1 fraction) smaller aggregates, compaction at 561.0 Nm, and an A/C ratio of 3.0. This combination ensures optimal aggregate packing, reduced porosity, and enhanced interlocking, resulting in superior compressive strength.

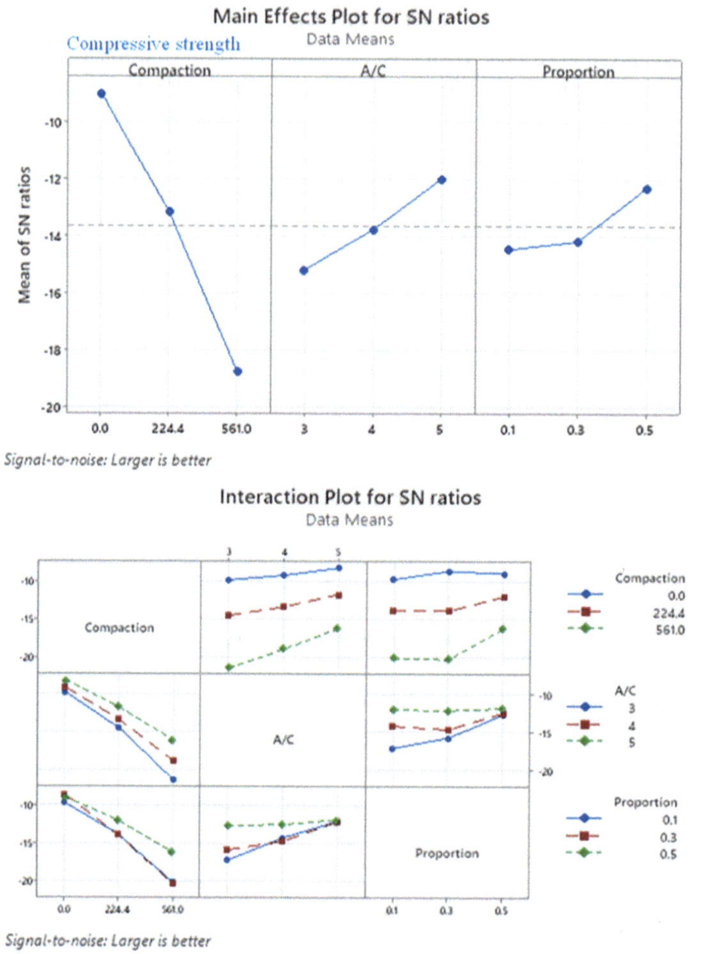

Fig. 7. Taguchi optimized S/N graph for compressive strength larger is better

Interaction graphs, which illustrate the relationship between packing efficiency and aggregate gradation, demonstrate that increasing aggregate proportions work best at higher compaction levels.

However, striking a balance between the performances is inevitable for practical implementations. Therefore, an approach of desirability function is adopted to optimize the mix design in terms of porosity and compressive strength by substituting compressive strength and porosity separately in the place of Y in Eq. (2).

$$S_i = \frac{Y_1 - Y_{Min}}{Y_{Max} - Y_{Min}} \tag{2}$$

Once the S_i is separately computed for compressive strength (S_c) and porosity (S_p), the desirability factor (D) is computed by providing a weight for each performance, compressive strength (Wc), and porosity (Wp) by using Eq. (3). In this study equal weightage is provided and future studies can alter according to their needs. Then the factor was optimized using the Taguchi method and the results is shown if Fig. 8.

$$D = \left(S_c^{W_c} * S_p^{W_p}\right)^{\frac{1}{W_c + W_p}} \tag{3}$$

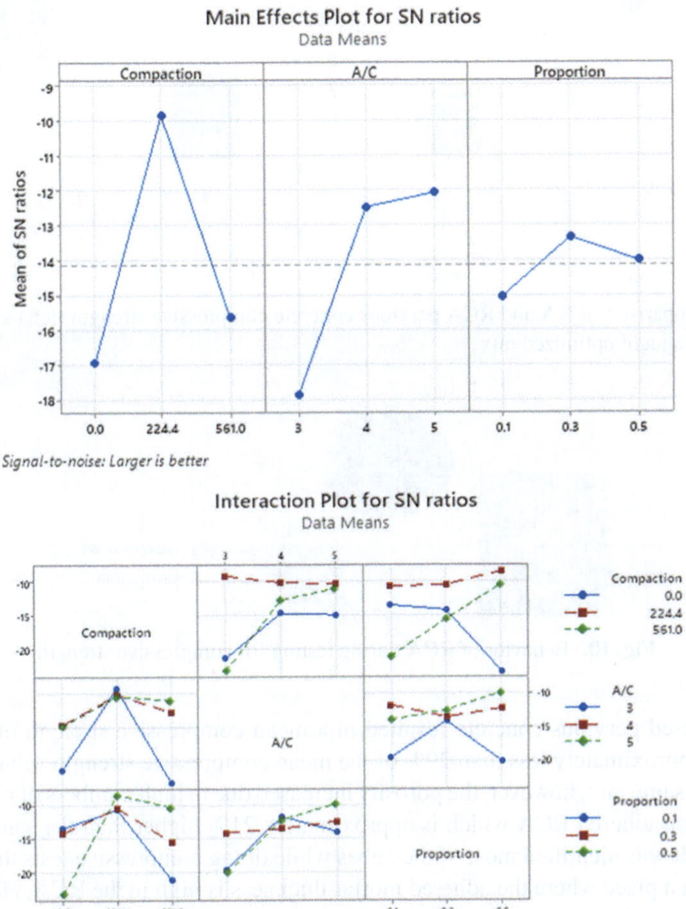

Fig. 8. Taguchi optimized S/N for desirable factors, considering both porosity and compressive strength

From the Fig. 8, optimized design is achieved with the mix design consisting of 30% smaller aggregates, compaction at 224.4 Nm, and an A/C ratio of 5.0. This combination ensures an optimal balance between compressive strength and porosity. Thereafter for the optimized mix again the concrete samples were cast using NA and RCA, each 6 cubes cast and compared the performances. Figure 9, illustrates the performance for both samples.

On the other hand, the main impacts plot shows that the best design for porosity, which is crucial for permeability, involves a lower compaction (0 compaction), larger proportions of smaller aggregates (50%), and an aggregate-to-cement ratio of 5.0.

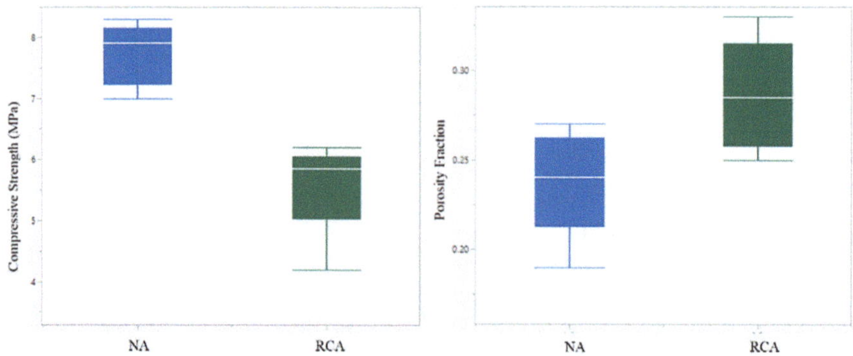

Fig. 9. Comparison of NA and RCA pervious concrete compressive strength (left) and porosity (right) for Taguchi optimized mix

Fig. 10. Behavior of RCA during testing for compressive strength

RCA used pervious concrete resulted in a mean compressive strength of 5.5 MPa which is approximately less than 29% of the mean compressive strength achieved from NA for the same mix, however, the porosity increased due to high numbers of micropores in the motor-adhered RCA which is approximately 21% higher than the sample made up of NA. It was identified most of the cubes while doing compressive tests the fracture occurred at a place where the adhered mortar thickness is high in the RCA, also bottom blind effect is observed due to the separation of fines during compaction as shown in Fig. 10. Low strength is because of the weaker bond between aggregates. Apart from that, several others including type of parent aggregates, age and the exposure condition of parent concrete, number of crushing stages, and amount of adhering mortar in RCA also affect the compressive strength of a concrete.

4 Conclusion

- This study comprehensively analyzed the influence of aggregate gradation, aggregate-to-cement ratio, and compaction energy on the performance of pervious concrete. Gradation significantly affects the parameters while compaction plays a major role in optimizing the compressive strength and porosity. A medium compressive strength of 224.4 J exhibited a favorable situation in all the cases.
- Taguchi method well explained the optimization of two objectives with three inputs with the help of the weights to the performances by using a desirable function, the mix designs were optimized for a balanced performance of porosity and compressive strength, and the prepared RCA pervious concrete samples using the optimized design showcased low compressive strength and high porosity compared to the NA pervious concrete. It can be concluded that the adhered motor thickness (Volume) plays a major role in determining the performances of RCA.
- Despite the advancements, the study identified a few limitations that the porosity is not the direct indicator of permeability (actual performance parameter), where permeability is not accounted for this analysis.
- Future research directions include testing various alternative sustainable aggregates and binders and examining the long-term durability of pervious concrete under varying environmental conditions will help to enhance the practical applicability of pervious concrete and promote its use in sustainable construction practices.

Acknowledgment. I sincerely thank the COWAM Plant – Municipal Council, Galle for their help in providing recycled concrete aggregate.

References

Anburuvel, A., Niruban Subramaniam, D.: Investigation of the effects of compaction on compressive strength and porosity characteristics of pervious concrete. Transp. Res. Rec. **2676**(9), 513–525 (2022)

BS-1881–124 Testing concrete methods for analysis of hardened concrete., British Standards Institution (BSI) (2015)

Liu, R., et al.: Influence of pore structure characteristics on the mechanical and durability behavior of pervious concrete material based on image analysis. Int. J. Concr. Struct. Mater. **14**(1), 29 (2020)

Meddah, M.S., Al-Harthy, A., Ismail, M.A.: Recycled concrete aggregates and their influences on performances of low and normal strength concretes. Buildings **10**(9), 167 (2020)

Mikami, R., Pieralisi, R., Pereira, E.: Influence of compaction energy on pervious concrete properties and vertical porosity distribution. Revista IBRACON de Estruturas e Materiais **17**(2), e17203 (2024)

Mitchell, C.J., Jayakaran, A.D.: Mitigating tire wear particles and tire additive chemicals in stormwater with permeable pavements. Sci. Total. Environ. **908**, 168236 (2024)

Nedeljković, M., Visser, J., Šavija, B., Valcke, S., Schlangen, E.: Use of fine recycled concrete aggregates in concrete: a critical review. J. Build. Eng. **38**, 102196 (2021)

Rasmussen, L.A., Lykkemark, J., Andersen, T.R., Vollertsen, J.: Permeable pavements: a possible sink for tyre wear particles and other microplastics? Sci. Total. Environ. **869**, 161770 (2023)

Sajeevan, M., Subramaniam, D.N.: Investigation of boundary layer impact on pervious concrete. Int. J. Pavement Eng. **24**(2), 2111423 (2023)

Shen, P., Zheng, H., Liu, S., Lu, J.-X., Poon, C.S.: Development of high-strength pervious concrete incorporated with high percentages of waste glass. Cement Concr. Compos. **114**, 103790 (2020)

Subramaniam, D., Hareindirasarma, S., Janarthanan, B.: An alternative approach to optimize aggregate-to-cement ratio and compaction in pervious concrete. Arab. J. Sci. Eng. **47**(10), 13063–13071 (2022)

Thomas, C., Setién, J., Polanco, J.A., Alaejos, P., Sánchez de Juan, M.: Durability of recycled aggregate concrete. Constr. Build. Mater. **40**, 1054–1065 (2013)

Wijekoon, S.H.B., Janarth, A., Dharmar, J., Vinojan, P., Sathiparan, N., Subramaniam, D.N.: Optimizing compressive strength prediction of pervious concrete using artificial neural network. Eng. Res. Express **7**(1), 015418 (2025)

Yavuz, D., Yazıcı, Ş: Experimental study of aggregate size and gradation on pervious concretes' mechanic, hydraulic, and surface properties. Struct. Concr. **24**(4), 5451–5464 (2023)

Yin, S., Chen, Y., Yang, Y., Huang, H., Lv, H.: Effect of supplementary cementitious materials on the resistance of cement paste to carbonic acid water. Struct. Concr. **19**(5), 1399–1408 (2018)

Yu, F., Sun, D., Wang, J., Hu, M.: Influence of aggregate size on compressive strength of pervious concrete. Constr. Build. Mater. **209**, 463–475 (2019)

Zaetang, Y., Wongsa, A., Sata, V., Chindaprasirt, P.: Use of lightweight aggregates in pervious concrete. Constr. Build. Mater. **48**, 585–591 (2013)

Zhang, P., Wang, W., Zheng, Y., Hu, S.: Effect of recycled aggregate and slag as substitutes for natural aggregate and cement on the properties of concrete: a review. J. Renew. Mater. **11**(4), 1853 (2023)

Zhang, Q., Feng, X., Chen, X., Lu, K.: Mix design for recycled aggregate pervious concrete based on response surface methodology. Constr. Build. Mater. **259**, 119776 (2020)

Zhang, X., Li, H., Harvey, J.T., Liang, X., Xie, N., Jia, M.: Purification effect on runoff pollution of porous concrete with nano-TiO2 photocatalytic coating. Transp. Res. Part D: Transp. Environ. **101**, 103101 (2021)

Building Sustainable Smart Cities for Energy Efficiency and CO_2 Emission Reduction: Insights from Europe

U. L. I. Udayantha[1](✉) and S. N. C. M. Dias[2]

[1] Department of Mechanical and Manufacturing Technology, Wayamba University of Sri Lanka, Kuliyapitiya, Sri Lanka
ishanudayantha8@gmail.com

[2] Centre for Environmental Studies and Sustainable Development, The Open University of Sri Lanka, Nugegoda, Sri Lanka

Abstract. Urbanization and climate change present two challenges to sustainable development, with more than 70% of global CO_2 emissions accounts for cities. This review explores the concept of Sustainable Smart Cities (SSC) as a solution to mitigate urban energy inefficiencies and reduce carbon footprints. Drawing lessons from Europe and explores how the integration of renewable energy, smart transportation, and innovative waste management systems could reduce the dependence on fossil fuels and enhance the quality of living. The smart cities with the use of potential technologies such as IoT (Internet of Things), AI (Artificial Intelligence), and smart grids achieve energy efficiency and environmental sustainability. Studying these strategies Sri Lanka (SL), could also contribute to meeting global climate goals. Focusing on stakeholder collaboration, policy reform, and research-driven solutions, this article presents a framework on how transition toward a sustainable urban future in SL can be achieved.

Keywords: Sustainable Smart Cities · Renewable energy integration · Urban sustainability · Climate change mitigation · Sri Lanka urban development

1 Introduction

Climate change and rapid urbanization are the two major challenges in the 21st Century; more than 70% of global CO_2 emissions rise from urban areas. Energy-intensive sectors like transport, industry, and construction fuels these emissions, and urban sprawl enhances biodiversity loss, energy consumption, and production of waste. Cities are prone to climate-related threats such as flooding and rising sea levels, which guide to damage or loss of urban infrastructure and incomes. With more than half of the world's population residing in cities, a number that will increase to 70% by 2030, the destruction of urban inefficiencies has become quite pressing (Luqman, Rayner and Gurney, 2023). SSC integrate resource-efficient advanced technologies into urban planning, thus helping optimize energy use and minimize emissions. Smart cities employ technology systems like IoT, AI, and renewable energy systems for the development of resource-efficient

R. Dissanayake et al. (Eds.): ICRES 2025, PTT, pp. 131–142, 2025.
https://doi.org/10.1007/978-981-96-9551-5_11

ecologies. Energy grids and demand-responsive systems, including renewable energy integrations, minimize energy inefficiencies. It also reduces CO_2 emissions through efficient building design, waste-to-energy conversion, and green public transportation (Bibri et al., 2023). Various tools for monitoring CO_2 levels and analytics help in tracking and managing emissions by the city. Sustainable urban planning fosters compact, mixed-use development and green spaces, and smart mobility solutions, such as electric vehicles and people-friendly infrastructure, decrease fossil fuel use. These measures enhance urban livability, reduce air pollution, and drive economic growth through green technologies (Winkler et al., 2023). SL faces significant challenges in rapid urbanization: increased energy consumption, congestion of traffic, and sprawl of cities have increased its dependence on fossil fuels and its vulnerability to climatic changes. The adoption of SSC models can help the country overcome these issues by exploiting abundant solar and wind resources, using intelligent transportation systems, and green building designs. Drawing from experiences in European cities like Amsterdam and Stockholm, SL can implement its context-specific strategy that balances economic growth and environmental sustainability with an improved quality of life for its citizens (Mahzouni, 2015; Adem Esmail et al., 2022).

2 The Concept of Sustainable Smart Cities

SSCs blend advanced technologies with sustainable development principles in order to address the challenges of rapid urbanization and climate change. These cities focus on economic efficiency, environmental sustainability, and social equity, enabled by digital innovation and active community participation. SSC make use of data and connectivity to improve resource efficiency, protect the environment, and improve the quality of life in urban areas. Key principles include: smart energy systems for optimized production and consumption, energy-efficient building stock, and sustainable urban mobility (Abaker et al., 2023; Pandiyan et al., 2023). The Circular Economy models further minimize the generation of waste by effective recycling and reuse. Citizens are at the center, and well-informed communities are quite important for the success of these initiatives. Examples of European cities such as Amsterdam illustrate effective CO_2 monitoring, integrating renewable energy, and waste-to-energy systems. This would involve, for SL, infrastructure designs that are energy-efficient, collaboration between the public-private sector, and stakeholders in decision-making processes. The backbone of SSC is modern technologies such as IoT, AI, renewable energy systems, and advanced energy storage. IoT enables real-time data collection and monitoring of energy, transport, and waste systems, hence facilitating informed decision-making (Ulpiani and Vetters, 2023). These are further developed through the processes of data analytics, aimed at resources optimization, consumption reduction of energy, and forecasting. Renewable energies are about solar, wind, decreasing dependence on fossils (Bibri et al., 2023; Mishra, 2023). Most European cities have managed to realize all these already. Stockholm smart grids and AI Amsterdam waste management system became bench-marks for their kind. Similar technologies can transform SL's urban systems into cleaner, more efficient, and resilient frameworks. Other essentials include urban planning and sustainable transportation. Compact city designs with integrated residential, commercial, and green spaces reduce energy and transport demand. Green building practices like passive solar design and efficient insulation further reduce energy consumption (Lee et al., 2023). The development of electrical vehicles, cycling networks, and public transport will, therefore, be cardinal in fulfilling the low-carbon ambitions in Amsterdam and Stockholm. Their integration with smart traffic management systems that minimize congestion levels and reduce air emissions will improve their livability quotient. In this connection, SL may adopt policies in prioritizing mass transit, providing incentives for electrical vehicles, and the integration of renewable energy sources within transport systems toward sustainable urbanization and responding locally to the present challenges (Kovačić, 2022).

3 Success Stories of Sustainable Smart Cities from Europe

European cities like Amsterdam and Stockholm are global litmus tests for SSC development, showcasing innovative approaches to urban sustainability. Amsterdam aims for carbon neutrality by 2025, leveraging renewable energy integration, efficient urban planning, and extensive cycling infrastructure. Its district heating systems powered by renewable energy considerably reduce emissions. Amsterdam merges smart energy systems with waste-to-energy technologies to power public transportation and urban areas. Intelligent urban systems provide, in real-time, monitoring data to support citizens in better energy management (Pandiyan et al., 2023). Stockholm has started the use of biogas-powered public transport and introduced an eco-district called Hammarby Sjöstad, integrating waste recycling, water treatment, and renewable energy in a closed-loop ecosystem. These cities show how the use of governance, innovation, and citizen participation brings successful models of SSCs (Leer et al., 2023). The most important thing that can be learned from European cities is the integration of renewable energy. Power in Stockholm is highly dependent on wind turbines, while biomass and waste-to-energy technologies are tapped for district heating systems in Stockholm. Amsterdam is leading in solar panel installations and micro-energy grids for distributed energy generation. These systems reduce dependence on fossil fuel and, consequently, CO_2 emissions-a strategy that SL could follow by leveraging its abundant solar and wind resources to meet growing energy demands in a sustainable manner. Smart grids and energy management systems are also at the heart of European smart cities (Menapace et al., 2020; Mishra, 2023). The "City-Zen" initiative in Stockholm optimizes energy use through smart meters, district heating, and cooling systems. The smart grids in Amsterdam can facilitate dynamic energy distribution, which will enable the consumer to feed excess renewable energy back into the grid. AI helps Amsterdam manage energy supply and demand for reduced energy wastage. Following such strategies can help SL make its urban centers more energy-efficient and reliable. CO_2 monitoring and circular economy practices contribute to urban sustainability (Mishra, 2023; Ratnasingam, 2023). While Copenhagen applies advanced data analytics to monitor and reduce its emissions, Stockholm integrates CO_2 monitoring in public transports and waste management systems. The model of Amsterdam's circular economy converts the municipal waste into electricity and heat, minimizing emissions and land filling. Those practices SL can only be followed by the introduction of CO_2 monitoring technologies and waste-to-energy systems to help build a self-sustaining urban ecosystem that works towards sustainability goals (Ulpiani and Vetters, 2023). Hamburg and Munich focus on district heating, while Helsinki and Espoo attach great importance to renewable energies and carbon neutrality. Public transportation is key to all cities: Hamburg and Munich have supported car-sharing, widespread cycling, whereas Helsinki and Espoo rely on integrated transit systems, autonomous shuttles, and efficient metro and bus combinations. In smart building initiatives, Hamburg leads with green roofs and retrofits; Munich maintains a very strict building code; Helsinki with energy metering and smart homes; and Espoo has applied smart grids to monitor energy consumption (Bouchard et al., 2023). Population density determines their focus on urban planning, where high-density cities like Hamburg and Munich focus on compact infrastructure and waste-to-energy initiatives, whereas Helsinki and Espoo have lower population densities and focus more on waste reduction and broader carbon-neutral goals. Despite their dissimilarities, all these cities drive sustainable urban development forward, with a common theme of energy, mobility, and environmental efficiency (Correia, Marques and Teixeira, 2022). Following Table 1 gives an overview of smart city features in few European cities in terms of energy, mobility, buildings and other main features.

4 Lessons for Sri Lanka

The European strategies for SSCs give a valuable understanding to SL in the field of renewable energy integrations, smart waste management, and smart mobility. Although differences in economic and technological improvements still exist, a common need remains between the two: optimizing energy use, reducing emissions, and making urban areas more livable. SL's abundant renewable energy resources naturally set with Europe's transition from fossil fuel-based energy systems to greener ones, such as solar and wind energies. Scalable solutions include wind farms in Copenhagen and district heating systems in Stockholm for the energy challenges of SL (Majeed et al., 2021). Localized energy systems in Amsterdam, such as solar grids and microgrids, are also supported by the geography of the country. Following a number of circular economy practices, such as waste-to-energy conversion, would solve some urban waste issues while generating energy. This can balance sustainability with socio-economic development in SL using low-cost, high-impact solutions. A strong policy and regulatory framework are needed to carry out SSC concepts. European cities highlight the need for local regulations to be aligned with national climate goals, such as carbon neutrality policies in Copenhagen and sustainable urban planning in Stockholm (Otamendi-irizar et al., 2022). SL needs to develop policies that will motivate and support investment in renewable energy, efficient infrastructure, and innovation in smart technologies. Various other regulatory measures, including feed-in tariffs and green technology subsidies, would ensure private sector participation. Introduction of zoning laws, making green building certification mandatory can be further assurance to sustainable urban development. As in European cities, transparent governance and stakeholder involvement are important in policy implementation. Capacity-building programs must be implemented to build a skilled manpower force that can continue smart city projects (Planning and Operation, 2018). The integration of renewable energy into European smart cities is a core lesson for SL. With the solar and wind potential of the country, these resources are of utmost importance in pursuing energy independence and sustainability. Examples such as wind farms in

Table 1. Overview of Smart City features in Amsterdam, Stockholm, Hamburg, Munich, Helsinki, and Espoo (Sanseverino *et al.*, 2017; Correia, Marques and Teixeira, 2022; Bouchard et al., 2023)

City	Smart Energy	Smart Mobility	Smart Building	Other Features
Amsterdam, The Netherlands	Renewable energy goals: 2,300 GWh by 2040 (solar: 1,000 GWh, wind: 740 GWh, waste: 560 GWh)	Reduction in private vehicle use: 20% in 5 years	Retrofitting historic buildings like De Groene Bocht with fuel cells (efficiency >85%)	Population: 810,084 Density: 3,699/km^2
	Energy from waste: Powers trams, city hall, public lighting	Goal: 200,000 electric vehicles by 2040; 10,000 EVs by 2015 achieved	Smart grid deployment to 10,000 homes in Nieuw-West district	CO_2 reduction target: 40% by 2025 (compared to 1990 levels)
	Biogas: 11 million m^3 from sewage sludge annually	Freight: City Cargo trams replaced 5,000 trucks daily (CO_2 cut by 16%)	Green roofs and walls supported via subsidies	Energy-saving measures reduced CO_2 emissions in Utrechtsestraat by 57%
Stockholm, Sweden	CO_2 emissions: Less than 4 tons per capita (half of Swedish average)	Public transport: 90% population within 300 m of a stop; 60% commuters use it (80% during rush hours)	Eco-district Hammarby Sjöstad: 22,000 residents; energy from renewables like biogas, solar, and district heating	Population: 843,139 Density: 3699/km^2
	69% of households on district heating (70% renewable)	Biogas from Bromma plant: 1.5 million liters/year	Royal Seaport district: 10,000 homes; building energy consumption <55 kWh/m^2/year; fossil-free by 2025	90% of residents live <300 m from green spaces
	Hammarby Sjöstad: Closed-loop system; waste produces 50% of energy needs	Congestion pricing cut private vehicle use by 20%	Waste: 73.5% incinerated for energy, 25% recycled, 1.5% biologically treated	Vision: Fossil-fuel-free city by 2050

(*continued*)

Table 1. *(continued)*

City	Smart Energy	Smart Mobility	Smart Building	Other Features
Hamburg, Germany	CO_2 emissions: 9.1 tons/capita (2022)	Public transport use: 92% of population <300 m from a stop	Green Roof Initiative: 3.6 million m^2 covered by 2030	Waste-to-energy: 70% of city waste
	District Heating: 20% renewable	Car-sharing: Momo project	Energy retrofitting: 15% energy savings	Population: 1.85 million
	Wind energy: 35% of power	Bikes: 2,500 public bikes available		Density: 2,400/km^2
Munich, Germany	Renewable electricity goal: 100% by 2025	Public transport use: 70% daily	Smart retrofitting: Reduced energy use by 30% in older districts	Population: 1.56 million
	Solar capacity: 600 MW installed	Cycling network: 1,200 km	Green building standards: Strict codes	Density: 4,700/km^2
	CO_2 emissions: <5 tons/capita	EVs: Over 1,000 public charging points		CO_2 goal: Carbon neutral by 2030
Helsinki, Finland	CO_2 emissions: 2.5 tons/capita (planned by 2035)	Public transport access: 98% of population within 300 m	Smart homes: Energy metering in Kalasatama	Population: 650,058 Density: 3,000/km^2
	District heating: 50% renewable	PRT: Jätkäsaari autonomous shuttles	Building retrofits: 15% annual energy savings	Vision: Carbon neutral by 2035
	Solar share: 10% planned	Cycling: 3,500 km lanes		Budget: €6 billion
Espoo, Finland	CO_2 emissions: 3.1 tons/capita	Public transport access: 95% of population	Smart grid: IoT monitoring in key districts	Population: 292,796

(continued)

Table 1. *(continued)*

City	Smart Energy	Smart Mobility	Smart Building	Other Features
	District heating: 50% renewable	Metro + buses: 10-min service intervals	Building retrofits: Smart energy systems implemented	Density: 1,500/km^2
	Energy goal: Carbon neutral by 2030	Cycling: Expanding network		Zero-waste goal: By 2030

Copenhagen and solar panel projects in Amsterdam shows how renewable energy can power urban infrastructure while decreasing emissions. This could include investment in decentralized solar grids, offshore wind projects, and energy storage systems such as batteries or pumped hydro storage to share out with intermittency issues and ensure a stable supply of power. This would include supportive policies toward rooftop solar installations and community-based renewable energy systems for encouraging inclusive energy access. This will support reduce dependence on fossil fuels, reduce energy costs, and improve resilience to climate change by combining renewables into the national grid and encouraging public-private partnerships (Enescu et al., 2023; Dobbelsteen, 2024). There are challenges in adapting these European tactics to the context of SL, such as financial limitations, resistance to change, and technological gaps. Inadequate funding for large-scale projects and a lack of digital infrastructure restrict to the adoption of high-cost technologies applied in Europe. For example, while smart grids in Stockholm and data-driven energy systems in Amsterdam present excellent examples, SL has a long way to go regarding investment in capacity building and technology transfer. Partnerships with international organizations, financial institutions, and technology providers will be required to overcome resource gaps. Community acceptance of sustainable smart city initiatives can be ensured only through education and awareness programs (Sharifi et al., 2024).

5 Proposed Framework for Sri Lanka

The urban development of SL should be able to focus more on energy-efficient infrastructure in the building of smart cities, as inspired by cities such as Stockholm and Amsterdam. Urban planning itself should be focused on renewable energy systems, which include solar panels and wind turbines. This would reduce travelling and energy use by mixing residences, commercial areas, and recreational spaces into one zone, an example of bioclimatic architecture similar to Helsinki (Mahzouni, 2015). Transportation is one of the most vital aspects that contribute to reducing carbon footprints. SL can adopt Amsterdam's electric vehicles approach by creating charging networks, offering incentives, and prioritizing electric buses and light rail. A PRT (Personal Rapid Transit) system could provide low-emission transit options, while the promotion of cycling and pedestrian infrastructure would reduce reliance on fossil fuels. Smart grids ensure better energy distribution and integrate renewable sources, as seen in Amsterdam's "City-ZEN" project, reducing energy wastage (Rivero, Oliva and Romero-ania, 2021). The Hammarby Sjöstad-inspired energy-efficient building should incorporate renewable technologies with intelligent automation systems. The public should be informed that, for example, Stockholm manufactures biogas out of household waste. A community-based energy cooperative will engage the residents in opportunities to produce clean energy themselves, just like what Amsterdam did. Participation and gamification tools provide residents with opportunities for ongoing commitment in relation to the fulfillment of the long-term SSC goals (Mahzouni, 2015; Johansson and Brandt, 2017).

6 Potential Benefits and Impacts

This is where integration with SSC technologies becomes so relevant in terms of energy efficiency and reduction of CO_2 emissions. European cities like Amsterdam and Copenhagen have already put across their models in the form of waste-to-energy plants and wind farms that show how clean energy systems can work efficiently. These, along with smart grids and real-time data monitoring, optimize energy distribution, reducing dependence on fossil fuels. The district heating systems run by renewable resources in Stockholm have considerably reduced greenhouse gas emissions, setting the global standard for energy-efficient urban environments (Juan et al., 2023). SL can also replicate such models, leveraging its rich solar and wind resources to implement solar-powered buildings and decentralized renewable energy grids. These systems could go a long way in reducing energy consumption and carbon footprints, thus contributing to the global climate change effort. SSCs take urban living into consideration for health and efficiency. Pedestrian-friendly streets, efficient transport systems, green building designs-offered partly by projects like Hammarby Sjöstad in Stockholm-reduce air pollution, improving both physical conditions and social harmony (Khalid, 2024). Amsterdam manages renewable-powered public transportation and intelligent traffic for better commutes with less congestion. These approaches will help SL transform Colombo into a cleaner and more livable city that attracts investment and tourism, ensuring sustainable urban development. SSCs align with international agreements on climate action, like the Paris Agreement. European cities have set challenging targets for carbon neutrality, thus proving that a reduction of emissions is achievable. As a signatory, SL can achieve its emission reduction targets by the use of renewable energy and enhancing energy efficiency to emerge as a regional leader in sustainable development with international partnerships and funding (Hoek, Struker and Danschutter, 2016; Lin et al., 2022).

7 Challenges and Limitations

The main challenges in carrying out SSC projects in SL are technology, finance, and governance. The technological infrastructure for IoT networks, smart grids, and renewable energy systems is still at a developing stage. Some of the major drawbacks include limited access to advanced technologies and the high cost of importing these systems. Financially, SL faces inadequate investments and dependence on government funding, which makes it difficult to scale smart city initiatives (Mora et al., 2023). While cities like Amsterdam and Stockholm have the backing of strong public-private participation, there are perceived risks and economic instability that may hamper SL's quest for similar investments. Discontinuity in policy and bureaucratic inefficiencies make implementation cumbersome. The lack of comprehensive strategies on urban development further weakens the efforts in making integrated and sustainable systems. These barriers can be overcome if supportive policies are put in place, private investments are invited, and international collaborations for both technical and financial resources are developed. There exists a big skill gap in the workforce (Ulpiani and Vetters, 2023). Advanced technologies such as smart grids, AI-based urban planning, and renewable energy systems need skilled workers, which cannot be offered by educational and vocational programs in SL. Other European cities are well equipped in research institutions and industry linkages. It is expected that developing this capacity, through specialized training, international collaboration, and on-the-job experience with regards to smart city projects, will provide the needed catalysts for this new era. For this, the plan is to build research and development Hubs on urban sustainability as part of driving innovation for self-reliance in achieving successful implementation (Gracias et al., 2023).

8 Future Perspectives

The pathway to SSCs for SL is, to a great extent, pegged onto the active engagement and collaboration of major stakeholders at policy, industry, and academic levels. The policymakers have to lead the way in laying down strong frameworks that align urban development with national goals of energy efficiency and carbon neutrality. These could include tax incentives for renewable energy adoption, stricter regulations toward energy-efficient buildings, and investments in public transportation like electric buses and cycling networks. Drawing inspiration from easy access to green spaces in Stockholm and the public-private partnerships of Amsterdam, SL can adapt these models to its socio-economic and environmental context (Yadav et al., 2019). Industries need to be encouraged toward cleaner modes of production with investments in green technologies and strategic alliances that would drive smart innovations. For example, renewable energy companies can partner with construction firms in the development of net-zero buildings, while transport firms can electrify their fleets. Academia is that important link connecting policy and practice, enabling research, innovation, and education. Interdisciplinary research centers on urban sustainability should be established at the universities, similar to the Masdar Institute in the UAE, for sharing knowledge and educating the future experts in the field. The strategic research and international collaborations will be very instrumental in developing smart city initiatives in SL. This will involve the adaptation of renewable energy technologies, including ultra-efficient solar panels and wind turbines for SL's tropical environment, while innovative smart grid technologies, emphasizing the use of low-cost, decentralized systems of energy production, could provide scalable solutions for urban and rural areas alike (Razmjoo et al., 2022; Tucunduva et al., 2022). A related priority involves the sustainability of transport systems, especially electric vehicles integrated within renewable energy-powered charging networks. This will be further enriched by collaborations with European institutions such as the KTH Royal Institute of Technology, Stockholm, or the Amsterdam City-ZEN project. Pilot projects accomplished in eco-districts like Hammarby Sjöstad or smart energy initiatives in Amsterdam will serve as a model for SL. Participation in global consortia and conferences on sustainability opens up avenues for funding, technology transfer, and skill development. Through research-driven collaborations, SL can innovate sustainable solutions, mitigate climate change impacts, and lead in regional smart city advancements (Wesseling, Faber and Hekkert, 2014; Generation, 2023).

9 Conclusion

The fundamental factor in solving the challenges brought by urbanization and climate change may stand with sustainable smart cities. Examples from Europe have shown that integrated technologies, renewable energy systems, and community engagement are drivers of urban sustainability merged with economic growth. Their adoption in Sri Lanka requires localized strategies that tap into abundant solar and wind resources, promote electrification of public transportation, and integrate waste-to-energy systems. Success shall be achieved by combined efforts on the part of policymakers, industries, and academia in making enabling policies, innovative technologies, and skilled human resources. While financial constraints and technological gaps exist, these challenges get better through collaborative partnership and capacity-building initiatives. By accepting sustainable smart city principles, Sri Lanka will take the lead in the region in aligning urban development with environmental protection and global climate dedications.

References

Abaker, I., et al.: Urban computing for sustainable smart cities: recent advances, taxonomy, and open research challenges (2023)

Adem Esmail, B., et al.: Greening cities through urban planning: a literature review on the uptake of concepts and methods in Stockholm. Urban For. Urban Green. **72** (2022). https://doi.org/10.1016/j.ufug.2022.127584

Bibri, S.E., et al.: Environmentally sustainable smart cities and their converging AI, IoT, and big data technologies and solutions: an integrated approach to an extensive literature review. Energy Inform. **6**(1) (2023). https://doi.org/10.1186/s42162-023-00259-2

Bouchard, P., et al.: A case study on smart grid technologies with renewable energy for central parts of Hamburg. Sustain. (Switz.) **15**(22), 1–29 (2023). https://doi.org/10.3390/su152215834

Correia, D., Marques, J.L., Teixeira, L.: The state-of-the-art of smart cities in the European union. Smart Cities **5**(4), 1776–1810 (2022). https://doi.org/10.3390/smartcities5040089

Van Den Dobbelsteen, A.: A parametric modelling approach for energy retrofitting heritage buildings: the case of amsterdam city centre, pp. 1–20 (2024)

Enescu, D., et al.: Energy tariff policies for renewable energy development: Sri Lanka (2023)

Generation, P.: A novel renewable smart grid model to sustain solar power generation (2023)

Gracias, J.S., et al.: Smart cities—a structured literature review, pp. 1719–1743 (2023)

Van Der Hoek, J.P., Struker, A., De Danschutter, J.E.M.: Amsterdam as a sustainable European metropolis: integration of water, energy and material flows. 9006 (2016). https://doi.org/10.1080/1573062X.2015.1076858

Johansson, S., Brandt, N.: The potential of the infrastructural system of Hammarby Sjöstad in Stockholm, Sweden. **59**(2013), 716–726 (2017). https://doi.org/10.1016/j.enpol.2013.04.027

Juan, A.A., et al.: Promoting energy efficiency and emissions reduction in urban areas with key performance indicators and data analytics, pp. 1–19 (2023)
Khalid, M.: Smart grids and renewable energy systems : Perspectives and grid integration challenges. Energy Strategy Rev. **51**(January), 101299 (2024). https://doi.org/10.1016/j.esr.2024.101299
Kovačić, M.: New paradigm of sustainable urban mobility : electric and autonomous vehicles—a review and bibliometric analysis (2022)
Lee, J., et al.: Smart city as a social transition towards inclusive development through technology: a tale of four smart cities. Int. J. Urban Sci. 75–100 (2023). https://doi.org/10.1080/12265934.2022.2074076
Van Der Leer, J., et al.: energy research & social science energy systems in sustainability-profiled districts in sweden: a literature review and a socio-technical ecology approach for future research. **101**(May) (2023). https://doi.org/10.1016/j.erss.2023.103118
Lin, H., et al.: After the epidemic, is the smart traffic management system a key factor in creating a green leisure and tourism environment in the move towards sustainable urban development ? (2022)
Luqman, M., Rayner, P.J., Gurney, K.R.: On the impact of urbanisation on CO2 emissions. NPJ Urban Sustain. **3**(1), 1–8 (2023). https://doi.org/10.1038/s42949-023-00084-2
Mahzouni, A.: The "policy mix" for sustainable urban transition: the city district of hammarby Sjöstad in Stockholm. Environ. Policy Govern. **25**(4), 288–302 (2015). https://doi.org/10.1002/eet.1688
Majeed, A., et al.: Municipal solid waste management practices and challenges in the southeastern coastal cities of Sri Lanka (2021)
Menapace, A., et al.: The design of 100% renewable smart urban energy systems: the case of Bozen-Bolzano. Energy 118198 (2020). https://doi.org/10.1016/j.energy.2020.118198
Mishra, P.: Energy management systems in sustainable smart cities based on the internet of energy: a technical review (2023)
Mora, L., et al.: Technovation Smart city governance from an innovation management perspective: theoretical framing , review of current practices , and future research agenda. Technovation **123**(February), 102717 (2023). https://doi.org/10.1016/j.technovation.2023.102717
Otamendi-irizar, I., et al.: Energy research & social science how can local energy communities promote sustainable development in European cities? **84** (2022). https://doi.org/10.1016/j.erss.2021.102363
Pandiyan, P., et al.: Technological advancements toward smart energy management in smart cities. Energy Rep. **10**, 648–677 (2023). https://doi.org/10.1016/j.egyr.2023.07.021
Planning, N., Operation, N.: Development of variable renewable energy policy in developing countries: a case study of Sri Lanka Xunpeng Shi * Hengsi Chen Fushuan Wen, vol. 14, pp. 10–29 (2018)
Ratnasingam, G.: JURISMA: jurnal riset bisnis dan manajemen volume XIII Nomor 2 (Oktober 2023) application of alternative energy sources as a sustainable strategy in Sri Lanka: cases review. **XIII**(Oktober), pp. 217–236 (2023)
Razmjoo, A., et al.: The key role of clean energy and technology in smart cities development. Energy Strategy Rev. **44**(September), 100943 (2022). https://doi.org/10.1016/j.esr.2022.100943
Rivero, L., Oliva, D.V., Romero-ania, A.: Managing sustainable urban public transport systems: an AHP multicriteria decision model, pp. 1–20 (2021)
Sanseverino, E.R., et al.: Smart cities: case studies (2017). https://doi.org/10.1007/978-3-319-47361-1
Sharifi, A., et al.: Smart cities and sustainable development goals (SDGs): a systematic literature review of co-benefits and trade-offs. Cities **146**(December 2023), 104659 (2024). https://doi.org/10.1016/j.cities.2023.104659

Tucunduva, T., et al.: Understanding sustainable energy in the context of smart (2022)
Ulpiani, G., Vetters, N.: Circular economy and waste management to empower a climate-neutral urban future. **421**(August) (2023). https://doi.org/10.1016/j.jclepro.2023.138454
Wesseling, J.H., Faber, J., Hekkert, M.P.: Technological forecasting & social change how competitive forces sustain electric vehicle development. Technol. Forecast. Soc. Change **81**, 154–164 (2014). https://doi.org/10.1016/j.techfore.2013.02.005
Winkler, L., et al.: The effect of sustainable mobility transition policies on cumulative urban transport emissions and energy demand. Nat. Commun. **14**(1) (2023). https://doi.org/10.1038/s41467-023-37728-x
Yadav, G., et al.: Developing a sustainable smart city framework for developing economies: an Indian context. Sustain. Cities Soc. **47**(January), 101462 (2019). https://doi.org/10.1016/j.scs.2019.101462

Dynamics of Total Solids Capture in Pervious Concrete Under Low Solids Loading Applications

K. Rishikeshan[1(✉)], A. Janarth[2], N. Sathiparan[1], and D. N. Subramaniam[1]

[1] Department of Civil Engineering, University of Jaffna, Ariviyal Nagar, Killinochchi, Sri Lanka
2022e068@eng.jfn.ac.lk

[2] Department of Multidisciplinary Studies, Eastern University Sri Lanka, Batticaloa, Sri Lanka

Abstract. Pervious concrete increases the permeability of paved surfaces, effectively mitigating flash flooding in urban catchments by allowing water to infiltrate the ground. However, runoff from urban areas often carries total solids, which can accumulate within the pores of pervious concrete, leading to clogging and reducing its effectiveness over time. Clogging can be assessed based on Event-Mean concentration (EMC), but this method provides limited information on the dynamics of solid retention. Specimens were cast with 2.36–5.0 mm and 5.0–10 mm aggregates (2 specimens each) with an aggregate-to-cement ratio of 4.0, a water-to-cement ratio of 0.3, and no compaction. With total solids of less than 75 µm and 75–150 µm particles, inflow designed under various hydraulic and solid loading configurations. Samples of outflow were taken at different times during an event and tested for turbidity and total solids. Specimens with larger aggregates, which creates larger pores captured solids less effectively, while specimens with smaller aggregates demonstrated better solid retention. Solid capture percentages remained consistent across multiple events, and none of the specimens reached clogging or saturation points.

Keywords: pervious concrete · total solids · pollutant removal · aggregate size · solids loading

1 Introduction

Urbanization is a major contributor to the unsustainable practices that are seen all across the world. Flash flooding is one of the major effects of urbanization. Urban catchment runoff is carrying pollutants ranging from micro-scale sand particles to nano-scale heavy metals (Kazemi and Hill, 2015). Although soil particles can potentially carry inorganic pollutants through sorption, it is advantageous to regulate the solid loading in water resources. Pervious concrete aims to increase permeability to retain the hydrology to predevelopment characteristics (Okaikue-Woodi et al., 2020). Coarse aggregates constitute a great proportion of the pervious concrete along with binder, while fine aggregate

R. Dissanayake et al. (Eds.): ICRES 2025, PTT, pp. 143–156, 2025.
https://doi.org/10.1007/978-981-96-9551-5_12

entirely lacks in pervious concrete. Permeability and strength characteristic are defined with aggregates size and their ratio to cement. To ensure the binder does not clog the pores at the bottom surface, pervious concrete requires optimized compaction and zero slump (Sathiparan et al., 2024b; Wijekoon et al., 2024). Therefore, the choice of aggregate size and aggregate-to-cement ratio affect designing the mixture elements of pervious concrete greatly. For a successful implementation in urban catchment, with reducing its impact particularly on the hydrologic characteristics of the catchment, permeability of the pervious concrete is crucial. The size and porosity distribution of the pores in pervious concrete influence its permeability. Additionally, the tortuosity of fluid flow depends on the width and size of the channels transmitting the fluid, which are determined by the size and shape distribution of the pores. Several studies tried and analyzed the permeability of pervious concrete with analyzing the relation to porosity or compressive strength. The consideration of even the same porosity can have different permeability owing to degree of vertical interconnection of pores and size and shape distribution, has been often neglected (Shobijan et al., 2024; Yao et al., 2018). These elements can also affect permeability to a higher extent comparing to the measure of porosity solely. For an accurate prediction of permeability, it would have to be predicted incorporating the pore characteristics. For subsequent management, planned application of pervious concrete in urban pavements would let infiltration of stormwater runoff to percolate through. The concentration of total solids in stormwater runoff, varying depending on the catchment characteristics and stormwater runoff speed (Subramaniam and Mather, 2016). The solids that are carried in water percolates through, could perhaps be captured in the pores of pervious concrete. This can occur due to either the process of straining or differential settling, interception or brownian motion (Subramaniam and Mather, 2016). Where size of the particle is approximately larger than 0.15 times if the size of the pore, retaining these larger particles on top of the pore is the process called retaining. Irrespective of flow conditions, capture of a solid particle due to process of straining would remain stable. However, particles smaller than 0.15 time the size of pore may also be captured in the pores caused by differential settling, interception and Brownian motion. Often solid particles attached to pervious concrete surfaces superficially, which is a temporary retention. These temporarily retained particles can be re-entrained and end up in the percolating water, eventually ending up in the outflow. Since total solids do not go through any physical or chemical transformation, by any other processes in captivity in the pore, they shall not be removed. Moreover, the effective pore area available for fluid flow gets progressively reduced, obstructing fluid flow and decreasing permeability when the particles are continually captured in the pores. Eventually this leads to clogging (Subramaniam et al., 2014). Due to the constriction in the fluid passage, increased sheer that in turn re-entrain trapped particles, guiding to flushing of solids in the outflow. Studies in contemporary literature have sought to track the clogging in laboratory-scale pervious concrete, typically by using an event-mean on the concentration of solids in both inflow and outflow. An event-man of loading and flushing of solids would fail to capture the dynamics in solids retention and re-entrainment of solids, significantly limiting the knowledge of solids capture mechanism (Subramaniam et al., 2018). Designing pervious concrete to maintain the hydrological characteristics of a catchment requires models that would incorporate solid loading characteristics, dynamics of capture and re-entrainment in relation to mix design parameters. This study examines the dynamics of solids retention and re-entrainment in pervious concrete optimized for mechanical properties and solids of comparable particle sizes observed in urban stormwater runoff.

2 Materials and Methods

Material Testing

This study was conducted on laboratory-scale experiments. Pervious concrete specimens were cast with Ordinary Portland cement (OPC, complies with BS-12–1996), crushed gneiss-based coarse aggregates sourced from local construction suppliers, and tap water (tested for negligible free chlorine). Aggregates were tested to verify that they satisfied the requirements to be used in concrete production. Aggregates used in the study conformed broad requirements specified in the standards. The deployed aggregate demonstrated significant resilience under high-impact, abrasive, and static pressure conditions, as evidenced by its Greater Aggregate Impact Value (AIV), Los Angeles Abrasion Value (LAAV), and Ten Percent Fines Value (TFV). Additionally, water absorption tests revealed minimal internal porosity within the aggregates.

Mix-Design and Casting of Specimens

The water-to-cement (W/C) ratio plays a critical role in influencing concrete's porosity, compressive strength, and workability. Low W/C ratios can lead to poor aggregate bonding due to insufficient cement paste, while high ratios may clog pores with excess paste. In this study, a constant W/C ratio of 0.3 was maintained, aligning with recent literature to balance strength and permeability while achieving zero slump without admixtures. To ensure consistent impact of the W/C ratio on permeability and strength, it was kept constant across all pervious concrete blends. Specimens were cast with an aggregate-to-cement ratio of 3.0 and a water-to-cement ratio of 0.3, using two aggregate sizes: 2.36–5.0 mm and 5.0–10.0 mm. Following guidelines, four fresh mix specimens with zero slump were prepared without compaction, apart from light tamping to level the top surface. During molding, 100 mm diameter, 50 cm long Perspex tubes were inserted to the bottom.

The pervious concrete specimens, along with the inserted Perspex tubes, were cured in a water tank for 28 days. A 120 L electrically powered concrete mixer was used to prepare the mix. The required aggregate-to-cement (A/C) and water-to-cement (W/C) ratios were calculated to determine the necessary quantities of coarse aggregate, cement, and water. Each component was precisely weighed using an electronic scale with a minimum count of 0.1 mg before being combined. The entire mixing process took 20 min to achieve a uniform mixture. According to the specifications for pervious concrete mix design, all mixes exhibited little to no slump (less than 10 mm). Following that, the mixture was poured into concrete molds measuring 150 mm × 150 mm × 150 mm. A tamping rod is used to compact the sample, where 10 tamping was given in equal three layers as of a total of 30 tamping, however, the blows were given uniformly across the cube's face as required in BS 1377–4:1990. After tamping the excess mix was leveled and removed using an iron rod. The cast pervious concrete cubes were left in the mold for one day at room temperature. The following day, the cubes were removed from the molds and fully submerged into a clean water bath for a 28-day curing period.

Solid Removal Experiment

This study aims to evaluate the capacity of pervious concrete specimens to retain solids. Experiments are designed to simulate rainfall events, where stormwater runoff contains suspended solids. For these simulations, native soil was collected from stormwater collection ponds following heavy rainfall, dried, and separated into solids. The solids were then sieved using 150 and 75-μm sieves to classify the suspended solid particles into two categories: (a) particles smaller than 75 μm and (b) particles between 75 and 150 μm.

The simulation events lasted 1.5 h (90 min) each, with hydraulic loading rates ranging from 50 to 300 mL/min (equivalent to approximately 6.37 to 38.20 mm/min). Solids concentrations in the inflow varied between 30 and 3200 mg/L. Each pervious concrete specimen underwent 13 events, with a constant 2-day dry period between successive simulated rainfall events. Two columns, each containing pervious concrete (PC) specimens with 2.36–5 mm and 5–10 mm aggregates, were exposed to simulated rainfall events. One set of specimens received suspended solids less than 75 μm in diameter, while the other set received suspended solids between 75–150 μm. All 13 events for each PC specimen were conducted consecutively without washing the specimens, to evaluate solids retention capacity over time. Outflow was collected at intervals of 5, 10, 20, 30, 40, 50, 60, and 90 min from the start of outflow. The collected samples were analyzed for turbidity and total solids (TS). Turbidity was measured using the desktop Hach DR6000, with three measurements per sample, and the mean value was recorded. For total solids, a dry petri dish was weighed, a known volume of the sample was added, and the dish was dried in an oven at 104 °C for 24 h before measuring the weight of the dry solids.

The measured weight of dry solids was recorded as the total solids (TS) concentration in the water samples, providing a time series of outflow turbidity and solids concentrations. Inflow concentrations and hydraulic conditions were kept constant throughout the experiments.

3 Results and Discussion

The results and discussion pertain to experiments conducted on four configurations of pervious concrete (PC) with varying aggregate sizes and total solids (TS). The configurations are as follows:

Column 01: 2.36–5.0 mm aggregates with 75–150 μm TS, Column 02: 5.0–10.0 mm aggregates with 75–150 μm TS, Column 03: 2.36–5.0 mm aggregates with 150–300 μm TS, Column 04: 5.0–10.0 mm aggregates with 150–300 μm TS. The experimental analysis included multiple runs to evaluate these configurations.

The analysis begins with a batch wise approach. The total mass of solids collected from the outflow channel, abbreviated as CTSOut, was weighed and recorded. Similarly, the total amount of solids loaded into the columns, abbreviated as CTSIn, was also documented. Figure 1 illustrates the variation of CTSOut versus CTSIn for: (a) individual columns, and, (b) all observations collectively. The data reveals significant variation in the mass of solids across observations. However, a general trend emerges, indicating a positive correlation between CTSOut and CTSIn, albeit with high uncertainty. Figure 1b demonstrates a linear trend for CTSIn versus CTSOut, combining all observations irrespective of columns or operational conditions. The model shows high accuracy, with an R^2 value of 0.89, and the trend is represented by a linear equation (intercept set to the origin) with a gradient of 0.5688.

The gradient reflects the general removal efficiency, approximately 56.88%. This indicates that the model effectively captures the total solids (TS) removal potential of about 57% across all events and conditions, with less than 10% variation. Therefore, the findings suggest that the pervious concrete system removes approximately 57% of the solids loaded, within the scope of the design and operational conditions used in this study. To generalize the model and assessment outcomes, a detailed evaluation of the uncertainty observed in this study is essential. The study involves two configurations based on: The aggregate size used in casting pervious concrete, and the particle size of the Total Solids (TS) in the inflow. To investigate the source of uncertainty, the observations were divided based on these two configurations, and the results are shown in Fig. 2. Figure 2a display the variation of CTSOut versus CTSIn for configurations based on TS particle size in the inflow. Linear trends for the 75–150 µm and 150–300 µm configurations yielded high accuracies, with R^2 values of 0.90 and 0.76, respectively. However, a noticeable difference in accuracy between the two configurations is evident, which will be discussed further in the study.

Another significant difference observed is in the gradient values: 0.4153 for the 75–150 µm configuration and 0.6325 for the 150–300 µm configurations. This corresponds to the removal efficiencies, with 41.53% efficiency for the smaller 75-150 µm and 63.25% for the larger 150–300 µm configuration. This suggests that pervious concrete, regardless of aggregate size (2.36–5.0 mm or 5.0–10.0 mm), is more effective at capturing larger solid particles than smaller ones.

The total mass of loading for 150–300 µmis significantly higher than that for 75–150 µm, primarily due to the settling of larger particles in the inflow channel and storage. The flow rates in this study were insufficient to flush larger particles, a trend also observed in natural stormwater runoff where larger particles are less common. The observed disparity in removal efficiency may be attributed to uncertainties in capturing and releasing larger particles, influenced by micro-flow properties in the pore channels of pervious concrete. Figure 2b shows the variation of CTSOut and CTSIn for different aggregate sizes used in casting pervious concrete. The trends for the 2.36–5.0 mm and 5.0–10.0 mm aggregate configurations are linear, with high accuracy (R^2 values of 0.83 and 0.97, respectively). A significant difference in accuracy is observed between the two aggregate size groups, especially between 75–150 µm and 1500–300 µm. The gradients for the 2.36–5.0 mm and 5.0–10.0 mm specimens differ notably, with values of 0.6037 and 0.3075, respectively. This corresponds to approximately 60% solids removal in 2.36–5.0 mm specimens and 30% removal in 5.0–10.0 mm specimens, indicating that smaller aggregates retain more solids (almost twice the retention efficiency) than larger ones. These findings hold regardless of the TS particle size supplied. If all aggregates are assumed to be perfectly spherical, mono-sized, and packed in configurations like Face Centred Cubic: FCC,

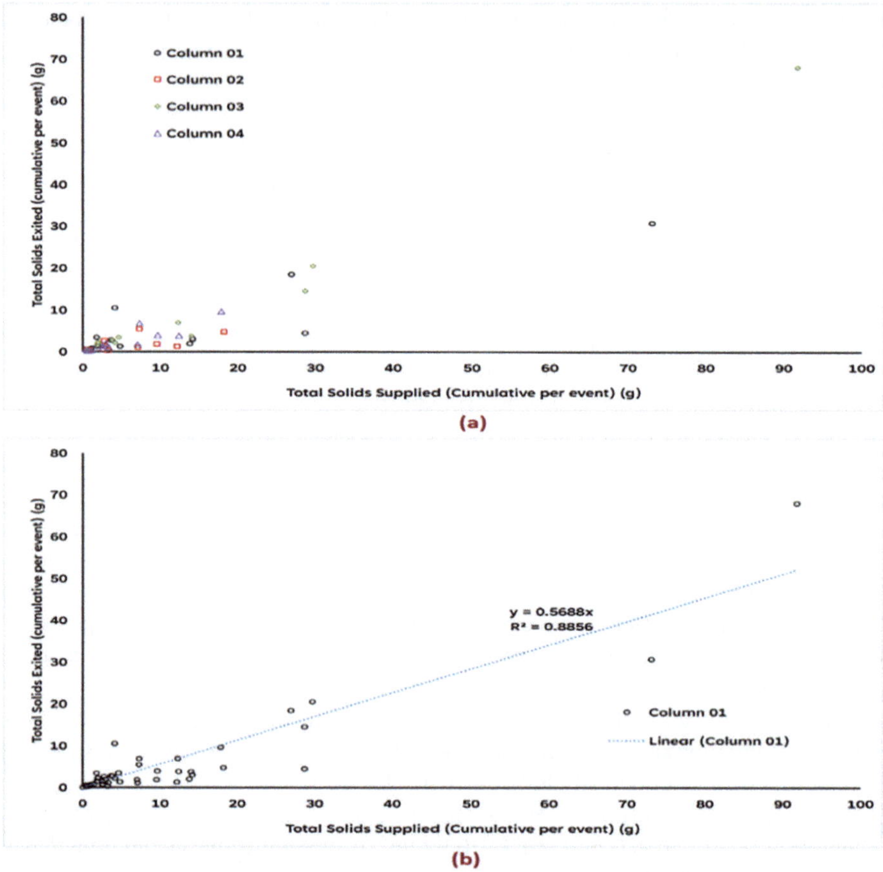

Fig. 1. Variation of Total Solids in the outflow per event with Total Solids loaded (supplied) for (a) observations by individual columns (b) all observations collectively

Body Centred Cubic: BCC, or Simple Cubic: SC, they would exhibit consistent packing density regardless of the size. In pervious concrete, porosity remains consistent regardless of aggregate size; however, pore size varies significantly. Larger aggregates create larger pores, while smaller aggregates produce smaller pores. When aggregates have a range of sizes and shapes, larger aggregates statistically result in larger pores compared to smaller aggregates. Consequently, smaller aggregates are expected to capture more solids and potentially retain larger aggregates due to their smaller pores.

Examining the performance of individual columns in detail would enhance the understanding of the observations discussed, as illustrated in Fig. 3. Linear trends for all columns, except Column 02, show high accuracy (R^2 above 0.8). Column 02 has an R^2 of 0.69. Column 01, with 2.36–5.0 mm aggregate specimens and 75–150 μm TS loading, achieves an average removal of 59.02% (gradient: 0.5902). In contrast, Column 02, with 5.0–10.0 mm aggregate specimens and 75–150 μm TS loading, achieves an average removal of 74.41% (gradient: 0.7441). This aligns with the earlier hypothesis

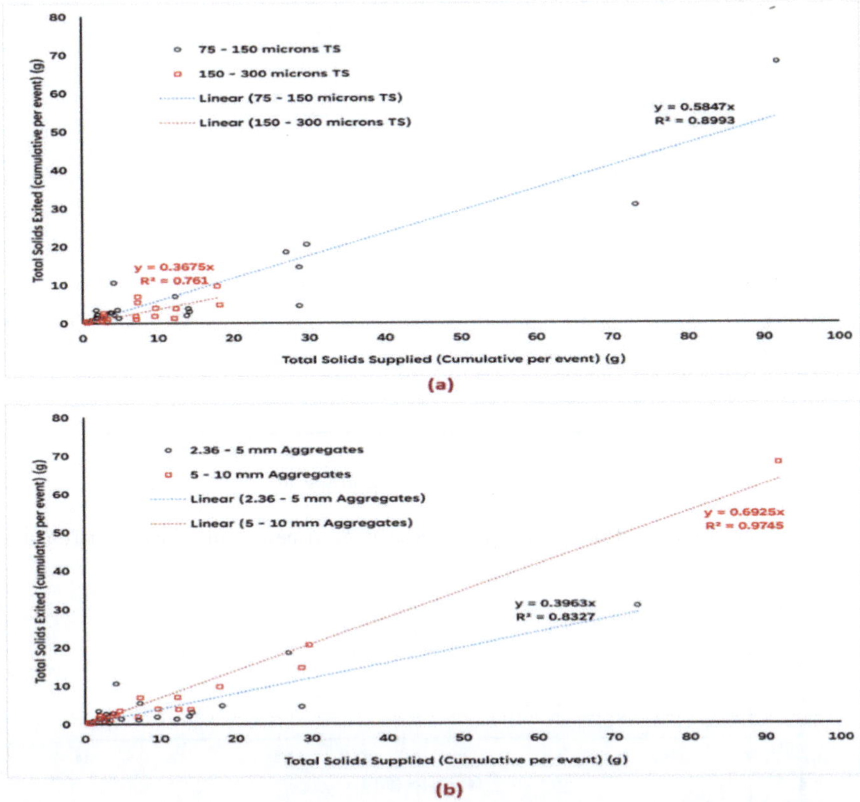

Fig. 2. Variation of Total Solids in the outflow (exited) per event with Totals Solids loaded (supplied) for (a) observations split by pervious concrete aggregate size (b) observations split by particle size of TS supplied

that larger pores are more effective at capturing smaller particles than smaller pores. For 150–300 μm TS loading, gradients indicate capture rates of 29.33% (0.2933) and 52.04% (0.5204) for 2.36–5.0 mm and 5.0–10.0 mm aggregate pervious concrete specimens, respectively. Comparing Column 01 (0.5902) and Column 03 (0.2933) shows the same specimen captures more smaller particles (59.02%) than larger particles (29.33%). Similarly, for 5–10 mm specimens, gradients of Column 02 (0.7441) and Column 04 (0.5204) reflect higher capture of smaller particles.

Most studies traditionally evaluate a system's performance in removing pollutants as a percentage of the outflow relative to the inflow of the specific pollutant (Eq. 1). It is also important to assess the performance of each column for individual events, as is commonly reported in conventional studies.

In this case, the governing system of equations can be written as follows:

$$\textit{Pollutant removal performance} = \frac{(\textit{total loading of pollutant} - \textit{total outflow of pollutant})}{\textit{Total loading of pollutant}} \times 100 \quad (1)$$

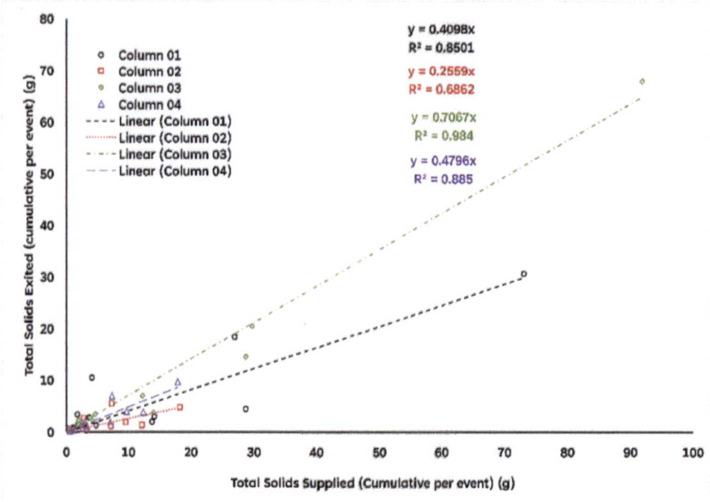

Fig. 3. Variation of TS exited the system against TS supplied to the system, for individual columns

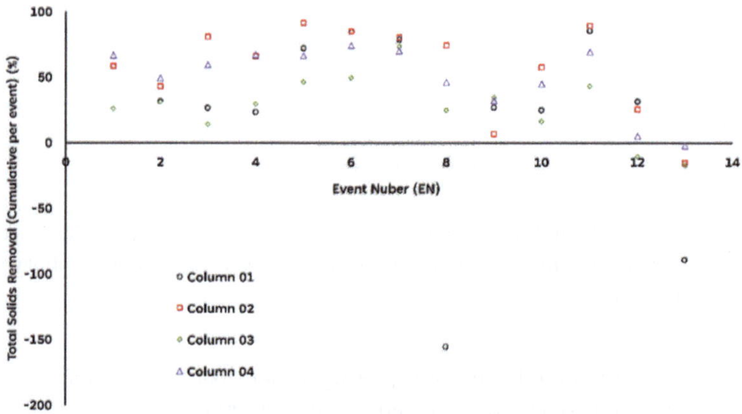

Fig. 4. Variations of TS removal (percentage) with event number (EN)

Figure 4 illustrates the variation in the percentage of total solids (TS) captured by pervious concrete specimens during individual events, plotted against event number (EN). The data shows significant fluctuation in the capture percentage across both columns and events, without any clear pattern suggesting the age of the specimen affects particle capture. Notably, all four columns show a negative removal percentage, indicating that total solids previously retained in the pores were flushed out during subsequent events. This flushing of solids was particularly evident in higher ENs, possibly suggesting a tolerance limit for the specimens in retaining particles as solid loading increases. Figure 5 shows the relationship between total solid removal and solid loading in the system. The data does not provide further clarity on solids retention and flushing, as the removal percentage fluctuates widely with different solid loadings. However, for events with more than 10 g of solid loading, the removal percentage ranged between 40% and 90%. In contrast, for events with less than 10 g of solid loading, the removal percentage varied from −160% to 90%. The negative or lower removal percentages in low total solids (TS) loading events can be primarily attributed to the sensitivity of the removal percentage calculation (Eq. 4) to the in-flow concentration (solid loading) into the system. Additionally, the removal percentage, as discussed, does not capture the dynamic nature of solid removal in pervious concrete, which changes over time during an individual event. For instance, a flush of solids may occur at different times during an event or at varying total loadings

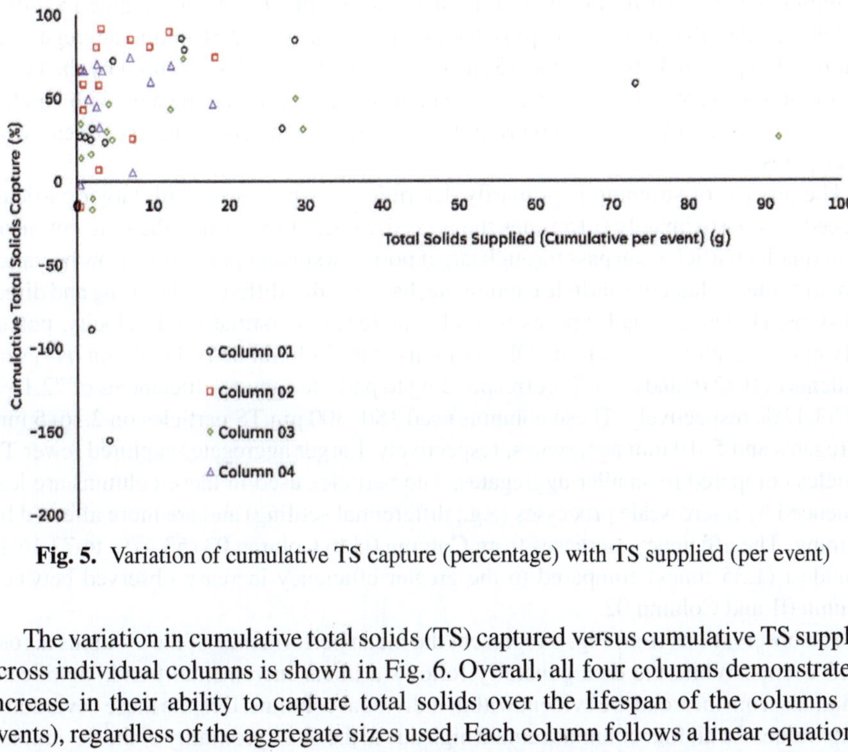

Fig. 5. Variation of cumulative TS capture (percentage) with TS supplied (per event)

The variation in cumulative total solids (TS) captured versus cumulative TS supplied across individual columns is shown in Fig. 6. Overall, all four columns demonstrate an increase in their ability to capture total solids over the lifespan of the columns (13 events), regardless of the aggregate sizes used. Each column follows a linear equation of the form y = mx, where 'm' represents the gradient, corresponding to the overall removal efficiency (m = y/x, with y as cumulative particles captured and x as cumulative particles loaded). The linear curves fitted with an R^2 above 0.99 for all columns. However, the gradients vary significantly between columns. Column 01, with a gradient of 0.52, and Column 03, with a gradient of 0.72, both used pervious concrete with the same aggregate size distribution (2.36–5.0 mm). Column 01 captured 75–150 µm TS, while Column 02 captured 150–300 µm TS. The gradient differences indicate that pervious concrete pores are more effective at capturing larger particles, with approximately 20% increased efficiency for 150–300 µm TS compared to 75–150 µm TS. A similar 22% increase in gradient was observed between Columns 02 and 04, which had 5–10 mm aggregates, capturing 75–150 µm (Column 02) and 150–300 µm (Column 04), with gradients of 0.29 and 0.53, respectively. Interestingly, the effectiveness of pervious concrete with

5–10 mm aggregates in capturing 150–300 μm particles (53.47%) was comparable to the effectiveness of pervious concrete with 2.36–5 mm aggregates in capturing 75–150 μm particles (52.08%). This suggests that the pore structures in pervious concrete with different aggregate sizes may capture specific loadings of total solids with similar efficiency.

The observed higher retention of larger particles (150–300 μm) compared to smaller ones (75–150 μm) can be attributed to the straining process and possibly differential settling. This phenomenon is consistent across PC specimens with different aggregate size distributions, supporting the hypothesis that these mechanisms contribute to the increased retention of larger total solids (TS).

Figure 6 shows that, despite both column 01 and column 02 having the same TS inflow particle size distribution (75–150 μm), the PC specimens with 2.36–5 mm aggregates in column 01 captured 52.08% of the TS, nearly double the 29.21% captured by the larger 5–10 mm aggregates in column 02. This suggest that smaller aggregates with smaller pores were more effective at capturing TS particles compared to the specimens with larger pores.

The process of straining is primarily determined by pore size, with larger particles (exceeding approximately 0.15 times the pore size) being retained at the pore entrance, while smaller particles can pass through larger pores. In smaller pores, fluid flow becomes more tortuous, enhancing particle capture mechanisms like differential settling and direct collisions. However, smaller pores may also increase interstitial fluid velocity, potentially opposing particle retention. Observations from Column 03 and Column 04 reveal gradients of 0.7216 and 0.5347, corresponding to particle capture efficiencies of 72.16% and 53.47%, respectively. These columns used 150–300 μm TS particles on 2.36–5 mm aggregates and 5–10 mm aggregates, respectively. Larger aggregates captured fewer TS particles compared to smaller aggregates. The particles used in these columns are less influenced by micro-scale processes (e.g., differential settling) and are more affected by straining. The efficiency increases from Column 04 to Column 03 (53.47% to 72.16%) is modest (1.35 times) compared to the greater efficiency increase observed between Column 01 and Column 02.

The observations from Fig. 6 represent the mean performance of the columns across all 13 events. While the data generally follows a linear incremental trend, significant deviations from the trend are evident within individual columns. Despite these deviations, the R^2 values for all four columns remain high due to the close proximity of the deviations. The pattern of these deviations, which appears to evolve with the progression of loading, suggests dynamic particle capture during events, potentially including flushing or re-entrainment of previously captured TS particles. Further detailed analysis is needed to better understand the mechanisms of particle retainment and re-entrainment in pervious concrete specimens. Figure 7 illustrates the relationship between cumulative total solids retained and cumulative total solids loaded for each event across four columns. While a generally linear trend is observed, with consistent gradients indicating a steady fraction of solids retained over time, some deviations are evident. These deviations highlight rare instances of flushing, where retained solids are re-entrained. For example, in Column 1 (EN 8) and Column 3 (EN 9), a downward trend indicates flushing during those specific events. The cumulative solids flushed were relatively low compared to the total

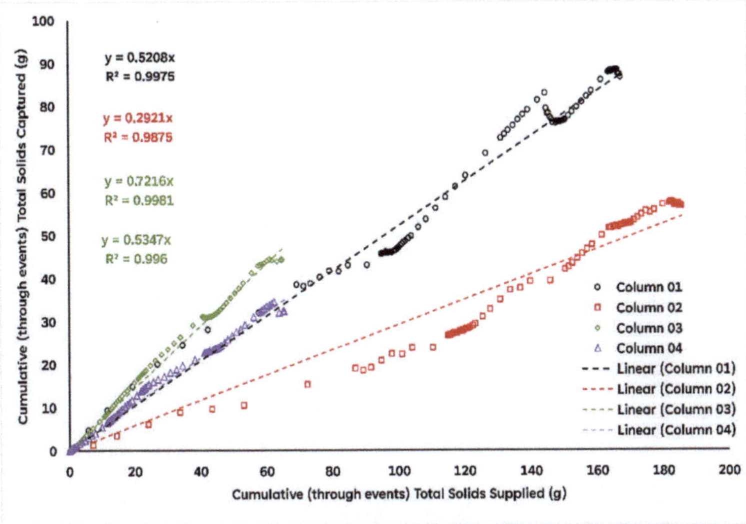

Fig. 6. TS output (cumulated through events) with TS supplied (cumulated through events)

loading, minimizing their impact on the overall trend accuracy. These columns, with 2.36–5 mm aggregates and total solids of 75–150 μm and 150–300 μm, respectively, experienced flushing at different loadings (140 g and 40 g). These occurrences do not suggest clogging, as the columns resumed normal capture trends afterward, indicating a spontaneous flush likely caused by flow condition changes.

In contrast, Columns 2 and 4 did not exhibit flushing throughout an entire event but showed significant within-event flushing. For instance, Column 4 (EN 12) experienced a notable flush after approximately one hour. These events suggest that spontaneous flushing is triggered by factors such as flow condition changes or uneven deposit distribution within pervious concrete pores. Importantly, none of the columns in this study reached a clogging potential or deposit ceiling, maintaining consistent performance across all events (Fig. 8).

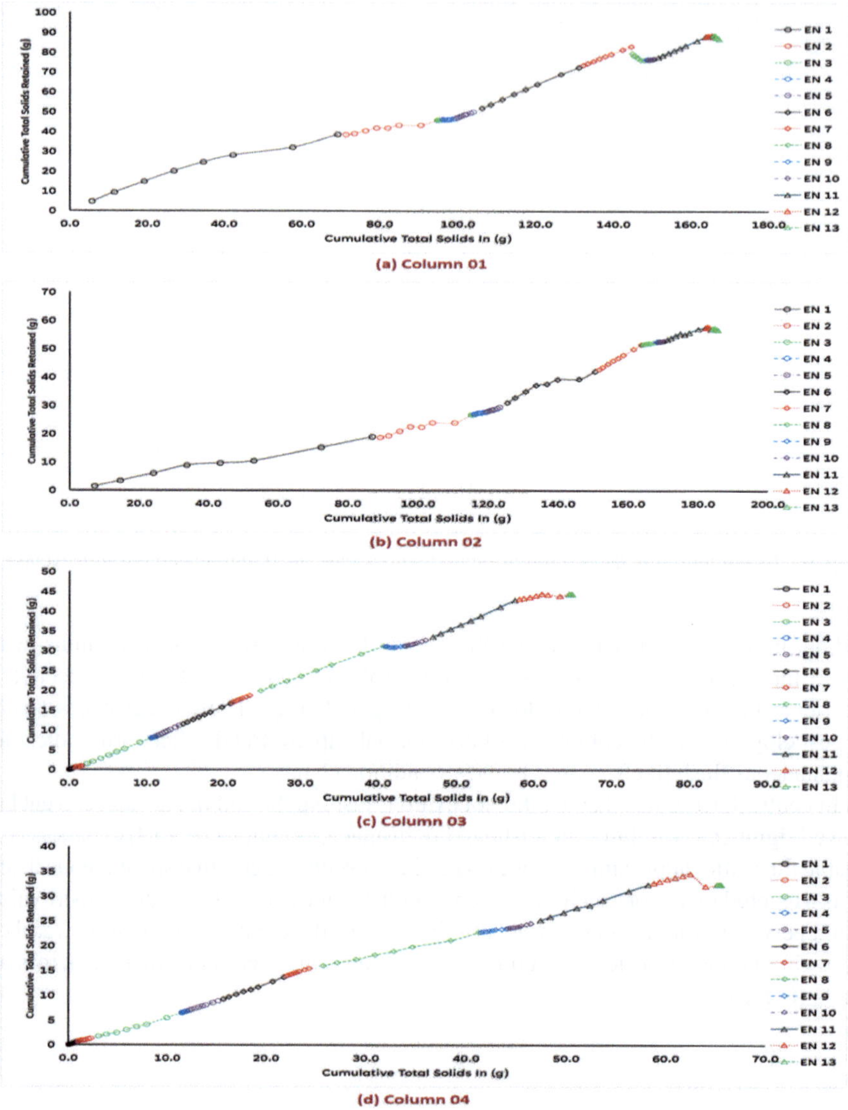

Fig. 7. TS output (cumulated through events) with TS supplied (cumulated through events) and event number (EN)

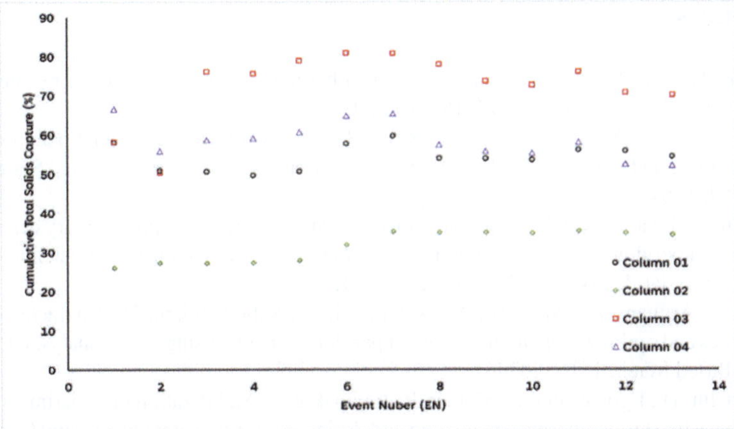

Fig. 8. Variation of cumulative TS capture with event numbers (EN) for individual columns

4 Conclusion

- Aggregate size plays a crucial role in the solids capture efficiency of pervious concrete. Smaller aggregates (2.36–5 mm) exhibited higher retention rates for fine particles, whereas larger aggregates (5–10 mm) allowed more particles to pass through due to increased pore sizes. This indicates that optimizing aggregate selection is essential for enhancing the filtration performance of pervious concrete in stormwater management applications.
- The consistent retention performance observed across multiple solid loading events suggests that pervious concrete maintains its filtration capacity over time without significant pore clogging. Even after exposure to 70–170 g of solid loading, the outflow rate remained stable, reinforcing the long-term viability of pervious concrete in capturing and retaining solids. These findings highlight its potential for sustainable urban drainage systems where clogging resistance is a key concern.
- The occurrence of spontaneous solids flushing indicates that flow conditions influence the retention efficiency of pervious concrete. Factors such as fluid velocity, turbulence, and uneven deposition of solids within the pore structure may contribute to the release of previously captured particles. Understanding these flow dynamics is crucial for improving pervious concrete design and ensuring consistent solids retention performance.
- Future research should focus on evaluating the impact of higher solid loading levels to determine the long-term clogging potential of pervious concrete. Additionally, investigating pore characteristics—such as size, shape, and tortuosity—will provide valuable insights for optimizing mix designs tailored to specific environmental conditions. Given the variability in pervious concrete applications, further studies are needed to assess its filtration efficiency under different mix configurations to enhance its effectiveness in real-world drainage systems.

References

Kazemi, F., Hill, K.: Effect of permeable pavement basecourse aggregates on stormwater quality for irrigation reuse. Ecol. Eng. **77**, 189–195 (2015)

Okaikue-Woodi, F.E.K., Cherukumilli, K., Ray, J.R.: A critical review of contaminant removal by conventional and emerging media for urban stormwater treatment in the United States. Water Res. **187**, 116434 (2020)

Sathiparan, N., Jeyananthan, P., Subramaniam, D.N.: Silica fume as a supplementary cementitious material in pervious concrete: prediction of compressive strength through a machine learning approach. Asian J. Civ. Eng. **25**, 2963–2977 (2024b)

Shobijan, J., Arunan, M., Pratheesh, S., Anburuvel, A., Subramaniam, D.: Characterization of porosities and optimization of mix design of pervious concrete using image analysis. J. Sustain. Cem.-Based Mater. 1–15 (2024)

Subramaniam, D., Egodawatta, P., Mather, P., Rajapakse, J.: Stabilization of experimental bioretention basins during intermittent wetting and drying. In: Stormwater Queensland Conference: Visions to Realities: Noosa Heads (2014)

Subramaniam, D.N., Mather, P.B.: Preliminary stabilisation of stormwater biofilters and loss of filter material. Water Sci. Technol. **74**(4), 787–795 (2016)

Subramaniam, D.N., Logeswaran, T., Tharshikka, V., Nilakshan, B.: Dynamics of clay particles in non-vegetated stormwater biofilters: water. Air Soil Pollut. **229**(9), 302 (2018)

Wijekoon, S.H., Shajeefpiranath, T., Subramaniam, D.N., Sathiparan, N.: A mathematical model to predict the porosity and compressive strength of pervious concrete based on the aggregate size, aggregate-to-cement ratio and compaction effort. Asian J. Civ. Eng. **25**(1), 67–79 (2024)

Potential of Improvement of Roof Runoff Quality Using Different Types of Biochar

D. G. Kivindya, N. P. Miguntanna(✉), and S. G. D. D. Rathnayake

Department of Civil Engineering, Faculty of Engineering, Sri Lanka Institute of Information Technology (SLIIT), Malabe 10115, Sri Lanka
nandika.m@sliit.lk

Abstract. Urbanization has significantly increased impervious surfaces, generating contaminated runoff that contributes to urban water pollution. Roof runoff is a major source of non-point pollution, carrying pollutants such as heavy metals, organic matter, and nutrients into water bodies. This study explores the potential of using biochar produced through biomass pyrolysis, as a sustainable solution for improving roof runoff quality. Specifically, three biochar types (rice husk, coconut shell, and wood) were tested for their ability to remove pollutants from runoff collected from three different types of roofs namely metal, clay, and asbestos roofs. Performance of biochar were assessed based on several key properties, including particle size, water holding capacity, bulk density, and pore structure, all critical to biochar's pollutant adsorption capacity. Improvement of the runoff quality was analysed based on laboratory-scale experiments conducted using filtration beds packed with uniform layers of biochar. Runoff samples were collected during rainfall events and analysed before and after treatment for various water quality parameters, including pH, Electrical Conductivity (EC), Turbidity, Total Solids (TS), Total Suspended Solids (TSS), and concentrations of heavy metals (Zn, Al). The results showed that coconut shell biochar performed the best across all parameters, exhibiting significant reductions in turbidity and TSS, making it ideal for stormwater treatment. Coconut shell biochar showed the overall best performance by effectively removing pollutants. Wood biochar demonstrated excellent performance in solids removal but resulted in higher EC and colour changes, which may limit its suitability for applications requiring clear water. Rice husk biochar maintained stable pH and EC but showed lower efficiency in removing solids. These findings underscore biochar's potential as a cost-effective and sustainable material for improving stormwater quality, particularly in rainwater harvesting and groundwater recharge applications. Further research, particularly field-scale trials, is recommended to optimise biochar use in real-world applications and ensure its long-term effectiveness.

Keywords: Roof runoff · Biochar · Water quality · Stormwater treatment · Turbidity reduction

R. Dissanayake et al. (Eds.): ICRES 2025, PTT, pp. 157–169, 2025.
https://doi.org/10.1007/978-981-96-9551-5_13

1 Introduction

The rapid growth of urban populations and widespread development have drastically altered landscapes, creating significant environmental challenges. Urbanization, coupled with increasing impervious surfaces such as asphalt, concrete, and metal, significantly contributes to the pollution of urban stormwater runoff. Urban runoff transports pollutants such as heavy metals, nutrients, pesticides, and microbial contaminants generated from residential, commercial and industrial activities and vehicle emissions into surface water bodies, leading to severe degradation of the water quality of receiving water bodies such as rivers, lakes, streams etc. (Herngren et al., 2005; Liu et al. 2016.; Liu et al., 2017). Furthermore, urban stormwater runoff significantly contributes to issues such as thermal pollution and eutrophication, which disrupts aquatic ecosystems by causing algae blooms, resulting in oxygen depletion of water bodies and habitat alterations (Smith et al., 1999). In particular, roof runoff forms a significant component of urban runoff, transporting pollutants accumulated on roof surfaces into water bodies during rainfall events.

The contamination of surface water quality complicates and increases the cost of treating water for human use, worsening water scarcity, particularly in areas already experiencing water shortages (Schwarzenbach et al., 2010). As urbanization increases, improving the quality of stormwater runoff becomes critical to conserving water resources and mitigating the stress on water supplies in water-scarce areas (Gleick, 2003). Traditional stormwater management practices, including detention ponds, vegetated swales, and permeable pavements, aim to reduce runoff and filter pollutants. Emerging materials like zeolites and alum-based flocculants target the removal of specific pollutants through ion exchange and flocculation mechanisms (Jusić et al., 2020).

However, recent research has turned toward biochar, a charcoal-like substance, as an alternative filter material, particularly in wastewater treatment due to its exceptional adsorption capacity making it effective in removing contaminants (Tareq et al., 2019). Biochar is produced by the thermal decomposition of organic material derived from agricultural and forestry wastes (also called biomass) in the absence of oxygen (pyrolysis) under the temperatures of 300 °C and 600 °C for 1 to 2 h (Chen et al., 2022). It is a carbon-rich (~85%) highly porous, complex material with a large surface area as shown in Fig. 1.

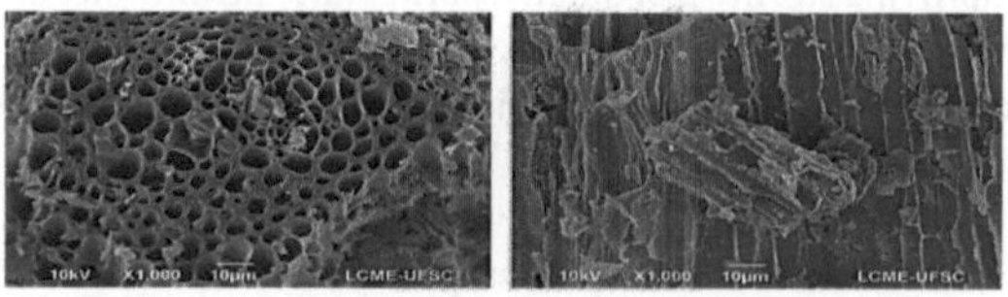

Fig. 1. Porous structure of biochar (Tomassoni et al., 2020)

Today, the use of biochar in the water sector has become increasingly popular due to its enhanced physicochemical properties which mainly include specific surface area, high porosity, high cation exchange capacity, pH, surface functional groups, surface reactivity, and mineral composition that affect its capability to adsorb organic, inorganic, and even pathogenic contaminants in water (Enaime et al., 2020).

According to Yaashikaa et al. (2020), several factors influence the properties of biochar, which include the type of biomass, pyrolysis temperature, rate of heating, retention time, and pressure. For instance, Enaime et al. (2020) concluded that high-temperature pyrolysis commonly results in biochar with a larger surface area and pore volume, making it ideal for the sorption of organic contaminants, while inorganic pollutants can be removed more effectively by biochar produced at low-temperature pyrolysis since it has more oxygen-containing functional groups. Furthermore, even though overall, the most common biochar size range is 1–5 mm and smaller and larger particle sizes of 0.25 mm and 9.5 mm respectively have also been reported in the research literature (Quispe et al., 2022, Emslie, 2019, Deepa et al., 2019). In thinner, porous media, the straining is greater, and adsorption sites are abundant, which can result in more reduction of pathogens (Molaei, 2014, Berger, 2012, Dalahmeh et al., 2016). The specific surface area is another property of great importance as it contributes to the removal of pollutants and determines the suitability of filtration material to allow biofilm growth and, consequently, biological degradation processes for organic pollutants. Quispe et al. (2022), conducting a comprehensive literature review on biochar-based column filtration systems, noted that the specific surface area of biochar varies in the range from 118 to 200 m^2/g, which is higher and lower than sand (0.152 m^2/g) and activated carbon (1000 m^2/g), respectively. Moreover, the total porosity varies from 42 to 74% and is higher than the typical porosity of sand particles (35%) which further implies the capacity of biochar for both water retention and biofilm development (Berger, 2012). Beyond water treatment, biochar also contributes to sustainability by promoting waste recycling and reducing greenhouse gas emissions (Ahmed et al., 2016). This makes biochar a promising, environmentally beneficial material for improving stormwater quality.

This study aimed to explore the potential of biochar to improve the quality of roof runoff, focusing on its ability to remove contaminants and enhance water quality. Therefore, this study specifically examined biochar derived from three locally and commonly available materials: rice husk, coconut shell, and wood. By leveraging locally sourced materials, this study aims to provide a cost-effective and sustainable solution for mitigating urban runoff pollution while promoting waste recycling.

2 Methodology

2.1 Biochar Characterization

The biochar used in this study was commercially sourced and derived from rice husk (RH), coconut shell (Coco), and wood (Wood). These three types of biochar were selected based on their local availability and economic viability in using them under real field conditions. The following properties of the biochar were characterized in this study according to the standard methods for activated carbon by the European Chemical Industry Council (CEFIC):

- **Particle Size Distribution:** Due to the considerable variation in particle sizes of Coconut shell and Wood biochar, samples of those were sieved through a series of standard sieves to determine their particle size distribution. Sieve analysis was not performed for rice husk biochar because rice husk particles typically have a more uniform and elongated shape compared to coconut shell and wood biochar.
- **Water Holding Capacity (WHC):** WHC was measured by soaking a known weight of biochar in water and calculating the percentage increase in weight.
- **Bulk Density:** Bulk density was determined by weighing a known volume of biochar and calculating the mass per unit volume.
- **Pore Structure:** The biochar's pore structure was observed under a microscope to evaluate surface morphology.

2.2 Experimental Setup

A laboratory-scale filtration system was designed using filtration beds filled with a 3 cm layer of biochar as shown in Fig. 2. The beds were arranged in triplicates to ensure reliability and reduce the variability of the experimental results.

Fig. 2. Experimental setup

2.3 Roof Runoff Collection

Roof runoff samples were collected from three common roofing materials (metal, clay, and asbestos) during rainfall events. These materials were selected based on their widespread use in urban and semi-urban settings, their distinct surface properties, and their recognition of generating contaminated stormwater runoff (Egodawatta et al., 2012, De Buyck et al., 2021, Tobiszewski et al., 2009). Metal roofs often contribute heavy metals, whereas clay roofs accumulate organic and inorganic particles, and asbestos roofs may release fine suspended solid (Farreny et al., 2011, Ojo, 2019). Additionally, these roof types were chosen for their easy accessibility, ensuring consistent sample collection and practical applicability of the study findings across a broader spectrum of field conditions. The runoff was collected using clean, airtight containers to avoid contamination. Each sample was filtered through the biochar filtration beds immediately after collection to assess its treatment potential. The samples were stored at 4 °C before analysis to prevent changes in water quality.

2.4 Water Quality Testing

Water quality tests were conducted for both pre- and post-treatment for each sample according to American Public Health Association (APHA) standard methods.

2.5 Data Analysis

Data was analysed using univariate data analysis techniques with the support of Microsoft Excel to compare the performance of different biochar types in terms of pollutant removal. Excel was utilized to organize, calculate, and visualize data, including pollutant levels before and after treatment. The results were interpreted to evaluate the effectiveness of each biochar type in improving stormwater quality. This methodology provides a practical and efficient approach for assessing biochar as a filtration medium for improving roof runoff quality, with the potential for scaling up to field applications.

3 Results and Discussion

3.1 Biochar Properties

- **Particle Size Distribution (PSD)**

Figure 3 shows distinct particle size characteristics for each biochar type. Coconut shell biochar has a coarser particle distribution, leading to fewer fine particles and promoting better water flow in filtration systems, with reduced clogging potential. Wood biochar has relatively finer particles, most of which fall below 6.3 mm, providing a higher surface area for pollutant adsorption but requiring slower flow rates to avoid clogging.

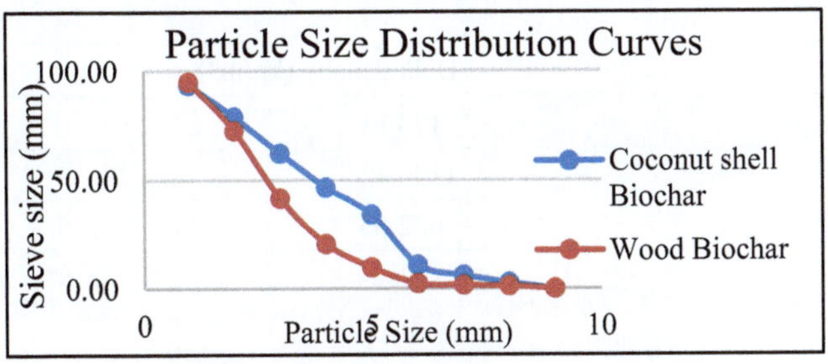

Fig. 3. Particle Size Distribution Curves

- **Water Holding Capacity (WHC)**

Table 1 presents the water-holding capacity of three types of biochar (rice husk, coconut shell, and wood), each with distinct properties suited for different water treatment applications.

Table 1. Water holding capacity

Biochar Type	Wet Weight (g)	Dry Weight (g)	WHC (%)
Rice husk	40.5	10	305
Coconut shell	30.5	21.1	44.55
Wood	49.55	20.74	138.91

As can be seen in Table 1, the rice husk biochar, shows the highest WHC (305%), implying a highly porous structure, which increases contact time with contaminants and enhances pollutant adsorption capacity. Coconut shell biochar, with the lowest WHC (44.55%), has a denser structure, promoting higher flow rates and reducing clogging risks. Its lower WHC makes it well-suited for high-flow filtration systems, where rapid water treatment is essential, as noted by Mohan et al. (2014) and Liu et al. (2017). Wood biochar, with a moderate WHC (138.91%), strikes a balance between water retention and structural stability, offering effective adsorption while maintaining manageable water flow. This makes it versatile for general filtration purposes, where both adsorption capacity and consistent flow rates are required, as highlighted by Choudhary et al. (2019) and Inyang et al. (2016). Each biochar type thus provides specific advantages, depending on the water treatment requirements.

- **Bulk Density**

Table 2 presents the measured bulk density values for each type of biochar, providing a comparison of their physical characteristics based on material composition.

Table 2. Bulk Density

Biochar Type	Bulk density (kg/m^3)
Rice husk	134.81
Coconut shell	1136.43
Wood	652.66

As shown in Table 2, rice husk biochar has the lowest density (134.81 kg/m^3), making it highly porous and effective for high-adsorption applications, though less stable in high-flow systems (Choudhary et al., 2019). Coconut shell biochar, with the highest density (1136.43 kg/m^3), offers excellent stability and is ideal for high-flow, long-term filtration setups, minimizing clogging (Mohan et al., 2014). Wood biochar, with a moderate density (652.66 kg/m^3), provides a balance between adsorption capacity and structural stability, making it versatile for general filtration. These findings align with previous studies, confirming each biochar's specific strengths for water treatment (Liu et al., 2017; Piscitelli et al., 2018).

- **Surface Pore Structure**

Figure 4, Fig. 5 and Fig. 6 show the microscopic images of the surface pore structure of the rice husk, coconut shell and wood biochar, respectively.

Fig. 4. Microscopic image of rice husk biochar

Figure 4 shows that rice husk biochar has a highly porous structure, with small, interconnected pores that enhance water retention, consistent with its high-water holding capacity. The extensive porosity increases its surface area, making it effective at adsorbing dissolved pollutants like heavy metals and organic compounds. These findings align with (Choudhary et al., 2019), who reported that higher porosity improves adsorption efficiency.

Fig. 5. Microscopic image of coconut shell biochar

The microscopic image of coconut shell biochar shown in Fig. 5 reveals a denser, less porous structure with fewer, larger pores compared to rice husk biochar. This lower porosity is consistent with its low water-holding capacity and high bulk density, making it suitable for high-flow filtration systems where clogging resistance is important. These characteristics align with findings from (Mohan et al., 2014), who noted that denser biochar is better suited for stable, high-capacity filtration systems.

Fig. 6. Surface pore structure of wood biochar

Figure 6 shows the wood biochar with moderate porosity, featuring a mix of small and medium-sized pores. This balanced pore structure provides a compromise between adsorption capacity and flow stability. Its moderate porosity, along with intermediate water holding capacity and bulk density, makes it versatile for general water treatment. These characteristics align with (Liu et al., 2017), who found that biochar with balanced porosity is effective for pollutant adsorption while maintaining manageable flow rates.

3.2 Water Quality Parameters

The analysis of water quality parameters aimed to determine how the water quality of filtrate differs from the raw water collected from each roof surface. This involved determining the change in each water quality parameter in the filtrates and expressing these changes as a percentage compared to the raw water for each type of roofing material as shown in Figs. 7, 8, 9, 10, 11 and 12.

- **pH**

Figure 7 shows that wood biochar causes the highest pH increase, suitable for applications needing strong alkalization. Coconut shell biochar results in a moderate pH increase, ideal for controlled alkalization. Rice husk biochar has the smallest effect, maintaining near-neutral pH, suitable for applications where minimal pH change is desired. Biochar selection should align with the desired level of alkalization.

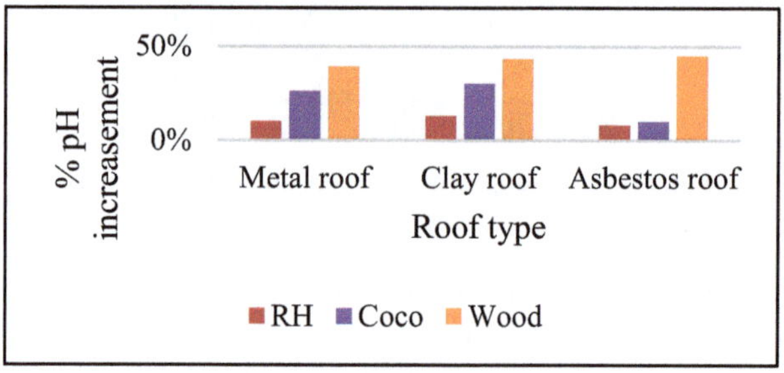

Fig. 7. pH Increment

- **Electrical Conductivity**

Figure 8 shows that wood biochar causes the highest increase in Electrical Conductivity, particularly for metal and clay roof runoff, due to its high ash content, which releases more soluble ions. Coconut shell biochar results in moderate EC increases, particularly for clay and metal roofs, offering a balanced release of ions. Rice Husk biochar has the smallest impact on EC, with minimal or negative changes, due to its low mineral (Abdiyev et al., 2023). These differences highlight that wood biochar is suited for nutrient-rich applications, coconut shell for moderate adjustments, and Rice Husk for maintaining low EC in sensitive environments.

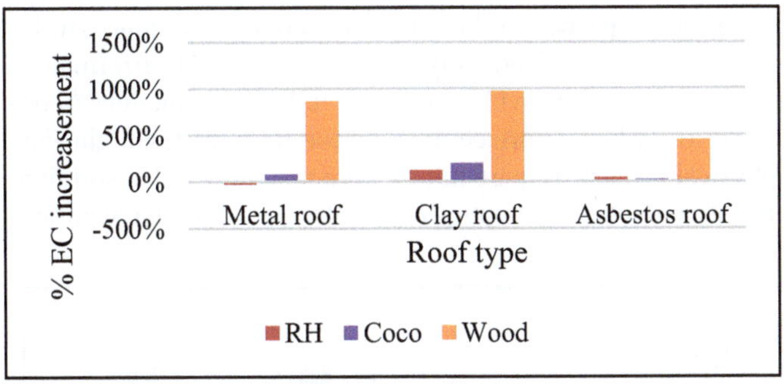

Fig. 8. EC Increment

- **Colour**

Figure 9 shows that wood biochar causes the most significant colour increase, particularly for clay and asbestos roof runoff, likely due to the leaching of organic compounds like tannins or lignin (Gusiatin & Rouhani, 2023). Coconut shell biochar generally reduces colour, especially for clay and asbestos roof runoff, making it suitable for maintaining water clarity (Ahmed et al., 2016). Rice Husk biochar has a minimal or slight reduction in colour, demonstrating stability and making it ideal for applications where minimal impact on water colour is desired (Bhatnagar et al., 2013). These findings highlight that wood biochar may not be suitable for water clarity-sensitive applications, while coconut shell and rice husk biochar offer better options for colour control.

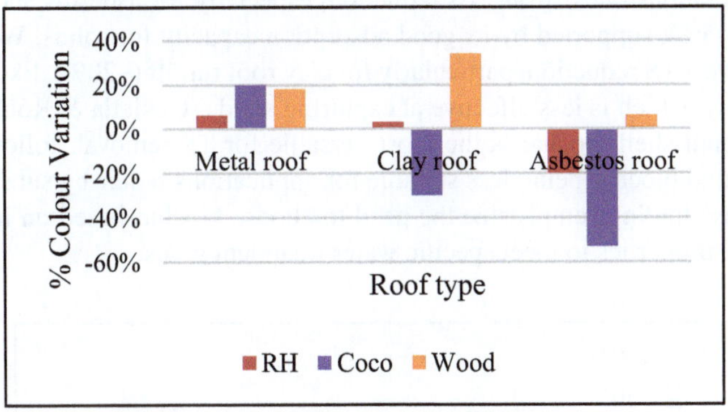

Fig. 9. Colour Variation

- **Turbidity**

Figure 10 illustrates turbidity reduction in runoff water from metal, clay, and asbestos roofs treated with rice husk, coconut shell, and wood biochar. Coconut shell biochar consistently achieves the highest turbidity reduction across all roof types, with reductions of up to ~ 56% in clay and asbestos roof runoff, which can be attributed to its balanced pore structure and surface properties. Wood biochar effectively reduces turbidity in metal roof runoff (~38%) due to its high porosity but exhibits variable performance across other roof types. Rice husk biochar provides moderate turbidity reductions (~10–40%) but is less effective due to its lower surface area. These findings highlight the critical role of biochar's physical properties, particularly pore structure and adsorption capacity, in enhancing turbidity reduction and optimizing water treatment applications.

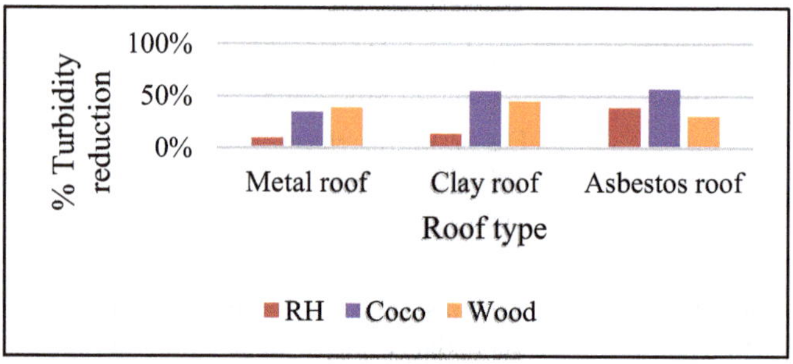

Fig. 10. Turbidity Reduction

- **Total Solids (TS)**

Figure 11 shows that coconut shell biochar is the most effective for total solids reduction, achieving significant reductions across all roof types, particularly in asbestos runoff (~50%), due to its balanced pore structure and high surface area. Rice husk biochar also demonstrates strong TS reduction, especially for asbestos and metal roof runoff (~40–45%), supported by its good adsorption capacity for solids. Wood biochar shows the lowest TS reduction, particularly for clay roof runoff (~20%), likely due to its higher porosity, which is less effective at capturing solids (Gusiatin & Rouhani, 2023). Overall, coconut shell biochar is the most versatile for TS removal, followed by rice husk, with wood biochar being less suitable for applications requiring substantial solid removal. These findings emphasize the need to choose biochar based on its feedstock and structural properties to meet specific water treatment goals.

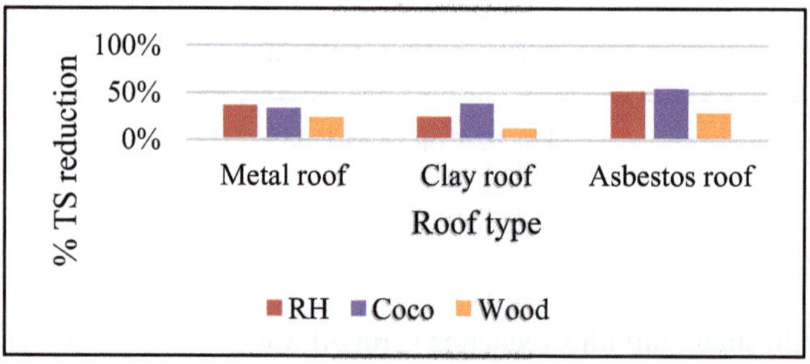

Fig. 11. TS Reduction

- **Total Suspended Solids (TSS)**

Figure 12 illustrates that wood biochar is most effective for TSS reduction, particularly in metal roof runoff (~80%), due to its high surface area (Inyang et al., 2016). Coconut shell biochar achieves strong TSS reduction in asbestos (~70%) and metal runoff (~60%) with balanced pores. Rice husk biochar offers moderate TSS reduction, especially in asbestos runoff (~60%) (Bhatnagar et al., 2013). Overall, wood biochar excels in high TSS removal whereas coconut shell provides consistent reduction, and rice husk is suitable for moderate TSS removal.

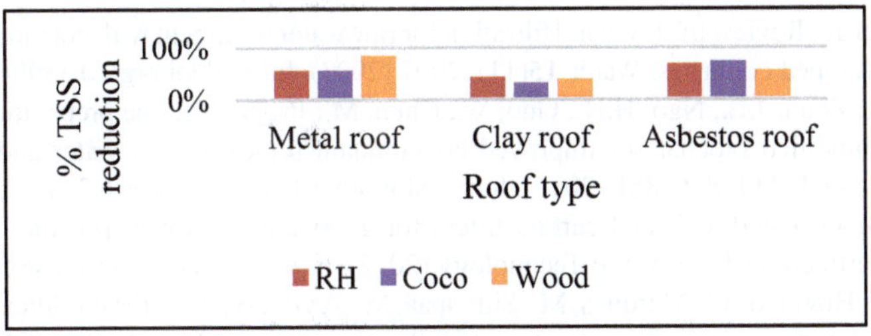

Fig. 12. TSS Reduction

4 Conclusion

This research demonstrates the effectiveness of biochar, derived from rice husk, coconut shell, and wood, in improving roof runoff quality. Laboratory experiments showed that biochar properties like surface area, porosity, and organic content significantly affect water quality parameters such as pH, EC, Colour, Turbidity, TS, and TSS. Coconut shell biochar was the most versatile in reducing turbidity (up to 50%) and TSS (up to 70%) with minimal impact on EC and colour, making it ideal for improving the quality of the runoff generated from asbestos and clay roofs.

Wood biochar excelled in removing particulates, achieving up to 80% TSS reduction and 40% turbidity reduction, but caused significant increases in colour and EC, limiting its use where clarity and low ionic content are important. Rice husk biochar, while less effective in particulate removal, maintained stable EC and colour, making it suitable for applications where a minimal impact on water quality is needed. The study highlights the need to select biochar based on specific treatment goals. Coconut shell biochar is recommended for its overall balanced performance, wood biochar for managing heavy particulate runoff, and rice husk biochar for stable water chemistry.

Further research is needed to validate these findings in real-world field conditions, optimizing biochar production, and minimizing organic leaching from wood biochar. However, large-scale implementation may present challenges, including sourcing sufficient amounts of biomass, maintaining consistent pyrolysis conditions, and ensuring long-term sustainability in continuous usage. It is also required to determine the saturation time of biochar to optimize its replacement cycle and long-term efficiency. Moreover, economic factors such as production costs, transportation, and market feasibility must also be considered.

Acknowledgments. The authors would like to express their heartfelt gratitude to the Faculty of Engineering at the Sri Lanka Institute of Information Technology (SLIIT) for providing invaluable resources and access to laboratory facilities essential for this research.

References

Abdiyev, K., et al.: Review of slow sand filtration for raw water treatment with potential application in less-developed countries. Water **15**(11), 2007 (2023). https://doi.org/10.3390/w15112007

Ahmed, M.B., Zhou, J.L., Ngo, H.H., Guo, W., Chen, M.: Progress in the preparation and application of modified biochar for improved contaminant removal from water and wastewater. Biores. Technol. **214**, 836–851 (2016). https://doi.org/10.1016/j.biortech.2016.05.057

Berger, C.: Biochar and activated carbon filters for greywater treatment [Second cycle, A2E]. SLU, Department of Energy and Technology (2012). https://stud.epsilon.slu.se/5183/

Bhatnagar, A., Hogland, W., Marques, M., Sillanpää, M.: An overview of the modification methods of activated carbon for its water treatment applications. Chem. Eng. J. **219**, 499–511 (2013). https://doi.org/10.1016/j.cej.2012.12.038

Chen, Y., Wu, Q., Tang, Y., Liu, Z., Ye, L., Chen, R., Yuan, S.: Application of biochar as an innovative soil ameliorant in bioretention system for stormwater treatment: A review of performance and influencing factors. Water Science and Technology **86**, 1232–1252 (2022). https://doi.org/10.2166/wst.2022.245

Choudhary, T.K., Khan, K.S., Hussain, Q., Ahmad, M., Ashfaq, M.: Feedstock-induced changes in composition and stability of biochar derived from different agricultural wastes. Arab. J. Geosci. **12**(19), 617 (2019). https://doi.org/10.1007/s12517-019-4735-z

Dalahmeh, S.S., Lalander, C., Pell, M., Vinnerås, B., Jönsson, H.: Quality of greywater treated in biochar filter and risk assessment of gastroenteritis due to household exposure during maintenance and irrigation. J. Appl. Microbiol. **121**(5), 1427–1443 (2016). https://doi.org/10.1111/jam.13273

De Buyck, P.-J., Van Hulle, S.W.H., Dumoulin, A., Rousseau, D.P.L.: Roof runoff contamination: a review on pollutant nature, material leaching and deposition. Rev. Environ. Sci. Bio/Technol. **20**(2), 549–606 (2021). https://doi.org/10.1007/s11157-021-09567-z

Deepa, K., Varier, M., John, A.V., Jayakeerti, M., Balaji, R.: Multi Layered Soil Column analysis for wastewater (2019). https://www.semanticscholar.org/paper/MULTI-LAYERED-SOIL-COLUMN-ANALYSIS-FOR-WASTEWATER-Deepa-Varier/418300bddc27e169bab9e4da2262f1e5a6f0de3b

Egodawatta, P., Miguntanna, N.S., Goonetilleke, A.: Impact of roof surface runoff on urban water quality. Water Sci. Technol. **66**(7), 1527–1533 (2012). https://doi.org/10.2166/wst.2012.348

Emslie, D.K.: Analysis of a sand filtration system with common filter media and innovative materials [University of British Columbia] (2019). https://doi.org/10.14288/1.0377280

Enaime, G., Baçaoui, A., Yaacoubi, A., Lübken, M.: Biochar for wastewater treatment—conversion technologies and applications. Appl. Sci. **10**(10), Article 10 (2020). https://doi.org/10.3390/app10103492

Farreny, R., Morales-Pinzón, T., Guisasola, A., Tayà, C., Rieradevall, J., Gabarrell, X.: Roof selection for rainwater harvesting: quantity and quality assessments in Spain. Water Res. **45**(10), 3245–3254 (2011). https://doi.org/10.1016/j.watres.2011.03.036

Gleick, P.H.: Global freshwater resources: soft-path solutions for the 21st century. Science **302**(5650), 1524–1528 (2003). https://doi.org/10.1126/science.1089967

Gusiatin, M.Z., Rouhani, A.: Application of selected methods to modify pyrolyzed biochar for the immobilization of metals in soil: a review. Materials **16**(23), 7342 (2023). https://doi.org/10.3390/ma16237342

Herngren, L., Goonetilleke, A., Ayoko, G.A.: Understanding heavy metal and suspended solids relationships in urban stormwater using simulated rainfall. J. Environ. Manage. **76**(2), 149–158 (2005). https://doi.org/10.1016/j.jenvman.2005.01.013

Inyang, M.I., et al.: A review of biochar as a low-cost adsorbent for aqueous heavy metal removal. Crit. Rev. Environ. Sci. Technol. **46**(4), 406–433 (2016). https://doi.org/10.1080/10643389.2015.1096880

Jusić, S., Hadžić, E., Milišić, H.: Urban stormwater management – new technologies. In: Karabegović, I. (ed.) New Technologies, Development and Application II, vol. 76, pp. 790–797. Springer (2020). https://doi.org/10.1007/978-3-030-18072-0_90

Liu, A., Miguntanna, N.S., Miguntanna, N.P., Egodawatta, P., Goonetilleke, A.: Differentiating between pollutants build-up on roads and roofs: significance of roofs as a stormwater pollutant source. Clean-Soil Air Water **44**(5), 538–543 (2016)

Liu, Z., Dugan, B., Masiello, C.A., Gonnermann, H.M.: Biochar particle size, shape, and porosity act together to influence soil water properties. PLoS ONE **12**(6), e0179079 (2017). https://doi.org/10.1371/journal.pone.0179079

Mohan, D., Kumar, S., Srivastava, A.: Fluoride removal from ground water using magnetic and nonmagnetic corn stover biochars. Ecol. Eng. **73**, 798–808 (2014). https://doi.org/10.1016/j.ecoleng.2014.08.017

Molaei, R.: Pathogen and indicator organisms removal in artificial greywater subjected to aerobic treatment (2014)

Ojo, O.M.: Effects of roofing materials on harvested rain water quality. J. Appl. Sci. Environ. Manag. **23**(4), 735 (2019). https://doi.org/10.4314/jasem.v23i4.25

Piscitelli, L., Rivier, P.-A., Mondelli, D., Miano, T., Joner, E.J.: Assessment of addition of biochar to filtering mixtures for potential water pollutant removal. Environ. Sci. Pollut. Res. **25**(3), 2167–2174 (2018). https://doi.org/10.1007/s11356-017-0650-6

Quispe, J.I.B., Campos, L.C., Mašek, O., Bogush, A.: Use of biochar-based column filtration systems for greywater treatment: a systematic literature review. J. Water Process Eng. **48**, 102908 (2022). https://doi.org/10.1016/j.jwpe.2022.102908

Schwarzenbach, R.P., Egli, T., Hofstetter, T.B., Von Gunten, U., Wehrli, B.: Global water pollution and human health. Annu. Rev. Environ. Resour. **35**(1), 109–136 (2010). https://doi.org/10.1146/annurev-environ-100809-125342

Smith, V.H., Tilman, G.D., Nekola, J.C.: Eutrophication: impacts of excess nutrient inputs on freshwater, marine, and terrestrial ecosystems. Environ. Pollut. **100**(1–3), 179–196 (1999). https://doi.org/10.1016/S0269-7491(99)00091-3

Tareq, R., Akter, N., Azam, M.S.: Chapter 10 - Biochars and biochar composites: low-cost adsorbents for environmental remediation. In: Ok, Y.S., Tsang, D.C.W., Bolan, N., Novak, J.M. (eds.) Biochar from Biomass and Waste, pp. 169–209. Elsevier (2019). https://doi.org/10.1016/B978-0-12-811729-3.00010-8

Tobiszewski, M., Polkowska, Ż., Konieczka, P., Namieśnik, J.: Roofing materials as pollution emitters – concentration changes during runoff (2009)

Tomassoni, F., Schneider, E.E., Giroletti, C.L., Nagel-Hassemer, M.E., Recio, M.A.L., Lapolli, F. R.: Production and characterization of activated biochar of cássia fistula l. Leaves. Rev. Gestão Sustentabilidade Ambiental. 9, 800 (2020). https://doi.org/10.19177/rgsa.v9e02020800-815

Yaashikaa, P.R., Kumar, P.S., Varjani, S., Saravanan, A.: A critical review on the biochar production techniques, characterization, stability and applications for circular bioeconomy. Biotechnol. Rep. **28**, e00570 (2020). https://doi.org/10.1016/j.btre.2020.e00570

Impact of Urban Expansion on Air Quality: A Comparative Study of Kandy and Colombo Using Cloud-Based Geospatial Analysis

C. Tharshika(✉) and W. M. S. S. Dias

University of Peradeniya, Peradeniya, Sri Lanka
abitharshi99@gmail.com

Abstract. Urban expansion affects air quality, making it a critical concern in sustainable urban planning. Colombo and Kandy, major urban centres in Sri Lanka, face significant air pollution challenges due to vehicular emissions, urbanisation and industrial activities. This study employed cloud-based geospatial analysis to evaluate the impact of urban sprawl on air quality in Kandy and Colombo from 2001 to 2023. Data were sourced from NASA Land Processes Distributed Active Archive Centre (LP DAAC) at the USGS EROS Centre, including Terra & Aqua MAIAC Land Aerosol Optical Depth Daily 1 km (MCD19A2.061), Sentinel-5P Tropospheric Monitoring Instrument (TROPOMI) for analysing air quality and MODIS Land Cover Type Yearly Global 500 m (MCD12Q1.061) to assess urban growth. Analytical tools such as Google Earth Engine and ArcGIS Pro were used for time-series analysis and spatial visualisation of annual urban expansion, mean annual Aerosol Optical Depth (AOD) at 0.47 μm, and pollutant concentrations of Nitrogen Dioxide (NO_2), Sulphur Dioxide (SO_2), Carbon Monoxide (CO) and Ozone (O_3). The findings revealed a substantial relationship between urban growth and variations in air quality. Colombo urban area expanded from 159.39 km^2 in 2001 to 199.63 km^2 in 2023 (25.25% increase), predominantly towards the east and northeast. This urban expansion was accompanied by a sharp rise in mean AOD levels, from 9.64 to 18.45 (91.39% increase). In contrast, Kandy experienced slower urban growth, expanding from 8.27 km^2 in 2001 to 8.77 km^2 in 2023 (6.05% increase) towards the southwest, with mean AOD levels fluctuating from 7.05 to 11.48 (62.84% increase). Furthermore, air quality indicators such as NO_2, SO_2, CO and O_3 have shown increasing trends in both cities since 2019. The rapid urbanisation in Colombo has a more significant adverse impact on air quality than the slower, steadier growth in Kandy, emphasising the necessity of developing city-specific mitigation to ensure sustainable urban development.

Keywords: Air Quality · Urbanisation · Cloud-based Geospatial Analysis · Aerosol Optical Depth (AOD)

1 Introduction

Air quality monitoring is essential for assessing the health and environmental impacts of air pollution. Although Sri Lanka has an impressive portfolio of environmental legislation and a set of standards for the quality of air (Central Environmental Authority, 2023), poor

R. Dissanayake et al. (Eds.): ICRES 2025, PTT, pp. 170–182, 2025.
https://doi.org/10.1007/978-981-96-9551-5_14

implementation of the law results in poor air quality in urban areas. During peak traffic hours (between 6 AM to 8.30 AM and 4 PM to 6 PM), urban areas such as Colombo, Kandy, Gampaha, Negombo, Kegalle, Kurunegala and Jaffna face severe air pollution (Agalan, 2015).

Kandy & Colombo Districts, which are major urban centres in Sri Lanka, face significant air pollution challenges due to vehicular emissions, urbanisation and industrial activities that have contributed to the degradation of air quality in this region. Various air pollutants pose a significant threat due to their ability to penetrate deep into the respiratory system, leading to severe health problems such as asthma, bronchitis, and cardiovascular diseases (Ministry of Environment, 2025). This highlights the importance of developing effective air quality management strategies and contributing to the overall improvement of environmental health in the region. But in regions like Sri Lanka, access to spatial air quality data is limited, and only locational air quality is calculated in areas like Colombo by the National Building Research Organisation (n.d.). This lack of comprehensive spatial and temporal data on air quality creates concerns about spatial and temporal analysis by challenging an understanding of air pollution dynamics.

As a solution, cloud-based geospatial analysis has emerged as an innovative method for evaluating air quality within Geographic Information Systems (GIS). It enables real-time, large-scale analysis of air pollution trends with high temporal accuracy. Several global studies have demonstrated the potential of satellite-based air quality assessments using Remote Sensing (RS) data (Gupta and Christopher, 2009), yet most focus on short-term pollution levels or lack integration with urban expansion analysis. Cloud-based geospatial platforms like Google Earth Engine (GEE) provide the facilities to analyse primary air pollutants, including Land Aerosol Optical Depth – AOD (0.47 μm), Nitrogen Dioxide (NO_2), Sulfur Dioxide (SO_2), Carbon Monoxide (CO) and Ozone (O_3). These datasets offer a valuable alternative for assessing pollution levels in regions where ground-based monitoring is limited. However, in Sri Lanka, such analyses remain scarce, limiting comprehensive spatial and temporal evaluations of air quality.

Unlike conventional approaches (Abeyratne and Ileperuma, 2006; Senarathna et al., 2024), this study examines the comparative impact of urban expansion on air quality in Kandy and Colombo, two cities with distinct topographical and urbanisation patterns over a long-time period (2001–2023). By using cloud-based geospatial tools, this study integrates air quality monitoring with urban growth analysis, providing a more holistic understanding of pollution dynamics in urbanising environments.

2 Objectives

The primary objective of this study is to analyse the spatial and temporal trends of urban expansion in Colombo and Kandy from 2001 to 2023 and assess its impact on air quality using cloud-based geospatial analysis. By using GEE and RS datasets, this research examines the relationship between urban growth and air pollution levels, focusing on key pollutants such as AOD (0.47 μm), NO_2, SO_2, CO and O_3. This study aims to provide a comparative analysis of urbanisation patterns in both cities and their influence on air quality over time. Additionally, it seeks to highlight the potential of cloud-based GIS technologies in monitoring environmental changes, particularly in data-limited regions like Sri Lanka. The findings will contribute to the understanding of pollution dynamics in urbanising environments and support the development of sustainable urban planning and air quality management strategies to mitigate environmental and health risks associated with urban expansion.

3 Methodology

3.1 Study Area

This study focuses on Colombo and Kandy, two major urban centers in Sri Lanka. Colombo, located in the lowland coastal region, is the country's economic hub with rapid urban growth and significant air pollution. Kandy, situated in the central highlands, has slower urban expansion due to its hilly terrain and heritage preservation requirements.

3.2 Data Collection

Data for this study were obtained with coding from NASA's Land Processes Distributed Active Archive Centre (LP DAAC) at the USGS EROS Centre. Key datasets include the Terra & Aqua MAIAC Aerosol Optical Depth (AOD) Daily 1 km, the Sentinel-5P TROPOMI air quality data (NO_2, SO_2, CO, O_3) and MODIS Land Cover Type Yearly Global 500 m (Table 1).

Table 1. Data Sources

Satellite	Sensor	Launched Date	Derived Parameter	Spatial Resolution	Temporal Resolution
Terra Aqua	Moderate Resolution Imaging Spectroradiometer (MODIS)	24-02-2000	Total Column AOD	10 km, 3 km, 1 km	Daily
Sentinel 5P	Tropospheric Monitoring Instrument (TROPOMI)	From 2018 - Present	Total and Tropospheric NO_2, SO_2, CO and O_3	1113.2 m	Daily
Terra Aqua	Moderate Resolution Imaging Spectroradiometer (MODIS)	01-01-2001	LC_Type1	500 m	Yearly

3.3 Data Analysis

The study used GEE and ArcGIS Pro for data analysis. GEE facilitated the temporal analysis of air quality and urban growth by processing large-scale datasets efficiently.

ArcGIS Pro was used for spatial analysis and visualisation, allowing the integration of urban expansion and pollutant concentrations to generate maps and conduct statistical analyses.

4 Results and Discussion

4.1 Urban Expansion in Colombo and Kandy

Colombo exhibited significant urban expansion, growing from 159.39 km^2 in 2001 to 199.63 km^2 in 2023, a 25.25% increase. This growth predominantly occurred towards the east and northeast, encroaching on peri-urban and rural areas. Conversely, Kandy showed slower urban growth, with its urban area increasing from 8.27 km^2 to 8.77 km^2 (6.05% increase), mainly towards the southwest. The differences in urban expansion rates can be attributed to Colombo as a commercial hub and geographical constraints in Kandy, including its hilly terrain and heritage conservation requirements.

The urban growth of Kandy and Colombo between 2001 and 2023 reveals both commonalities and significant differences in their expansion patterns. Both cities have seen substantial increases in built-up areas, a sign of outward urban sprawl driven by population growth and economic development. However, Colombo's expansion is more intense and widespread, following a linear pattern along major transportation routes and coastal areas, particularly in the Colombo Metropolitan Area, coastal zones like Galle Face and Port City, and suburban regions such as Malabe, Battaramulla, and Maharagama. This rapid growth has been fueled by commercial activities, tourism, infrastructure improvements, and land reclamation projects (Fig. 1).

In contrast, Kandy's urban growth is more fragmented, due to its hilly terrain, with development concentrated in valleys, slopes and along major roads. Significant expansion has occurred in areas like Peradeniya, Katugastota, and Tennekumbura, as well as along roads such as the A1 and A9. Limited flat land and the city's cultural and tourism significance have driven residential and commercial growth, especially in areas like Gannoruwa and Kundasale (Fig. 2). While Colombo's growth is largely driven by its role as the economic and commercial hub of Sri Lanka, Kandy's development is shaped by its unique topography and cultural importance.

4.2 Air Quality Variations in Colombo and Kandy

Geospatial analysis of air quality in Colombo and Kandy, based on NO_2 concentration maps from 2019 and 2023. It reveals alarming trends of increasing pollution in both cities. In Colombo, areas such as Pettah, Fort and Borella stand out as major pollution hotspots, primarily due to high vehicular emissions, ongoing construction and a lack of green spaces (Fig. 3). These regions, marked by intense traffic and limited green areas, are at a high risk of deteriorating air quality, which poses serious health risks, especially to vulnerable populations like children, the elderly, and those with respiratory conditions.

Similarly, in Kandy, the city center and areas around the Kandy Lake show elevated pollution levels, especially in Gagawatakorale (Fig. 4), driven by traffic congestion and

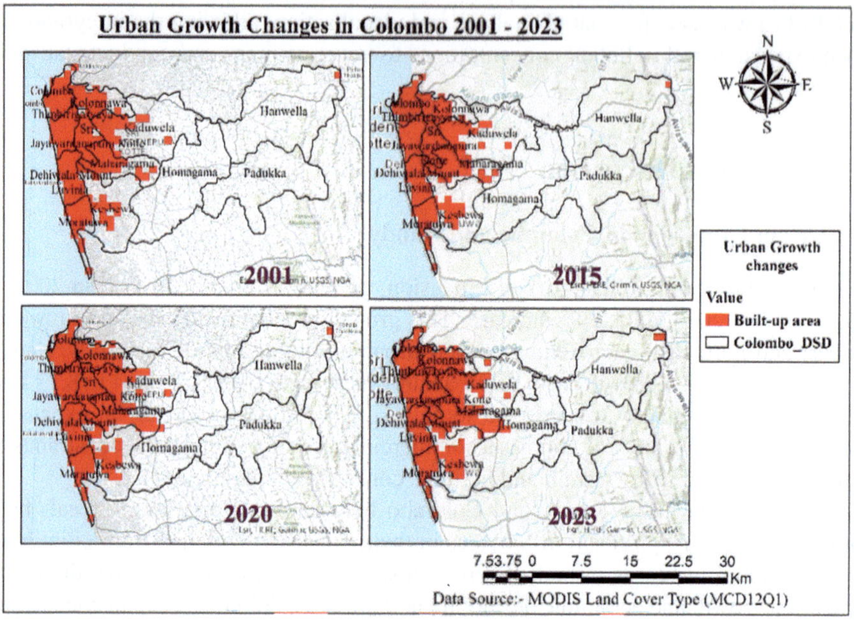

Fig. 1. Urban Expansion in Colombo

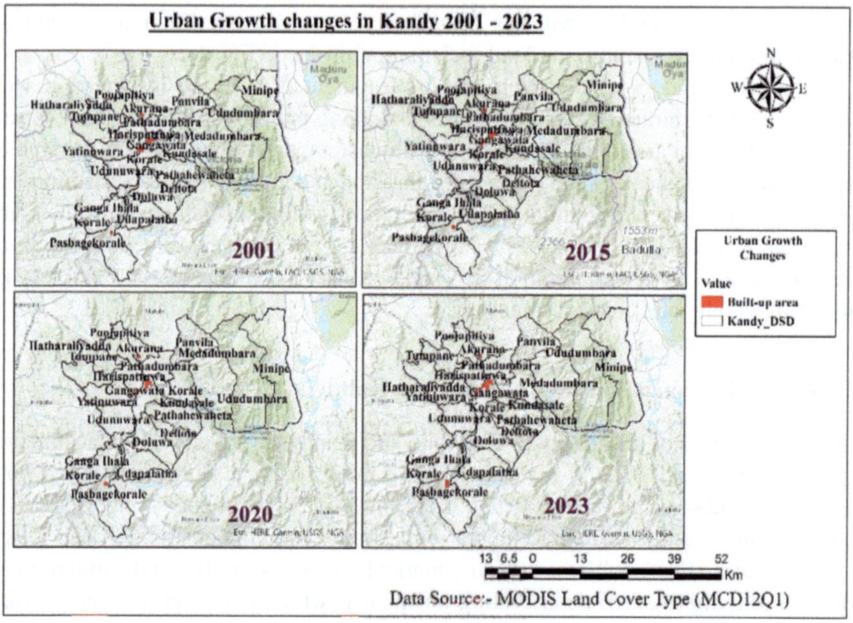

Fig. 2. Urban Expansion in Kandy

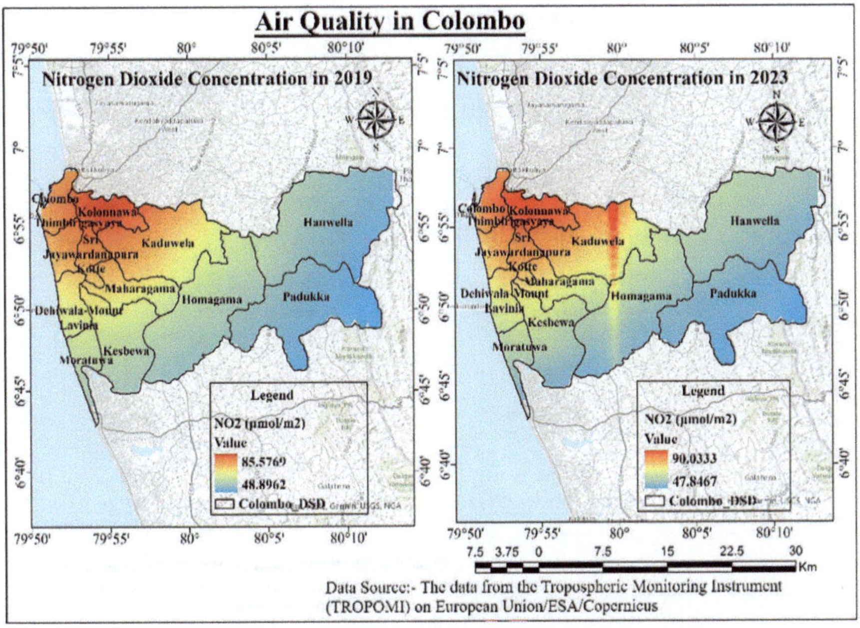

Fig. 3. TROPOMI NO_2 column number density in Colombo

construction activities. These red zones, representing the highest NO_2 concentrations, highlight the growing pollution challenges faced by both cities. While blue areas indicate relatively better air quality, the expanding red zones in both Colombo and Kandy signal an urgent need for targeted interventions to improve air quality, protect public health, and implement sustainable urban planning to reduce pollution in these key areas.

The analysis of CO concentrations in Colombo and Kandy, using satellite data from 2019 and 2023, reveals a concerning rise in pollution levels in both cities. In Colombo, the maximum CO concentration in 2023 reached 34,405.3 $\mu mol/m^2$, reflecting an 8.06% increase from 31,838.2 $\mu mol/m^2$ in 2019 (Fig. 5). The highest CO concentrations are primarily located in the central business district and along major transportation routes, areas heavily impacted by dense traffic, industrial activities, and residential fuel burning. This increase in CO pollution presents significant health risks, including headaches, dizziness, and, in severe cases, unconsciousness.

In Kandy, the maximum CO concentration increased from 31,516.8 $\mu mol/m^2$ in 2019 to 32,221.6 $\mu mol/m^2$ in 2023, representing a 2.24% rise (Fig. 6). The highest concentrations (indicated by red/orange) are found in the city center and along major roads, areas affected by traffic congestion, urban density and the topography of the valley, which can trap pollutants. While areas in the southern and western parts of Kandy show lower CO concentrations, the elevated levels in other regions still pose significant health risks, particularly for vulnerable populations such as children and the elderly.

To mitigate these risks in both cities, it is essential to promote public transportation, enforce stricter vehicle emission standards, improve traffic management and regulate industrial emissions. Additionally, raising awareness about the health hazards of CO

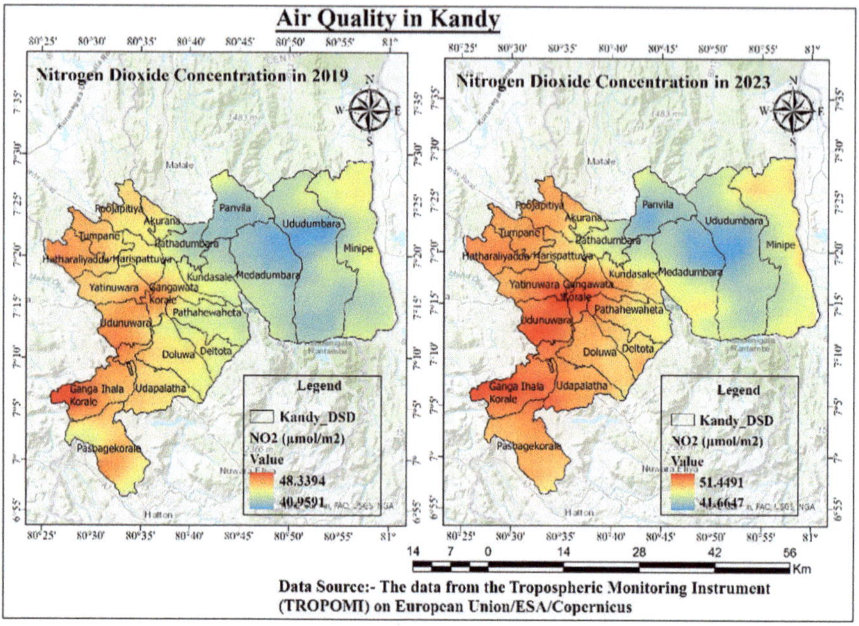

Fig. 4. TROPOMI NO_2 column number density in Kandy

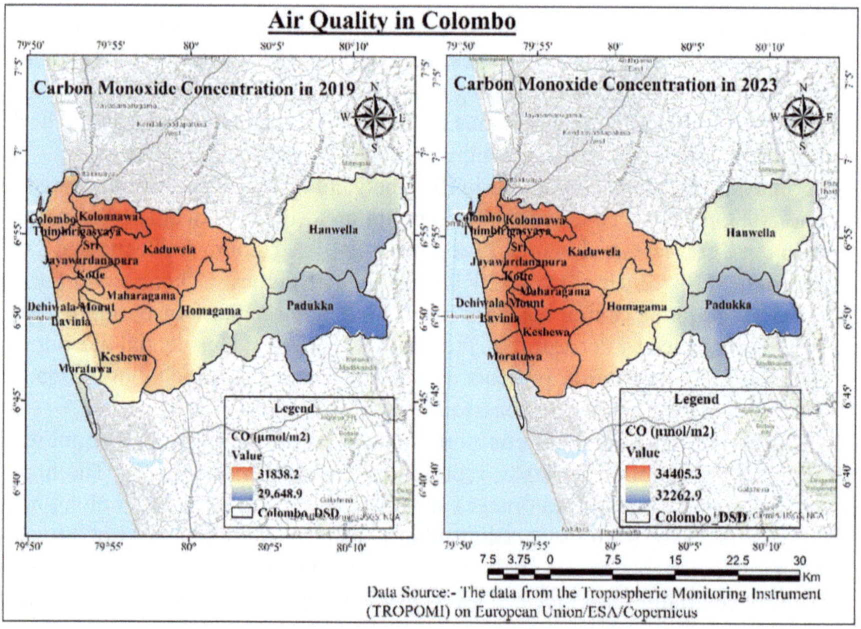

Fig. 5. TROPOMI CO column number density in Colombo

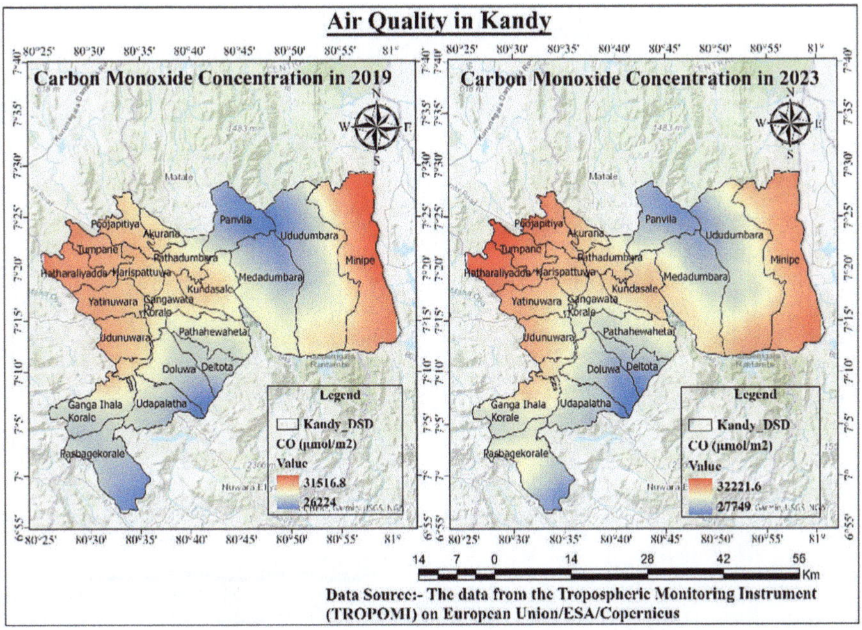

Fig. 6. TROPOMI CO column number density in Kandy

through public campaigns is crucial in encouraging citizens to adopt practices that reduce emissions, ultimately improving air quality and protecting public health.

The analysis of SO_2 concentrations in Colombo and Kandy from 2019 and 2023 reveals a significant rise in pollution levels in both cities. In Colombo, the maximum SO_2 concentration increased from 47.727 μmol/m^2 in 2019 to 61.9115 μmol/m^2 in 2023, marking a 29.72% rise. The highest concentrations are observed in the northern and western coastal areas, including Wattala, Ja-Ela, and Negombo, where industrial activities, power plants and shipping emissions are major contributors (Fig. 7).

Similarly, Kandy experienced a dramatic increase in SO_2 levels, with the maximum concentration rising from 59.7042 μmol/m^2 in 2019 to 80.5037 μmol/m^2 in 2023, reflecting a 35.33% rise. The central and eastern regions, particularly areas like Kundasale, Medadumbara and Pathadumbara, along with major roads and urban centers, recorded the highest concentrations due to industrial emissions, vehicular pollution, biomass burning and possible pollution transport from neighboring districts (Fig. 8).

The increasing SO_2 levels in both cities pose severe health risks, especially for vulnerable groups. To mitigate these risks, it is crucial to enforce stricter industrial emission regulations, promote cleaner fuels in industries and power plants, implement ship fuel sulfur content regulations and encourage the use of public transportation. Additionally, transitioning to renewable energy sources and controlling biomass burning are essential measures for reducing SO_2 emissions, improving air quality and protecting public health in both Colombo and Kandy.

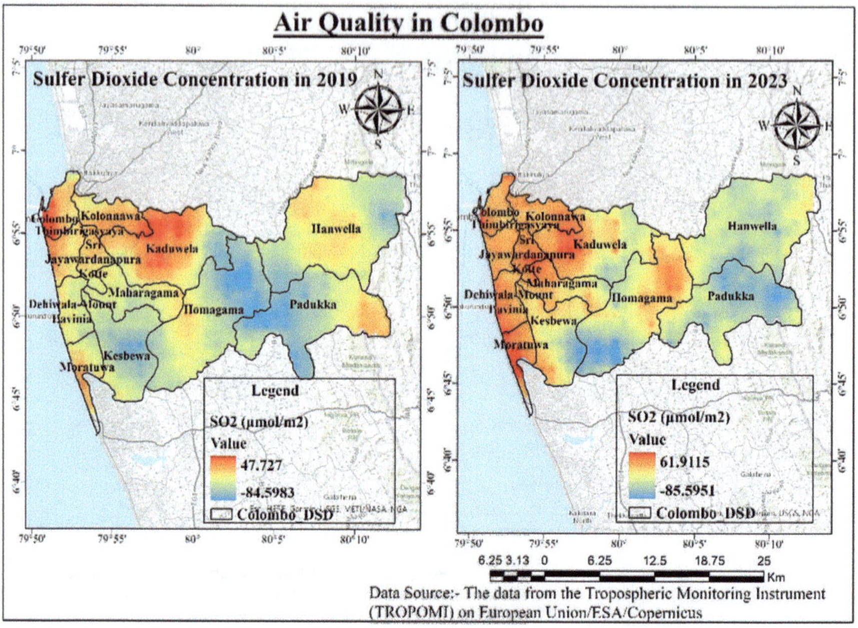

Fig. 7. TROPOMI SO_2 column number density in Colombo

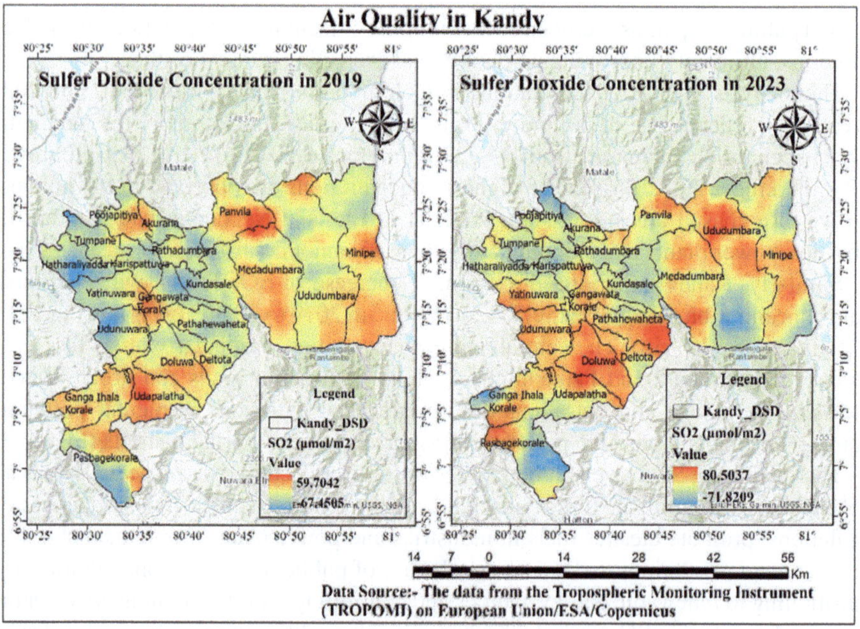

Fig. 8. TROPOMI SO_2 column number density in Kandy

Figures 9 and 10 represent the analysis of O_3 concentrations in Colombo and Kandy, using satellite data from 2019 and 2023 revealed a slight increase in pollution levels in both cities. In Colombo, the maximum O_3 concentration in 2023 reached 120,505 μmol/m^2, reflecting a 0.91% rise from 119,418 μmol/m^2 in 2019. Similarly, Kandy recorded a maximum O_3 concentration of 120,323 μmol/m^2 in 2023, marking a 0.61% increase from 119,588 μmol/m^2 in 2019. Higher O_3 concentrations in Colombo are primarily observed in the northern and western coastal areas (Fig. 9). In Kandy, elevated levels are concentrated in the central and eastern regions (Fig. 10). The increase in O_3 levels is driven by photochemical reactions involving nitrogen oxides (NOx) and volatile organic compounds (VOCs), emitted from vehicular traffic, industrial activities, shipping, and biomass burning. Hot and sunny weather conditions further accelerate ozone formation. Elevated O_3 levels pose significant health risks, including respiratory irritation, coughing, wheezing, and exacerbation of asthma, especially for vulnerable populations such as children and the elderly. To mitigate these risks, it is essential to enhance traffic management, promote cleaner fuels, improve public transportation, enforce industrial emission regulations, and implement stricter controls on shipping emissions in Colombo and biomass burning in Kandy.

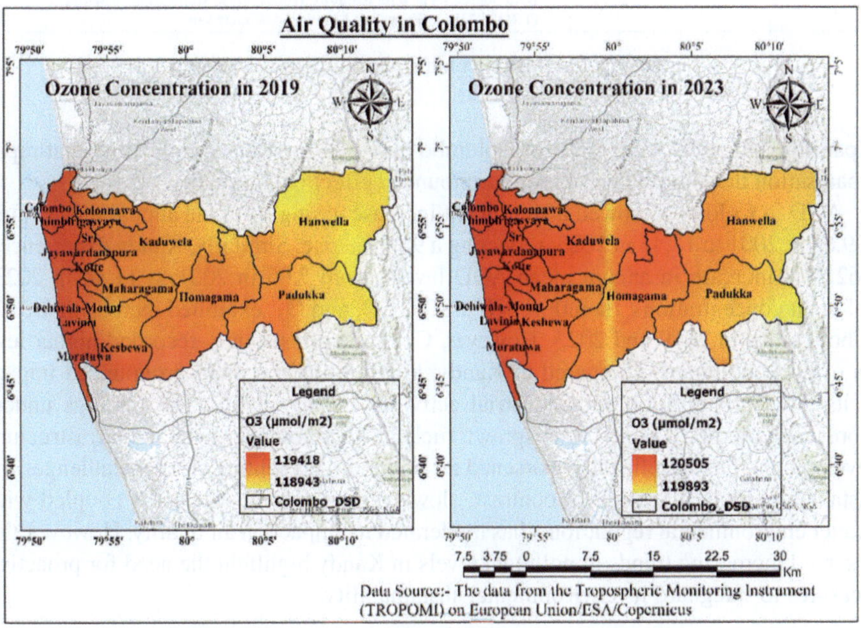

Fig. 9. TROPOMI O_3 column number density in Colombo

4.3 Relationship Between Urban Expansion and Air Quality

Statistical analysis revealed a positive relationship between urban growth and deteriorating air quality indicators in both cities (Fig. 11). The association between urban area

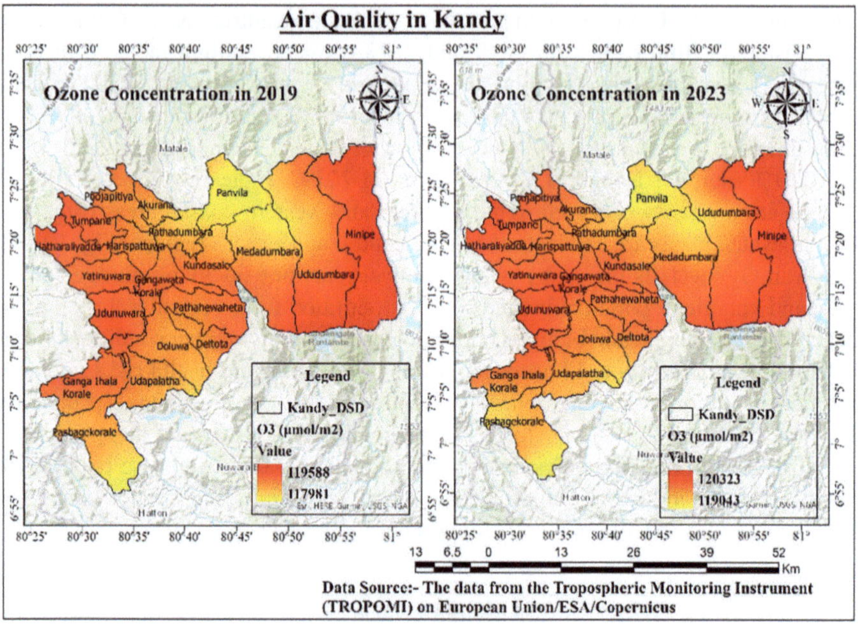

Fig.10. TROPOMI O_3 column number density in Kandy

expansion and AOD were 0.89 for Colombo and 0.74 for Kandy, indicating that rapid urbanisation in Colombo has a more pronounced effect on air quality.

AOD in Colombo showed a significant increase, rising from the annual mean value of 9.64 in 2001 to 18.45 in 2023, reflecting a 91.39% rise. Similarly, Kandy experienced a 62.84% increase in annual mean AOD levels, from 7.05 in 2001 to 11.48 in 2023. Pollutant concentrations, including NO_2, SO_2, CO, and O_3, exhibited increasing trends in both cities in 2019 and 2023. However, Colombo consistently recorded higher levels of these pollutants compared to Kandy, highlighting the more pronounced impact of its rapid urbanisation and industrial activities on air quality. The findings underscore that Colombo's rapid urban sprawl, fueled by economic growth and infrastructure development, has significantly worsened air pollution, presenting critical challenges for sustainable urban planning. In contrast, slower urban growth in Kandy, coupled with stricter environmental regulations, has moderated its impact on air quality. However, the observed increasing trends in pollutant levels in Kandy highlight the need for proactive measures to safeguard its environmental sustainability.

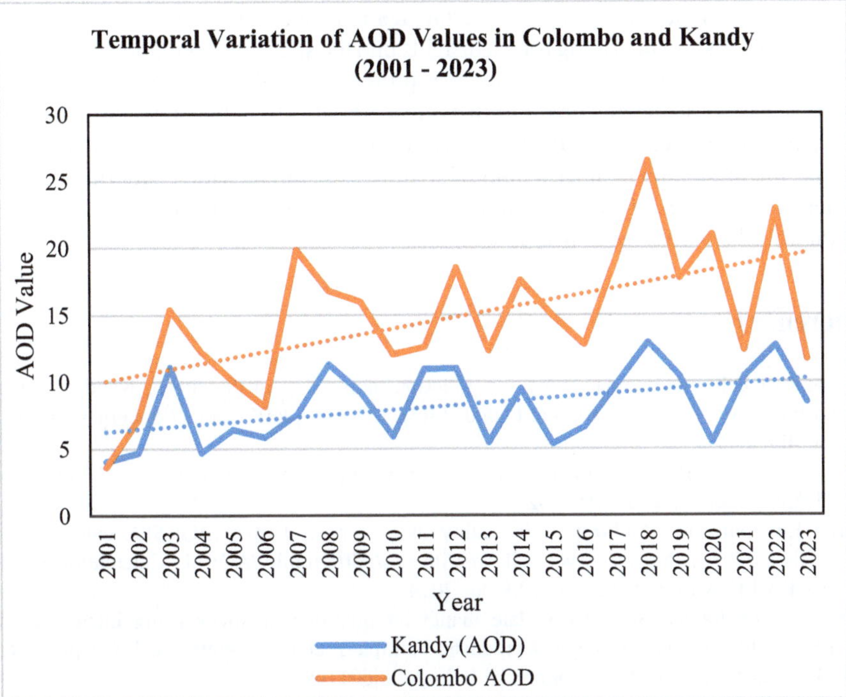

Fig. 11. Air Quality Variations of AOD (2001–2023)

5 Conclusion

This study underscores the link between urban expansion and air quality, offering a comparative analysis of Colombo and Kandy, two prominent urban centers in Sri Lanka. The rapid urbanisation in Colombo (25.25%), driven by accelerated economic growth and infrastructure development, has led to a substantial increase in AOD, rising from 9.64 in 2001 to 18.45 in 2023, an increase of 91.39%. Concurrently, concentrations of key air pollutants, including NO_2, SO_2, CO and O_3, exhibited significant upward trends, highlighting the environmental challenges associated with unregulated urban sprawl. In contrast, Kandy, characterised by slower urban growth and geographical constraints, recorded a more modest urban expansion of 6.05%, with AOD increasing from 7.05 to 11.48 over the same period, reflecting a 62.84% rise. While stricter environmental regulations in Kandy have moderated the adverse impacts on air quality, the observed upward trends in pollutant levels underscore the need for proactive measures to mitigate emerging environmental risks. The strong positive correlations identified between urban growth and declining air quality indicators (0.89 for Colombo and 0.74 for Kandy) emphasise the significant role of urbanisation in shaping environmental outcomes. These findings highlight the necessity of integrating air quality considerations into urban planning frameworks, particularly in Colombo, where rapid urban sprawl poses significant challenges for sustainable development. For Kandy, maintaining stringent regulatory measures while addressing increasing air pollution levels remains critical to preserving its environmental integrity. This study further demonstrates the potential of cloud-based geospatial analysis in providing robust insights into the spatial and temporal dynamics of urbanisation and air quality. By utilising platforms such as GEE, it is possible to perform comprehensive assessments that inform evidence-based urban planning and policy formulation. Future research should aim to refine these analytical approaches and develop city-specific mitigation strategies tailored to the unique urbanisation patterns and environmental challenges of each region, thereby fostering sustainable urban development and improved environmental health outcomes.

References

Abeyratne, V.D.K., Ileperuma, O.A.: Air pollution monitoring in the city of Kandy: possible transboundary effects. J. Natl. Sci. Found. **34**(3), 137–141 (2006). https://doi.org/10.4038/jnsfsr.v34i3.3644

Agalan, R.: Sri Lanka's Environmental Changes (2015). https://www.researchgate.net/publication/265453638. Accessed: 25 Dec 2024

Central Environmental Authority. Air quality index for Sri Lanka (AQI-SL), calculation & guideline (2023). https://www.cea.lk/web/images/pdf/airqulity/AQI-SL_Calculation_Guideline_CEA.LK_V1.0.pdf. Accessed 15 Dec 2024

Gupta, P., Christopher, S.A.: Particulate matter air quality assessment using integrated surface, satellite, and meteorological products: multiple regression approach. J. Geophys. Res. **114**(D14205) (2009), https://doi.org/10.1029/2008jd011496

Ministry of Environment. Clean Air 2025 Action Plan: Challenges in Implementation (2025). https://lpr.adb.org/sites/default/files/resource/434/sri-lanka-clean-air-2025-action-plan.pdf. Accessed 15 Jan 2025

National Building Research Organisation. Air quality monitoring and assessment in Sri Lanka (n.d.). https://www.nbro.gov.lk/index.php?option=com_content&view=article&id=85&Itemid=378&lan. Accessed 15 Dec 2024

Senarathna, M., et al.: PM2.5 air pollution trends and patterns in Kandy, Sri Lanka. Ceylon J. Sci. **53**(2), 197–206 (2024). https://doi.org/10.4038/cjs.v53i2.8403

Driving Sustainability through AI, Circular Economy, and Governance for Resource Efficiency and Climate Resilience

Biomimetic Fog Harvester Design: Enhancing Durability and Wind Resistance for Sustainable Water Collection in Harsh Environments

K. S. Dawson(✉) and L. N. Wickramarathna

Department of Mechanical Engineering, Faculty of Engineering, University of Peradeniya, Peradeniya, Sri Lanka
e18057@eng.pdn.ac.lk

Abstract. This research presents the design of a biomimicry-inspired fog harvester, aimed at optimizing water collection in arid and semi-arid regions while ensuring durability in extreme weather conditions. Drawing inspiration from natural systems such as spider webs, the Namib desert beetle, and pine needles, the harvester incorporates a triple-layered nylon mesh structure to enhance water collection efficiency and withstand high wind forces. To ensure long-term structural integrity, the harvester's ability to minimize turbulence and resist wind damage was evaluated through Computational Fluid Dynamics (CFD) simulations and physical testing. Simulations and calculations indicated that the optimized design could sustain wind speeds of up to 44 m/s while achieving an average water collection efficiency of 60%. The results demonstrate the potential for scalable and sustainable water harvesting solutions, focusing on wind resistance and material resilience.

Keywords: Biomimicry · Fog harvesting · CFD (Computational Fluid Dynamics) · Hydrophobic materials · Material resilience

1 Introduction

Fog harvesters are innovative tools designed to collect water from fog which provides a sustainable solution to water scarcity, which most probably would be applicable in arid and semi-arid regions, as shown in Fig. 1. The global water crisis is a growing issue gradually increasing because of population growth, climate change, pollution, inefficient management and improper usage. Around 2.3 billion people live in water-stressed areas, with climate change, worsening droughts and altering rainfall patterns. Pollution from industry and agriculture contaminates freshwater, leading to over 485,000 diarrheal deaths annually. Agriculture consumes 70% of freshwater, putting further pressure on limited resources, and rapid urbanization strains water infrastructure. The UN estimates that over 2 billion people lack access to safe drinking water while shrinking wetlands and depleted rivers threaten biodiversity. As demand rises, geopolitical tensions may escalate into "water wars", particularly in regions like the Middle East and Africa. The unequal distribution of water resources not only exacerbates poverty but also threatens food security, public health, and ecosystem stability.

R. Dissanayake et al. (Eds.): ICRES 2025, PTT, pp. 185–198, 2025.
https://doi.org/10.1007/978-981-96-9551-5_15

Fig. 1. Fog harvesters installed in Morocco copyright: aqualonis

The whole concept of fog harvesting is based upon biomimicry. Biomimicry in engineering design (Pohl and Nachtigall, 2015) is the practice of drawing inspiration from nature's time-tested patterns, structures, and processes to develop innovative and efficient solutions to engineering challenges. Its importance and relevance lie in several key areas such as sustainable solutions, innovation and efficiency, enhanced performance, problem solving in complex systems, interdisciplinary collaboration, resilience and adaptability. In engineering design, biomimicry is becoming recognized as a powerful tool for driving innovation, sustainability, and efficiency while providing a pathway to solving modern engineering challenges by learning from nature's wisdom. There are a few categories of fog collectors. In brief mainly it can be described as follows.

- Mesh based fog harvesters
- Fog fences
- Hybrid systems with active fog collection
- Electrically charged meshes

Despite proving to be effective in water collection the provided meshes and structures remain to unsustainable in high wind (>100 kph) and harsh weather conditions. Because most of these structures have to be mounted in mountainous areas or coastal areas exposing them to such conditions and in the process damaging the design. As it is evident in Fig. 2 (Rivera and Lopez-Garcia, 2015) and Fig. 3 (Olivier et al., 2015) many tend to be neglected afterwards due to the lack of technical knowledge by the residing communities which in turn threatens the environment and local fauna with artificial materials being deposited in the vicinity.

The objective of this project was to design a durable mesh-based structure which could be used in the process to be durable and efficient. The objectives include overcoming mesh clogging issues due to obstructive mesh designs, producing a passive structure with minimal maintenance work, structure build-up with durable materials while minimizing the cost, optimization of fog harvesting mesh to withstand high wind and reduce turbulence, development of a robust anchoring system for any terrain, evaluation and optimizing water collection efficiency and to asses feasibility and scalability of the design.

Fig. 2. Failure of large fog collectors: (top; Peña Blanca, Chile) failure of the supporting structure; (bottom, left; Anagua project—Canary Island, with permit by Carlos Sanchez Recio) failure of the foundation; (bottom, right; Peña Blanca, Chile) breakage of the mesh.

Fig. 3. Wind damage to the Whirly Mk II (left) and to the aluminium drive ring (right).

2 Methodology

2.1 Conceptual Design Progression

The concept was inspired by the Namib desert beetle's outer shell as in Fig. 5 but considering the cost of chemicals and materials needed for such a design it had to be discarded. Though this might prove to be more effective in the future incorporated to a tile of a house which could be built in an arid or semi-arid geographical area.

Then the interest was shifted to existing work (Klemm et al., 2012), most of them utilizing a 2-dimensional or a 3-dimensional mesh a model was created to test the most common Raschel mesh for force. Initial force calculations were done using the following equation (Eric Bank and Finance, 2018) for the SolidWorks model in Fig. 4.

$$\boldsymbol{F} = v^2 \times (0.00483) \tag{1}$$

- $\boldsymbol{F}$ is force per square meter in *N/m*^{2}
- v is the wind speed in *kph*
- *0.00483* being the coefficient that converts wind speed squared into force per unit.

Even in a slow wind, a 3-D mesh structure would exert a significant amount of force when subjected to a high wind. Not taking turbulence effects into consideration. For a thin and tall 10 m^2 structure, a force of 48.3 kgf/m^2 (4732.23 N for 10m^2) is significant and could pose a risk of structural instability if not properly designed for wind loads. The structure would likely need to be carefully engineered, considering material strength, anchoring, and possible bracing or reinforcement to withstand such forces.

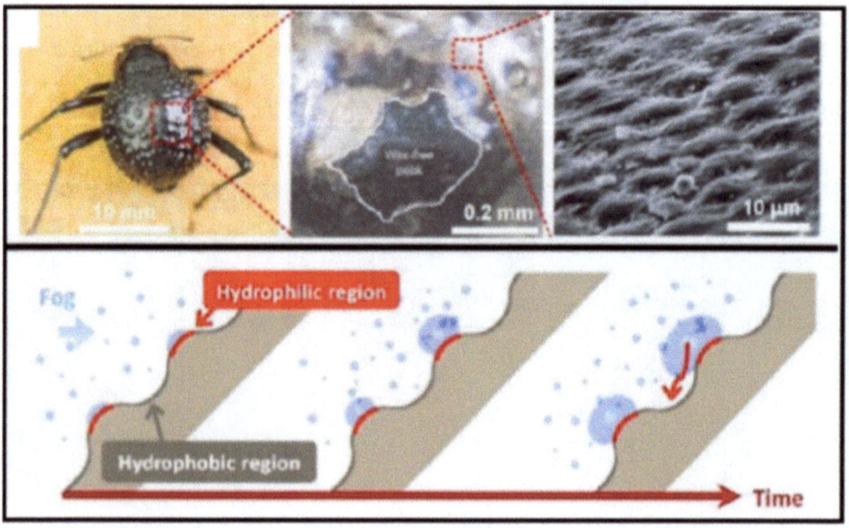

Fig. 4. *Stenocara gracilipes* also known as the Namib desert beetle.

Next a spider web design was infused with an EPDM mesh in between for more stability and durability in high wind conditions. Here the model was limited to a 1 m^2 model which is a small-scale fog collector as it was much feasible when in the fabrication phase. As seen in Fig. 6 the model was developed with four main detachable components to make traversing and installation much easier.

After careful consideration of the cost-effectiveness and availability of materials the rubber mesh had to be dropped due to its high cost and unavailability. Being an elastic material, this would also oscillate in high wind conditions making the rest of

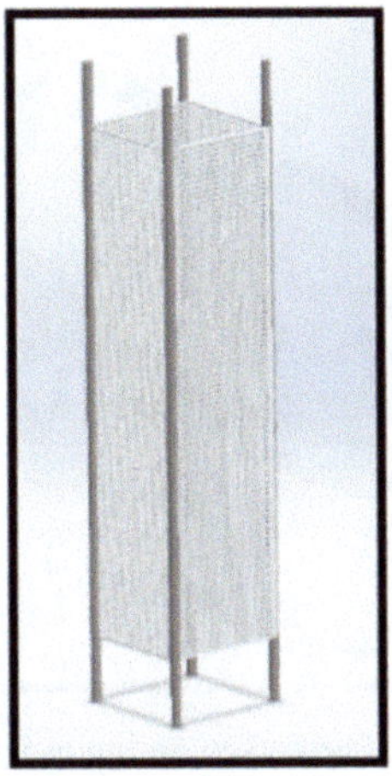

Fig. 5. A moderate fog collector model each side covered in a 10 m^2 Raschel mesh

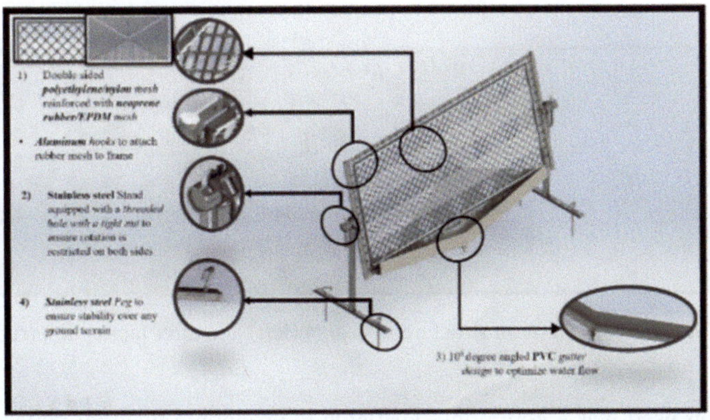

Fig. 6. A much more feasible small fog collector design model 1 m^2 equipped with the novel mesh

the structured mesh useless in its sole purpose. Utilizing these design considerations, a new model was finalized to fit the description of the job inspired by the spider web design and its structure. For fabrication, a convenient model was developed under the available conditions for demonstration and data collection purposes. It was scaled down to make a functional but easily manoeuvrable prototype. (As seen in Fig. 7) The design was intended as shown in the figure to be built using GI pipes, steel or *Al* box bars, sheet metal strips, nylon strings, PVC gutters and a reducer which are commonly available locally. Thus, rendering the design cost effective.

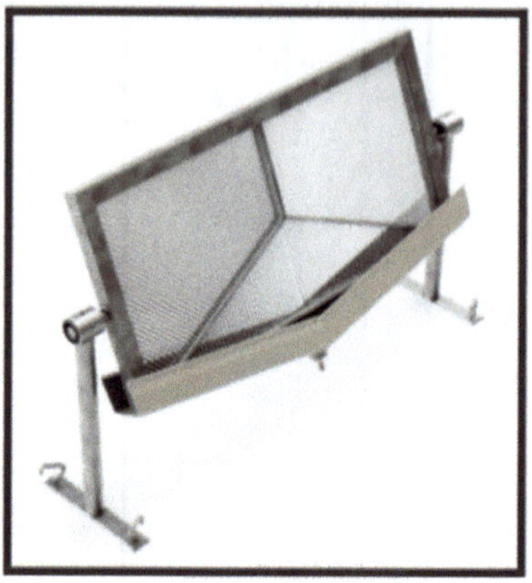

Fig. 7. Final prototype design in SolidWorks

2.2 Material Selection

Fig. 8. Triple nylon mesh structure in design (left) and after fabrication (right)

The nylon used was a commonly available hydrophobic material which is usually used as fishing line. The specifications of the line are that it has a diameter of 0.75 mm (see Fig. 8) and only has one fibre rather than being a braided line. This was aimed to improve the dripping mechanism which was already utilized in the design. Considering the strength of the spool lines it was more conveniently durable than other materials. The hollow box bars provided a large enough surface area (1 1/2 inches of width) for 3 rows of 5mm holes with sufficient distance between them. The mild steel stirps were of 1 1/2 inch width providing the same distance and structural rigidity for the frame. They were less in weight after the whole assembly (Fig. 9) approximately 5 kg which was much more convenient for one person to carry around and simultaneously serve as a strong structure in extreme weather conditions. The connections between the steel members were done using manual metal arc welding.

The PVC gutter as shown in Fig. 10 was to evenly let the water drip towards a trough where the water could be collected. By using PVC as the material, it made it considerably lighter and stronger proving to be ideal for the design and cost-effective considering the general availability. The connections were done with a metal wire mesh covered in metal epoxy to ensure rigidity and smooth flow of water.

Fig. 9. Structure being fabricated using steel box bar and mild steel strips with 3 rows of 5 mm drilled holes

Fig. 10. Angled PVC gutter design fabricated for collection of accumulated water in the mesh

After the process all components were coated with an anti-corrosive paint to increase longevity in exposed weather conditions while testing (Fig. 11).

2.3 Structural Design

All the above components in Table 1 can be observed in Fig. 12 in the physical prototype after the assembly. An anti-corrosive paint coating was applied to ensure safe water collection and longevity of the structure in varying weather conditions.

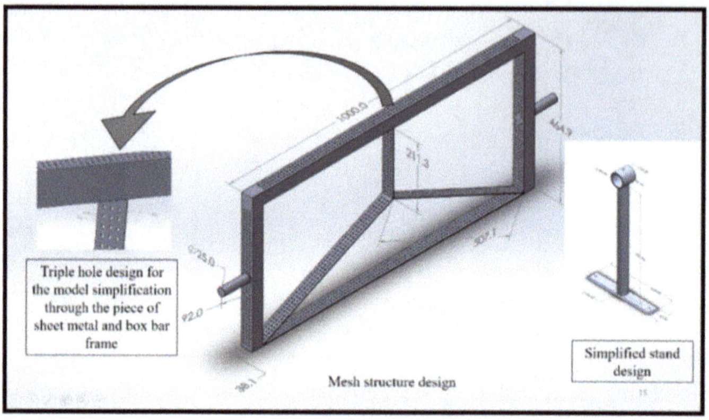

Fig. 11. Dimensions of the simplified model structure with 1 1/2 inch box bar

Table 1. Components and description

Component	Description
Nylon Mesh Structure	Mesh Structure: Three parallel nylon fishing string meshes with a circular cross-section (diameter 0.75 mm). Functionality: Maximizes fog harvesting efficiency with multiple layers for droplets to condense upon. Turbulence Reduction: Minimizes turbulence for consistent airflow and better droplet adherence. Support: Robust framework withstands wind pressure
Box Bar Steel Frame	Material: Durable box bar steel provides high strength and rigidity for outdoor conditions (wind speeds 25–44 m/s). Design: Serves as the structure's skeleton, supporting meshes and maintaining stability
Triangular Support	Location: Positioned centrally in the structure. Functionality: Enhances stability with lateral reinforcement and efficient load distribution. Purpose: Assists in weaving the nylon pattern, forming a spider web-like structure

2.4 Design Optimization

The following equations were used to find the optimum parameters which were obtained using MATLAB and can be observed in Figs. 13 and 14. There were a few assumptions to make the modelling process more convenient and accurate. Then after the representation of data, an optimum value was found for each changeable condition of the model as shown in Table 1.

$$F_d = \frac{1}{2} C_d \rho_{air} v_{mist}^2 A_f \tag{2}$$

- C_d (Drag coefficient) is constant (typically 1.2 for cylindrical objects in airflow)
- v_{mist} (mist speed) is uniform across the frontal area.

Fig. 12. Finished structure after fabrication (Courtesy of the Workshop, Department of Mechanical Engineering UOP)

- ρ_{air} (Air density) is a constant (typically 1.225 kg/m^3: standard air density at sea level)
- A_f frontal contact area

$$\eta = 1 - exp\left(-\frac{d}{d_0}\right) \tag{3}$$

- $d_0 = 0.75$ mm (Based on effective minimum diameter for water collection)

$$\eta_c = \frac{d_{drop}}{d_{string}}.C_s^2 \tag{4}$$

- C_s (Slip coefficient) 0.5 for nylon strings
- d_{string} often 0.75 mm
- $d_{drop} = 10$ µm

$$A_{surface} = \pi d_{string}.L \tag{5}$$

- $L = 1\ m$: Considering a uniform length

$$F_{detach} = \gamma.\pi d_{string}.cos(\theta) \tag{6}$$

- $\gamma = 0.0728$ *N/m (Average surface tension)* does not vary
- $70^0 \leq \theta \leq 80^0$ (Contact angle between water and nylon)

$$M_{collected} = \rho_{water}.A_{surface}.\eta_c.v_{mist} \tag{7}$$

- $\rho_{water} = 1000$ kg/m^3; standard density of water
- $0\text{m/s} \leq v_{mist} \leq 44\text{m/s}$.

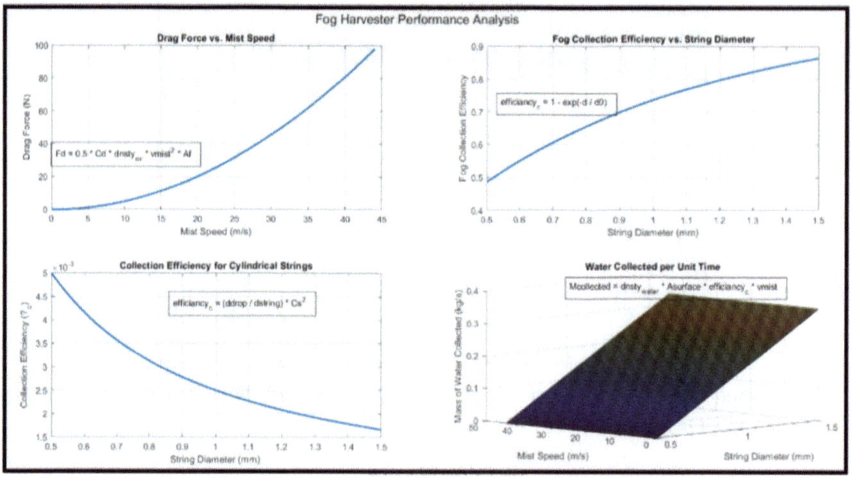

Fig. 13. Variation of Drag force vs Mist speed, Efficiency vs String diameter, Efficiency vs String diameter and Mass of water collected against variable mist speed and string diameter.

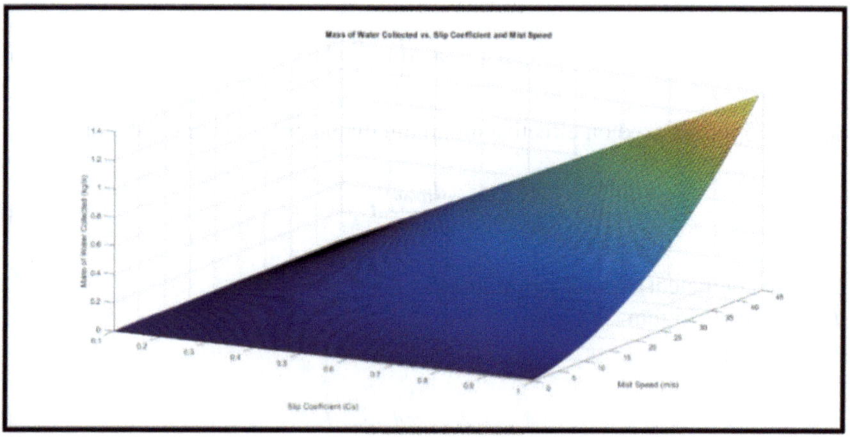

Fig. 14. Mass of water collected against varying parameters for 0.75 mm string diameter

3 Results

3.1 CFD Simulations

The results were obtained using the simulation platform SimScale. The initial boundary conditions used for the triple mesh are shown in Fig. 15.

Initially the velocity magnitude contours were obtained using the K-Omega turbulence model present in the CFD analysis tool which can be seen in Fig. 16. Here the fluid used was observed to have a velocity drop of 5m/s and increment of 13 m/s yielding a minimum velocity = 18.7 m/s and a maximum velocity = 46.64 m/s.

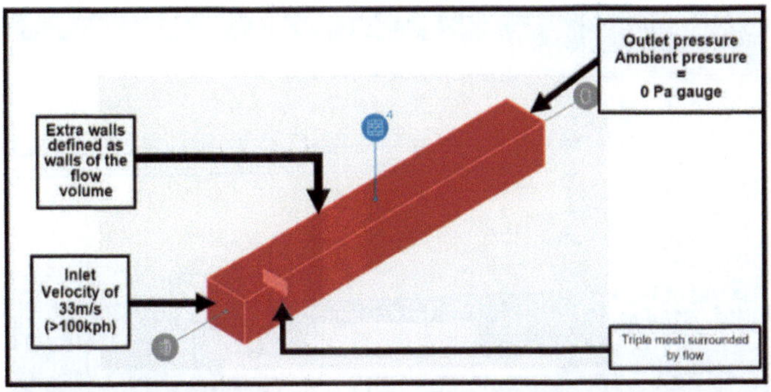

Fig. 15. Boundary conditions for the CFD analysis (For triple mesh)

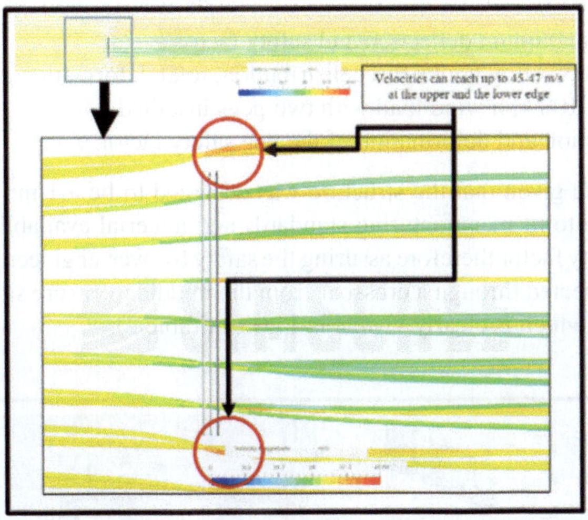

Fig. 16. Velocity magnitude contours streamline using particle tracing

Figure 16 shows the cross-sectional side view midplane of the mesh and according to Bernoulli's principle, there is a velocity increment taking place at the upper and lower edges which in turn amounts to a pressure loss creating a pressure difference within the mesh and thus generating a force on the structure. The turbulent kinetic energy results were also obtained to observe the disruption in energy through the mesh. The pressure contours along the streamlines can be observed in Fig. 17. The simulation yield results for the pressure drop to be around 323 Pa ranging below 793 Pa to just below 470 Pa which is the minimum pressure.

3.2 Structural Simulations

The initial design's structural analysis was done using ANSYS structural (See Fig. 18).

- The mesh and frame under 100 kph wind force.

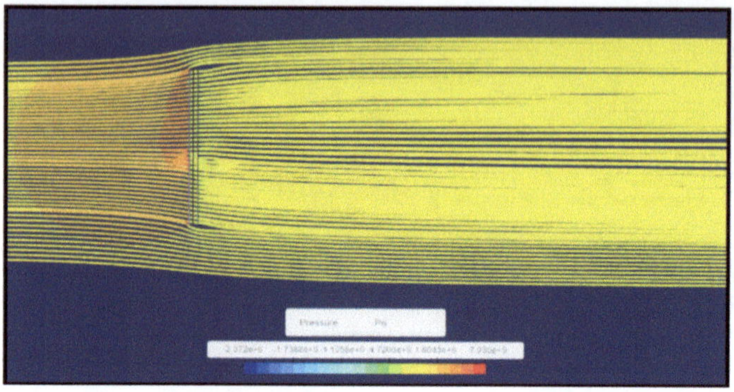

Fig. 17. Pressure contour streamlines using particle tracing

- EPDM mesh maximum deflection and safety factor.
- Aluminium 1069 alloy hooks force distribution, total deformation and safety factor.
- Stand under a 100 kph wind load with two pegs installed.
- Force distribution and deformation of the peg safety factor.

As it was safe given that the structure was designed to be a 1 m^2 model and later was scaled down to fit manufacturing standards and material availability the structure increased its safety factor therefore assuring the safety by over-engineering. The two side stands were connected through a crossbar from the middle to ensure structural integrity over flat terrains which is clearly visible in Fig. 12 (Table 2).

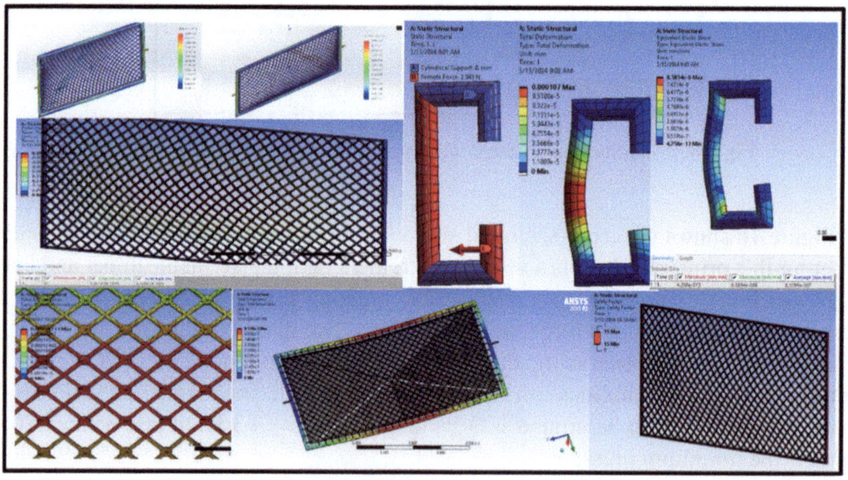

Fig. 18. ANSYS structural simulation

Table 2. Optimum parameters found using MATLAB

Parameter	Optimum Value
String Diameter	1.5 mm
String Spacing (Mesh Density)	5 mm to 10 mm
Wind Speed (Optimum)	16 m/s
Droplet Size (Ideal Range)	10 μm to 30 μm (10 μm considered for least optimum conditions)
Mesh Wettability	70° < Contact Angle < 90°

4 Conclusion

The design process of the fog harvester began with an initial concept inspired by the Namib desert beetle, but it was abandoned due to cost considerations. The focus was then shifted to existing mesh-based designs, culminating in a spider web-inspired prototype. Several iterations were explored, including Raschel mesh and EPDM mesh, however the final design utilized a triple-layered nylon mesh supported by a robust box bar steel frame. ANSYS simulations confirmed the structure's stability under high wind loads, while material choices such as PVC gutters and steel bars ensured durability and cost-effectiveness. Simulations conducted with MATLAB helped optimize the design, identifying key parameters like string diameter (1.5 mm) and ideal wind speed (16 m/s) for maximum water collection efficiency. Field testing revealed challenges with inconsistent mist conditions, but the second round of testing—despite some rainfall interference—demonstrated successful water collection. The structure proved the capability of withstanding high wind speeds, and the data gathered provided valuable insights for further refinement.

Acknowledgements. I sincerely thank Dr. Lalith N. Wickramarathna, my supervisor, for his guidance and encouragement, and Professor Asanga Rathnaweera and Ms. Lekha Bakmeedeniya for their support. Special thanks to Dr. Janaka Rajaguru for providing materials and to Professor Leelananda Rajapaksha and other evaluators for their feedback.

I am grateful to the technical staff, including Mr. K.V.A.T.L. Kekulawala and the team, for their assistance during fabrication. My heartfelt thanks to my parents for their unwavering support both financially and mentally, and to God for the strength to complete this project.

References

Eric Bank, B., Finance, M.: How to calculate wind loads from wind speeds (2018)

IntoAction 5_Fog nets (2015)

Klemm, O., et al.: Fog as a fresh-water resource: overview and perspectives. Ambio (2012). https://doi.org/10.1007/s13280-012-0247-8

Olivier, J., Van Heerden, J., Rautenbach, H., Pretorius, E.: Optimising fog water harvesting in South Africa. Water Research Commission (2015)

Pohl, G., Nachtigall, W.: Biomimetics for Architecture & Design. Springer (2015). https://doi.org/10.1007/978-3-319-19120-1

Rivera, J.D.D., Lopez-Garcia, D.: Mechanical characteristics of Raschel mesh and their application to the design of large fog collectors. Atmos. Res. **151**, 250–258 (2015). https://doi.org/10.1016/j.atmosres.2014.06.011

Optimizing a Heel-Strike Energy Harvester for Off-Grid Applications Using Piezoelectric and Electromagnetic Technologies

P. C. N. Fernando(✉) and L. N. Wickramarathna

Department of Mechanical Engineering, Faculty of Engineering, University of Peradeniya, Peradeniya, Sri Lanka
e18101@eng.pdn.ac.lk

Abstract. This study focuses on resource efficiency in the energy sector by exploring biomechanical energy harvesting technologies for off-grid applications. Various energy harvesters, including knee joint and backpack systems, have been evaluated. Still, heel-strike harvesters have emerged as the most efficient portable solution, offering energy generation with minimal disruption to the human gait cycle. To maximize energy output while ensuring a lightweight and ergonomic design, this research leveraged piezoelectric and electromagnetic technologies. The optimization procedure was done under three concepts. They are selecting an optimized piezo plate configuration on the heel area (concept I), increasing the rate of piezoelectric pulse generation within a stair gait cycle with a newly designed mechanism (concept II) and using a hybrid technology method with piezo-electricity and electro-magnetism (concept III). A hiker scenario was chosen to evaluate the feasibility of this mechanism in off-grid areas with temporary power outages. Power generation data from testing was used to estimate the system's ability to meet the energy consumption needs of hikers. A prototype was designed only using four piezoelectric plates integrated into a shoe insole to test concept I. It resulted in a peak AC voltage of 3.05 V; an estimated power of 29 mW; and an estimated current of 7.8 mA for a 57.10 kg person whose hiking speed was 0.5 m/s. It was only sufficient to cover the minimum power consumption range of a hiker (100 mAh) within 12.81 h. The system's performance was suggested to be optimized using concepts II and III to achieve the higher power consumption ranges (100 mAh<). A specific mechanism was designed for this optimization and simulated considering mechanical strength over foot pressure and electromagnetism. The research concludes that combining piezoelectric and electromagnetic harvesting technologies while implementing the conceptual mechanism developed provides a reliable, practical energy solution for hikers, with the potential for future energy efficiency and user comfort improvements.

Keywords: biomechanical energy harvesting · resource efficiency · off-grid applications · piezoelectric technology · electromagnetic technology · heel-strike harvesters

R. Dissanayake et al. (Eds.): ICRES 2025, PTT, pp. 199–214, 2025.
https://doi.org/10.1007/978-981-96-9551-5_16

1 Introduction

The global energy sector faces a growing need for resource-efficient solutions to meet the increasing demand for electricity while minimizing environmental impact. In off-grid scenarios, such as rural areas, temporary urban power outages and remote exploration, access to reliable power becomes critical for essential needs like communication and navigation. Among these, hiking provides a compelling example, with over 118 million participants annually requiring daily electricity for devices ranging from low (100–200 mAh) to high (>2000 mAh) consumption. (Globally, how many people practice hiking, and what is the global market size for this sport?, n.d.). However, conventional energy sources like solar panels are often impractical due to weight and dependency on external conditions like weather and terrain. (Andreas Holzinger). Besides, external power banks add weight and require recharging from external sources, often unavailable on long hikes. Therefore, off-grid communities like hikers need an energy source that is highly reliable, consistent and low weight to minimize the metabolic cost.

Biomechanical energy harvesting as a renewable energy method offers a groundbreaking approach to this matter by converting human motion into electricity and recharging the only external power bank of the hiker consistently as long as there is movement (Huang, 2023-05). According to a recent review, piezoelectric energy harvesting from human legs and footwear represents a higher output power and voltage than electromagnetic and triboelectric harvesters (Mohankumar, 2021), (V, 2021). Out of them, the motion of the knee harvests more than the ankle motion and heel strike. However, knee harvesters significantly hinder the gait cycle. That means it increases the metabolic cost of the human by adding extra positive energy to the climbing-up phase of the stair gait cycle as shown in Fig. 2 (Huang et al., 2023). Among these methods, energy recovery from heel strikes using piezoelectricity during walking or climbing is particularly promising due to its consistency and minimal impact on natural gait. Researchers have explored various methods for harvesting energy from human walking using shoe-based piezoelectric harvesters. By 2023, the highest electric power harvested by a piezoelectric footwear energy harvester is 23.9 mW (Zhao, 2023). Therefore, this study has focused on optimizing an existing flat piezo plate-based design (Fig. 1), selected from (Zhao, 2023) due to its minimal gait disruption.

Fig. 1. Existing piezoceramic discs attached to the insole of the ball and heel areas (Zhao, 2023)

While this study highlights hiking as an illustrative use case, the underlying technology has broad applicability for low-power electricity usage across diverse off-grid environments where portability and reliability are essential. This study tries to fill the energy gap between the power generation of a selected heel strike harvester (Fig. 1) and the power consumption of a hiker just before the hiker reaches the mountain peak. In this scenario, the body with a backpack exerts a higher net weight force on the heel strike, resulting in greater energy generation.

With the above purpose, key features of this study include the use of hybrid piezoelectric and electromagnetic technologies with an optimized piezo plate configuration to maximize power generation, ergonomic design to minimize metabolic cost, and rigorous simulation and testing to ensure system durability and efficiency. By leveraging these innovations, this research contributes to the broader theme of resource efficiency in the energy sector, demonstrating a scalable and user-friendly approach to sustainable energy harvesting.

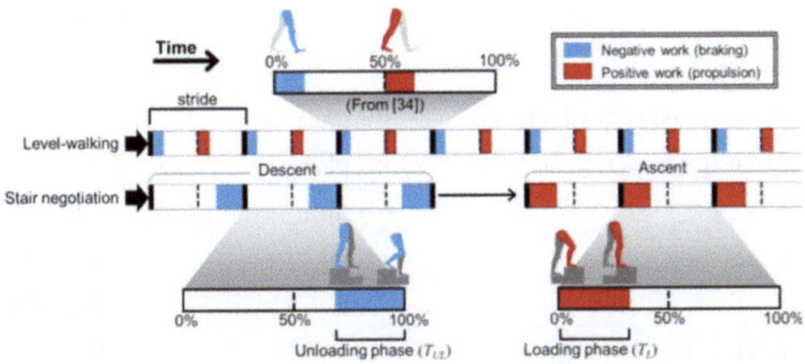

Fig. 2. Distribution of negative and positive energy during the stair gait cycle and walking gait cycle (Yun Seong Song, 2017)

2 Methodology

The mechanism developed combines three concepts and is constrained by the 3D space shown in Fig. 3, as it must fit within the limited space of a shoe heel. The corresponding concepts are selecting an optimized piezo plate configuration on the heel area (concept I), increasing the rate of piezoelectric pulse generation within a stair gait cycle with a newly designed mechanism (concept II), and using a hybrid technology method with piezo-electricity and electro-magnetism (concept III).

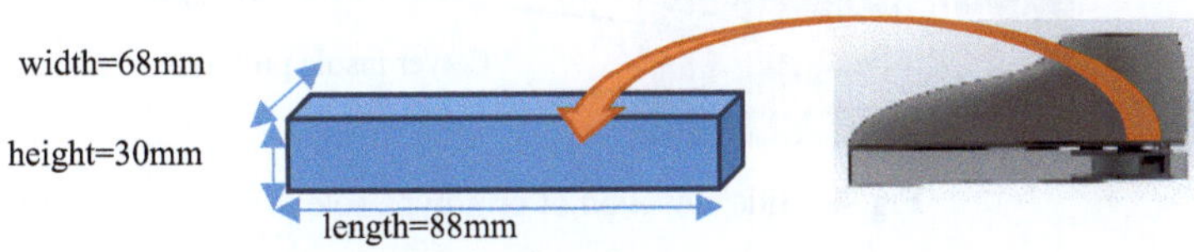

Fig. 3. Exterior dimensions of the design

2.1 Selection of Optimal Configuration of Piezo Plates -Concept I

For concept I, the piezo plate type (Table 1) was chosen based on its diameter and the space constraints of the shoe insole, assessed through visual inspection. The feasible number of piezo plates within the constant area (68 × 88 mm) and the feasible configuration of piezo plates were selected. Besides, these plates were connected in parallel to increase the charge.

Table 1. Data related to selected piezo plate type

Parameters of The Selected Piezo Plate	Units	Value
Type	–	27D
Relative permittivity (ε_r.)	–	1200
Piezoelectric charge sensitivity (d_{31})	pF/n	110
Permittivity of free space (ε_o)	pF/m	8.85
Thickness	m	5.10E−04
Crystal area radius	m	1.00E−02
Capacitance	F	1.60E−08
Crystal voltage sensitivity factor(g)	V.m/N	0.010

There were previous experiments only done considering the whole foot area, but in this study, only the heel area has been considered to find the configuration of piezo plates that gives the most stable analog voltage output. Four specific configurations were tested, as shown in Fig. 6. These configurations were designed to evaluate the effects of increasing the number of piezo plates and adjusting their positions. A physical setup, as illustrated in Fig. 5, was tested by a 57.10 kg person walking on the prototype (Fig. 4) with random stepping frequency. The prototype shoe sole, made from a PVC hard sheet, had holes cut to accommodate the piezo plates; piezo plates were glued, and rubber supports were attached.

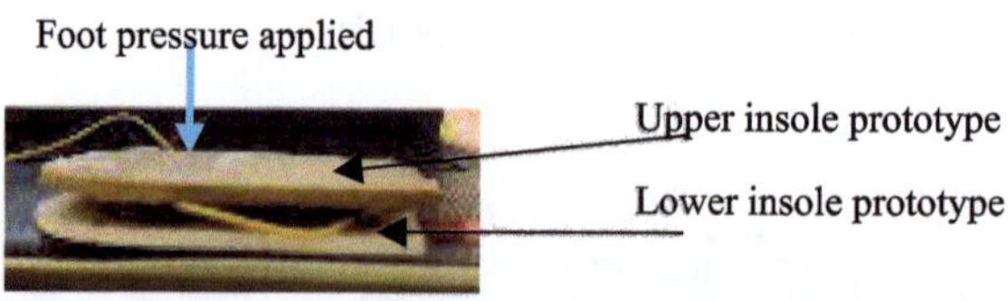

Fig. 4. Side elevation of prototype sole

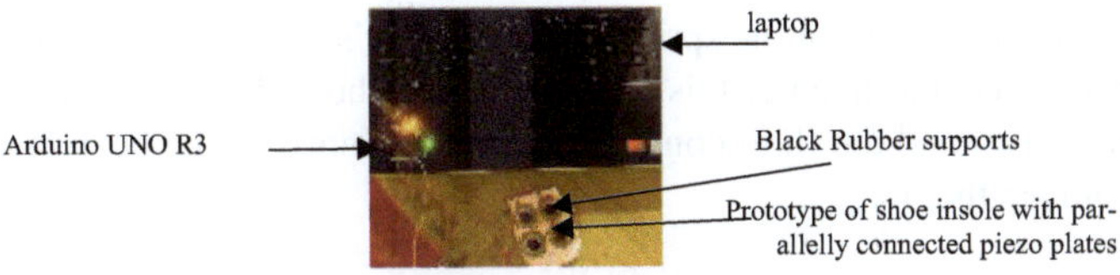

Fig. 5. Setup implemented to select the optimal configuration

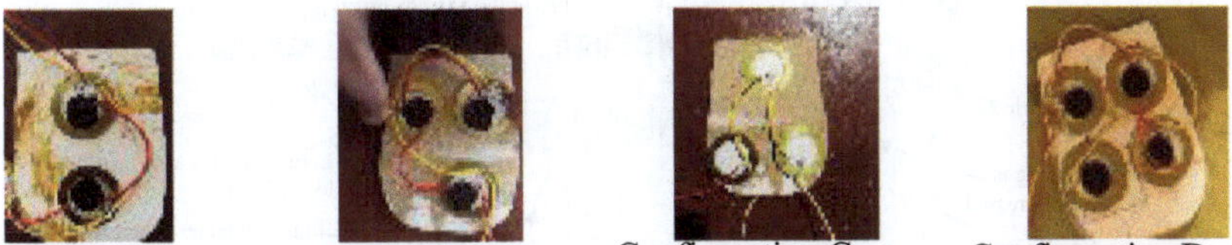

Configuration A (a) | Configuration B (b) | Configuration C (c) | Configuration D (d)

Fig. 6. Tested four configurations that were attached to the prototype of the lower insole

2.2 Increasing the Rate of Piezoelectric Pulse Generation Within a Stair Gait Cycle- Concept II

For concept II, a conceptual mechanism was designed to maximize power generation by increasing the rate of piezoelectric pulse generation during a stair gait cycle. The main components of this mechanism are shown in Fig. 7. In the existing method (Fig. 1), a voltage signal is generated only by the heel strike. However, with this new mechanism, the piezo plates are compressed twice within a stair gait cycle: first by the direct impact of the heel strike, and second by the impact of a large piston in the fluid container at the beginning of the swing phase. The fluid container transfers energy from the see-saw mechanism to the large plate, which compresses the piezo plates during the swing phase. This design ensures that the piezo plates are compressed when the heel is lifting off the ground, thereby optimizing energy harvesting.

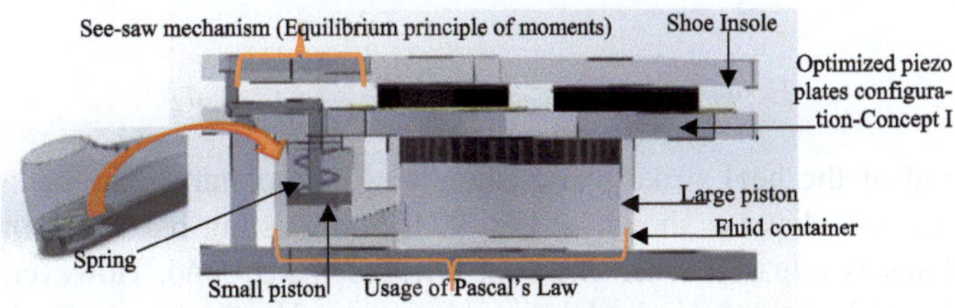

Fig. 7. Side view of the conceptual mechanism

The mechanism within one stride(gait cycle) has been described below in two phases. In the heel strike phase, piezo plates are compressed by direct foot pressure to generate energy as the shoe upper insole touches the cylindrical supporter attached to the piezo plates. The see-saw mechanism is in a state as shown in Fig. 8. Therefore, the spring has compressed, storing energy in the spring. For the hiker's comfort, the upper insole and the lower insole become fixed at this phase, keeping the stability of the footstep. The see-saw mechanism and the fluid container were developed as the heel strike pressure would not damage them.

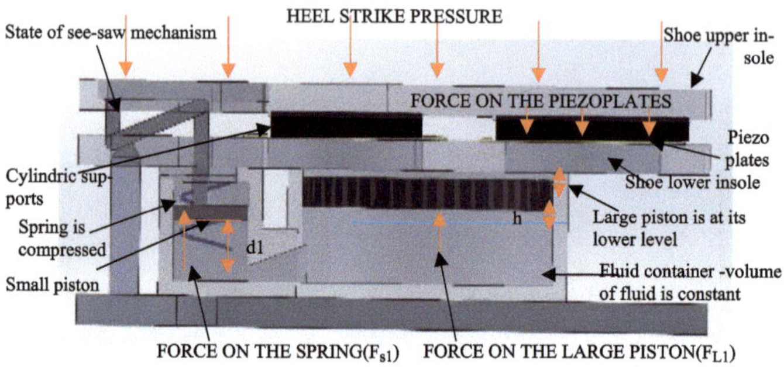

Fig. 8. Overview of the mechanism at heel strike phase

In a U-tube system, the pressure generated by a force on one side must be balanced by the combination of hydrostatic pressure due to fluid level difference and the applied pressure from the other side. By applying Pascal's Law to the U-shaped fluid transferring system in Fig. 8, Eq. (1) has been derived. On the left side, the small piston is applying a force F_{s1} on area A_s (surface area of the large piston). On the right side, a force F_{L1} is occurring on area A_L (surface area of the large piston). Between the two sides, there's a fluid of density ρ, and there's a height difference h in the fluid levels.

$$\frac{F_{s1}}{A_s} = h\rho g + \frac{F_{L1}}{A_L} \tag{1}$$

Assuming the spring has compressed d_1 distance from its initial position (Fs = 0), Eq. (2) has derived for the energy stored in a spring (E_{SPRING}). F_{s-avg} is the average force on the fluid by the small piston and F_{L-avg} is the average force on the large piston by the fluid.

$$E_{SPRING} = F_{s-avg}d_1 = \frac{kd_1{}^2}{2} = F_{L-avg}d_2 \tag{2}$$

At the end of the heel strike phase, the swing phase starts. The behavior of the mechanism in this phase is shown in Fig. 9. In the existing harvester, the stress on the piezo plates is released when the foot releases the ground. However, when this mechanism is implemented, piezo plates are compressed back again by the large piston. The seesaw mechanism is activated by foot releasing; the spring begins to go to its initial position while converting spring energy into work (work is moving the fluid); this work compresses the piezo plates and converts them into electric energy.

If there is no energy loss in the fluid container, the work done by the small piston will be equal to the work done by the large piston. Therefore, Eq. (3) can be derived.

$$d_1{*}F_{S-avg} = d_2 * F_{L-avg} \tag{3}$$

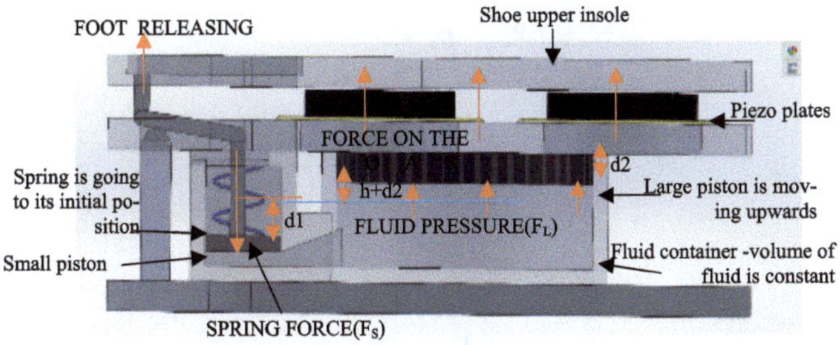

Fig. 9. Overview of the mechanism at swing phase

Using Fig. 9 and Pascal's Law Eq. (4) has derived same as the Eq. (1). (Fs = 0);

$$\frac{F_{s2}}{A_s} = (h + d_1 + d_2)\rho g + \frac{F_{L2}}{A_L} \tag{4}$$

Equation (4)–Eq. (1);

$$\frac{F_{s2} - F_{s1}}{A_s} = (d1 + d_2)\rho g + \frac{F_{L2} - F_{L1}}{A_L} \tag{5}$$

Equation (5) has been derived for the fluid transferring system and for the See Saw mechanism, and Eq. (6) has been derived by using the Equilibrium Principle of Moments.

$$F_{applied\ by\ foot} {*} X_1 = F_{s1} * X_2 \tag{6}$$

When designing this mechanism, small pistons, large pistons, see-saw bars, fluid containers, and the spring had to be designed as they could satisfy the above equations.

According to Eq. (7)–Eq. (9), foot force (F_{foot}) is directly proportional to the electricity generated and the voltage in the heel strike phase. Using Eq. (10)–Eq. (12), some design calculations were conducted to optimize the design to maximize the electricity generated and the voltage using concept II by changing the variables d_1, d_2, A_L, A_{piezo} in Eq. (14) and Eq. (15). Piezoelectric charge sensitivity, crystal voltage sensitivity factor, thickness and spring constant are represented by d,g,t and k in order.

$$Electrical\ energy\ generated\ by\ a\ piezo\ plate = \frac{QV}{2} \tag{7}$$

$$Electric\ energy\ generated = \frac{dgt}{2A_{piezo}} x \left[\frac{F_{foot} A_{piezo}}{A_{foot}}\right]^2 \tag{8}$$

$$Voltage\ generated = \frac{gtF_{foot}}{A_{piezo}} \tag{9}$$

Fluid transferring mechanism is used at the beginning of the swinging phase, and the energy equation has derived below, and energy is represented by E.

$$E_{spring} = E_{loss} + E_{transfer} \tag{10}$$

$$\frac{kd_1{}^2}{2} = \frac{F_L d_2}{1} + Eloss \tag{11}$$

$$\sigma_{piezo} = \frac{F_L}{A_L} = \frac{F_{piezo}}{A_{piezo}} \tag{12}$$

$$E.Energy = \frac{dgt}{2A_{piezo}} x \left[\frac{F_L A_{piezo}}{A_L} \right]^2 \tag{13}$$

$$E.Energy = \frac{dgt}{2A_{piezo}} x \left[\frac{(kd1^2/2 - Eloss)A_{piezo}}{d_2 A_L} \right]^2 \tag{14}$$

$$V = \frac{gt \frac{(kd1^2/2 - Eloss)A_{piezo}}{d_2 A_L}}{A_{piezo}} \tag{15}$$

This whole mechanism was modeled in SOLIDWORKS, and simulations were done using ANSYS for the fluid container and the seesaw mechanism, which are the critical parts of the mechanism considering the mechanical strength. The deformation and stress simulations were performed to confirm the structural integrity of the fluid container and see-saw mechanism under a maximum force of 9408 N, representing the weight of a 100 kg person with a 20kg loaded backpack (How Much Should Your Pack Weigh?, n.d.). Then a convergence test was done to approach the true solution with a refined mesh. For the final design of the fluid container (Fig. 10), simulations were conducted using Polycarbonate, Polyurethane, and PVC foam due to their direct contact with the electric circuit.

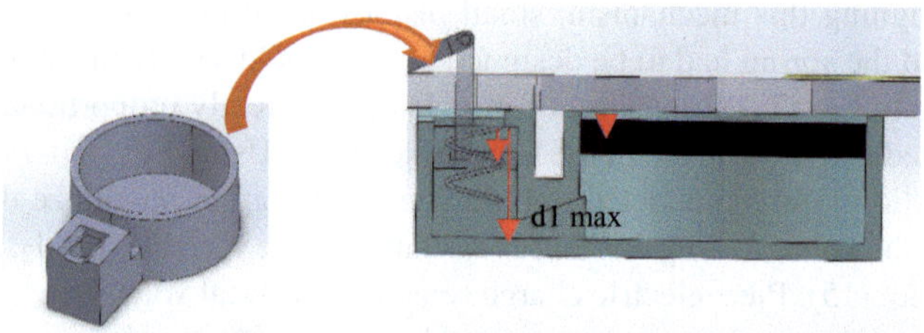

Fig. 10. Final design of the fluid container (3D view and the side view)

For the see-saw mechanism, simulations were conducted using Structural Steel as it did not come into direct contact with the circuit. The see-saw mechanism shown in Fig. 11 converts the foot's upward motion into the downward motion of the small piston during the swing phase. This is achieved using a rotational joint and a translational joint, ensuring the smooth vertical motion of the piston without rotational interference, as in Fig. 12. Additionally, a spring was incorporated to assist the downward motion of link 3, ensuring continuous functionality without hindrance. A predesign calculation was used to determine the optimal position of the rotational joint, minimizing the vertical motion Δd3 of the upper insole to less than 2 mm, thereby maintaining the hiker's comfort. Then the simulation was conducted.

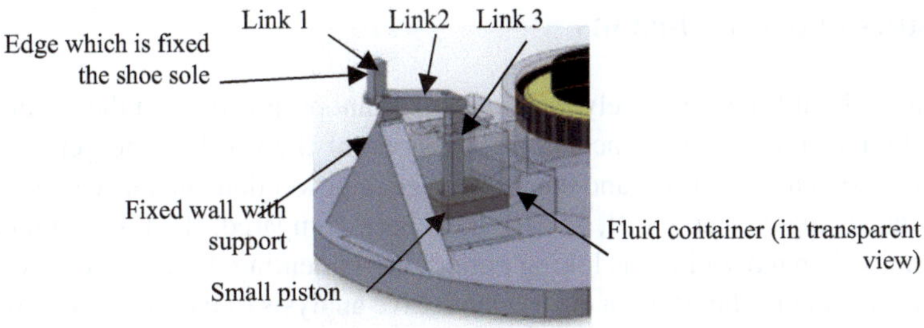

Fig. 11. Final design of the see-saw mechanism (3D view)

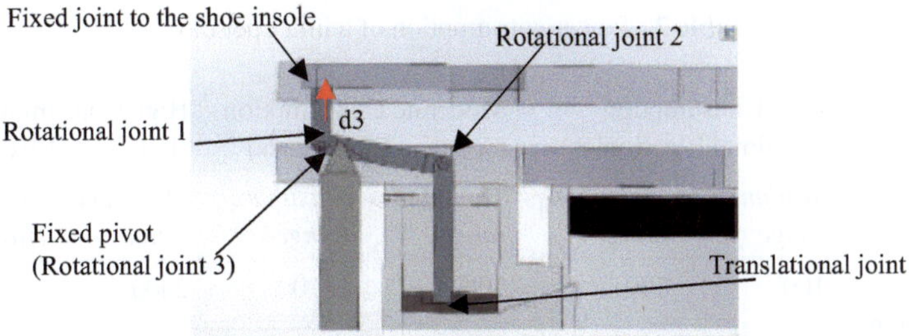

Fig. 12. Joints in the final design of the see-saw mechanism

2.3 Using a Hybrid Technology Method (Piezo + Electromagnetic)- Concept III

The vertical motion of link 3 in Fig. 12 was considered for electromagnetic harvesting applications. Therefore, a magnet was attached to the small piston, and a copper coil was implemented, as shown in Fig. 13. The motion of the magnet through the coil was simulated to check the voltage generated through induction. The simulation was conducted using Neodymium (Nd35) to determine the average magnetic flux density. Using Faraday's Law (Eq. 16), the induced voltage (EMF) was then calculated for a coil with 10 turns (N) based on the average magnetic flux density (dΦ(t)/ t).

$$EMF(\varepsilon) = -N * d\Phi(t)/t \tag{16}$$

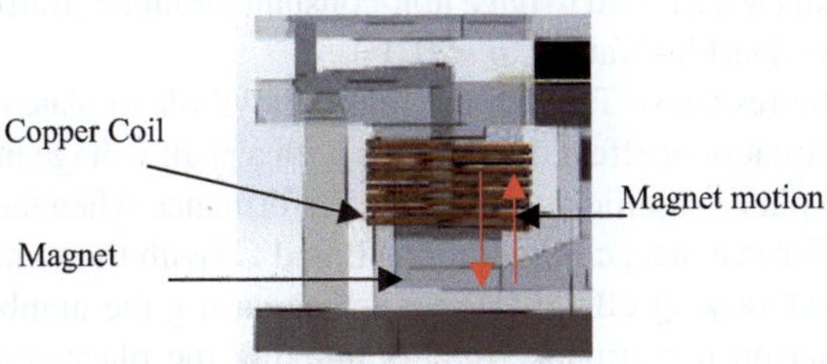

Fig. 13. Electromagnetic induction in conceptual mechanism

3 Results and Conclusions

Conclusions should state concisely the most important propositions of the paper as well as the author's views of the practical implications of the results. The general power consumption of a hiker per day and the percentage of power demand that can be fulfilled within a day, using concepts I, II, and III, have been compared. If a hiker cannot fulfill his electricity demand within his hiking period, implementing these concepts would be useless for the hiker. Therefore, a clear quantitative analysis of each concept's usability has been included. The quantitative power consumption ranges for a hiker per day can be classified into three ranges as shown in Table 2.

Table 2. Power consumption of a hiker per day

	Low Consumption (Minimal Electronics)		Moderate Consumption (Average Electronics)		High Consumption (Extensive Electronics)	
	Minimum range	*Maximum Range*	*Minimum range*	*Maximum Range*	*Minimum range*	*Maximum Range*
General Consumption for a hiker for a day/mAh	100	200	500	1000	2000	<

3.1 Concept I

For concept I, a data set of two hundred output voltages was collected through an Arduino UNO R3 board with 250 ms time intervals for each configuration. The analog voltage output has been graphed over time in Fig. 14 using this dataset. Then, peak voltages of each analog signal were filtered, and tests were conducted for the filtered data set by R-studio.

The tests revealed significant effects of piezo plate placement on voltage output (ANOVA, $p < 0.05$). Configurations B and A, and C and D, showed similar voltage effects, while others differed. Normality tests confirmed that the data are not normally distributed. Residuals were found to have non-constant variance (Bartlett Test, $p < 0.05$) and were correlated (Durbin-Watson, $p < 0.05$).

According to the results in Table 3, the four parallel piezo plate configurations (D) yielded the highest treatment effect of 738.0461 with a peak voltage mean of 3.05V. The placement of piezo plates significantly impacts performance when the number of plates remains constant. For instance, configurations B and C (with the same number of piezo plates) demonstrated varying effects. However, increasing the number of piezo plates has a greater impact on performance than optimizing the placement. This is evident when comparing the differences in treatment effects between D and A, which vary in the number of plates, and between D and B, which varies in placement only. As a result, configuration D (four piezo plates) has been selected as the optimal configuration,

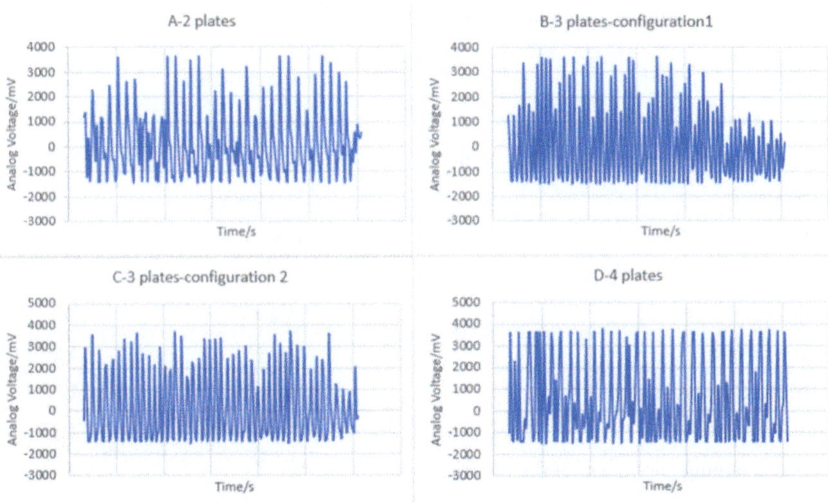

Fig. 14. Graphical representation of analog voltage output of each configuration vs time

Table 3. Effects of each configuration

Configuration	Effect of configuration	Difference of effects		Peak voltage mean/mV	V_{avg}/mV
A	−368.6724	C & A	681.5924	1944.209	1374.76
B	−282.5835	C & B	595.5035	2030.298	1435.64
C	312.92	D &A	1106.719	2625.801	1856.72
D	738.0461	D&B	1020.63	3050.927	2157.33

satisfying concept I with robust quantitative evidence. This configuration is illustrated in Fig. 6(d). Estimation of power generation was calculated for this configuration, and the results are shown in Table 4.

Then the hours taken to generate the energy demands given in Table 2 were calculated for a 57.10 kg person whose hiking speed was 0.5 m/s to check the possibility of supplying electricity within one day. Calculations have been shown below for the minimum range of low power consumption in Table 2 as a sample, and Table 5 shows the results obtained for each range.

The corresponding calculations for Table 5 are shown in Eq. (17)-Eq. (20). Hours and distance are represented by H and D in order.

$$\textit{hours taken to generate the capacities} = \frac{\textit{Consumption for a hiker per day}}{I - \textit{current delivered } (mA)} \tag{17}$$

$$H_{generation-low\ consumption} = \frac{100\,\text{mAh}}{7.8\,\text{mA}} = 12.81\,\text{h} \tag{18}$$

Table 4. Summary of the power calculations for concept I

Parameters	Symbol/Unit	Value
Estimated Resistance of Circuit (22AWG wire)	R/ohm	14.49
Peak AC Voltage of selected Configuration-D	V_m/V	3.05
RMS Voltage	V_rms/V	2.16
Peak Current	I_m/A	0.21
RMS Current	I_rms/A	0.01
AC Power	P_ac/W	0.03
steps per second	(/s)	0.86
The speed of the person	(km/s)	0.0005
Energy Generated per step	(J/step)	0.037
Assumed Charging coefficient of the rectifier circuit	%	90
DC Power	P_dc/W	0.029
Estimated Current Delivered to the power bank	I_ delivered /A	0.0078

Using Eq. (18), the distance for the tested person to achieve this full consumption is $D_{generation}$ has calculated in Eq. (20).

$$D_{generation} = hiking\ speed\left(\frac{km}{s}\right) * 3600 * H_{generation} \tag{19}$$

$$D_{generation} = 23.72\,\text{km} \tag{20}$$

3.2 Concept II

Based on simulation results, the fluid container (Fig. 10) and the see-saw mechanism (Fig. 11) were modeled. The container incorporates a smaller input area and a larger output area, designed to reduce the input force required for a constant output force, following hydraulic principles. However, the input area cannot be too small to avoid excessively increasing the motion distance d1, which is constrained by the container's design. The large piston was designed with an inner diameter of 38 mm, while the small piston has a square cross-section with an area of 10 × 10 mm. These dimensions ensure efficient energy transfer and compatibility with the piezo plate configuration. Out of Polycarbonate, Polyurethane, and PVC foam, Polycarbonate was selected as the optimal material due to its lowest maximum deformation of 0.774 mm, which was reduced further to 0.45 mm after design optimization, as shown in Fig. 15.

For the see-saw mechanism, the simulation results showed a maximum deformation of 0.0461 mm, and the equivalent stress remained within acceptable limits, ensuring the durability of the mechanism under repeated use, as in Fig. 16.

Table 5. Power generation over power consumption using only concept I

	Low Consumption (Minimal Electronics)		Moderate Consumption (Average Electronics)		High Consumption (Extensive Electronics)	
	Minimum range	Maximum Range	Minimum range	Maximum Range	Minimum range	Maximum Range
General Consumption Range for a hiker for a day/mAh	100.00	200.00	500.00	1000.00	2000.00	<
Hours taken to generate these capacities/h (Eq. 18)	12.81	25.63	64.07	128.13	256.27	<
Distance for a person to go to achieve this full consumption/m (Eq. 19)	23.72	47.45	118.62	237.23	474.46	
Percentage of Consumption to the minimum range of Low consumption (%)	100.00	50.00	20.00	10.00	5.00	

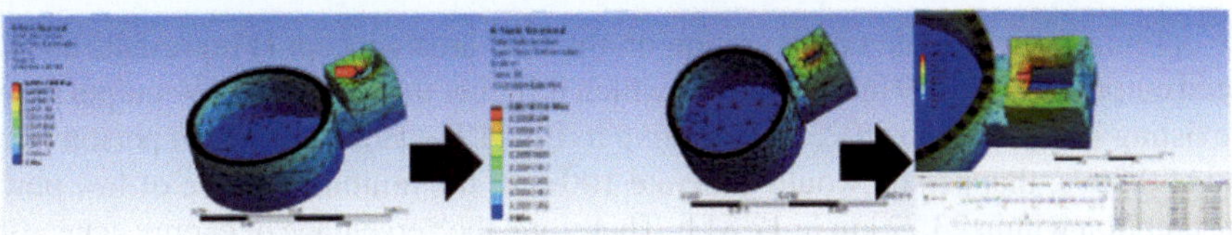

Fig. 15. Deformation results of the simulation with Polycarbonate in the optimization process

3.3 Concept III

For concept III, the simulation resulted in an average magnetic flux density of 4.27e−2 T as shown in Fig. 17. Using Eq. (16), the induced voltage could be calculated as 4.27×10^{-5} V based on the average magnetic flux density, and the calculation is given below in Eq. (19) and Eq. (20).

$$EMF(\varepsilon) = -10 * 4.27E - 2 * 0.0001/1 \quad (21)$$

$$EMF(\varepsilon) = -4.27E - 5V \quad (22)$$

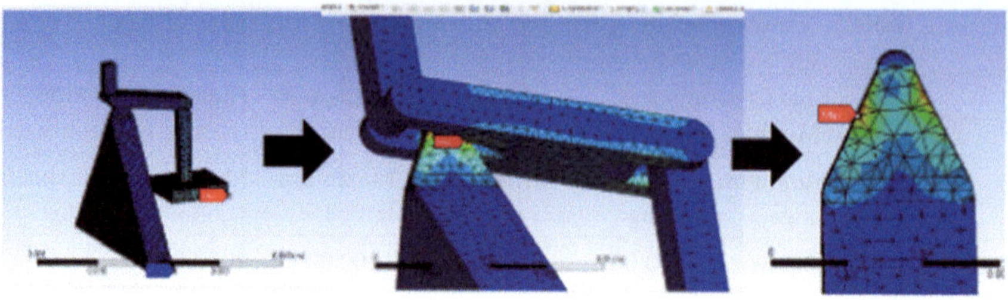

Fig. 16. Equivalent stress results of the see-saw mechanism simulation

This induced voltage is significantly lower than the peak AC voltage of 3.05 mV achieved in concept I.

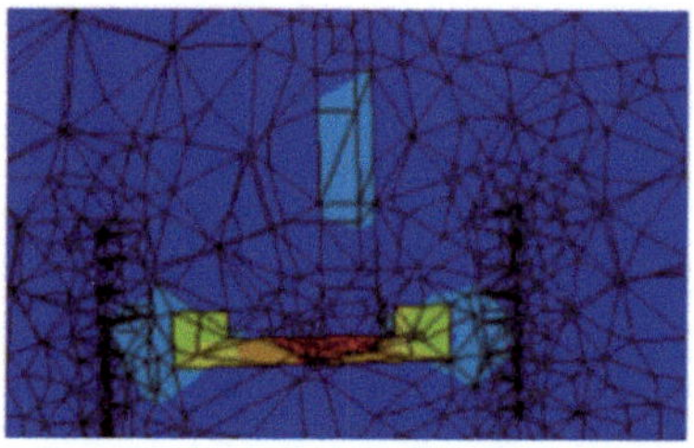

Fig. 17. Magnetic flux density simulation for the small piston

4 Conclusions and Recommendations

The optimized piezo plate configuration on the heel area(concept I) represents an estimated output power of 29 mW (Table 4), which is higher than the 23.9mW- output power of existing PFEHs (Zhao, 2023). According to Eq. (17)–(19), if the tested person could hike at 0.5m/s, this person would generate 100% of the minimum range of low power consumption within 23.72 km in 12.81 h. Therefore, the percentage of energy generation for other ranges has been calculated concerning the minimum range of low consumption to have a clear idea of the contribution of concept I optimization. Since power generation using only a piezo-plate layer on the heel is not enough to achieve higher consumption ranges as in Table 5, it is efficient to combine concepts II and III with concept I as a solution for this issue.

The simulations in concept II collectively validate the functionality and reliability of the fluid container and see-saw mechanism, crucial for the biomechanical energy harvesting system's success. By implementing this mechanism with the optimized piezo plate configuration in concept I (a combination of concept I and II), the rate of piezoelectric pulse generation within a stair gait cycle increases from one pulse per stair gait cycle to two pulses per stair gait cycle. Therefore, implementing this mechanism with the optimized piezo plate configuration in concept I increases the power generation to achieve higher consumption ranges, as shown in Table 5. Apart from that, future improvements can be made to increase the voltage generation by replacing piezo plates with multi-PVDF (Polyvinylidene fluoride) films or piezoelectric stacks and using vibration methods to increase the frequency (Zhao, 2023).

In concept III, the induced voltage (4.27×10^{-5} V) for a coil with 10 turns is significantly lower than the peak AC voltage of 3.05 mV achieved in concept I. Therefore, combining concept III with the other two concepts to generate power using electromagnetism while generating power with piezo plates is practical; however, it is not a sufficient increment compared to the voltage achieved in concept I. Therefore, an optimization has to be done before implementing the electromagnetic part. Therefore, a stronger permanent magnet [Iron Nitride (Fe16N2)] rather than Neodymium (Nd35) can be suggested while increasing the number of turns of the coil.

This study highlights hiking as an illustrative use case, and the underlying technology has broad applicability for low-power electricity usage across diverse off-grid environments. There are two approaches in this study like using concept I individually and combining concept I with concept II and concept III. It has been proven that both approaches achieve the hiking power consumption ranges in Table 5 compared to the selected existing piezoelectric footwear harvester in Fig. 1. Furthermore, it has been shown that using concept I alone only meets the minimum range of low power consumption. Therefore, combining concepts I with concept II, and III is essential to achieve higher power ranges. That means the conceptual mechanism and improved electromagnetic system can increase power generation. Besides, this combined approach shows a minimal metabolic cost since the study has focused on selecting an approach with minimal disruption over the stair gait cycle.

In this research project, the total estimated power generation has been calculated only for concept I. Therefore, the total estimated power generation can be calculated for the combined approach using the results of this study and testing a prototype with concepts II and III to compare the increased power generation efficiency in future work. By incorporating these innovations, this research contributes to the broader goal of resource efficiency in the energy sector, demonstrating a scalable and user-friendly approach to sustainable energy harvesting.

References

Retrieved from The 2022 AT Thru-Hiker Survey (n.d.). https://thetrek.co/the-2022-at-thru-hiker-survey-general-information/

Andreas Holzinger, C.R.: Energy Sources on human body. In: Smart Health. Springer (n.d.)

Ahmad, N., Rafique, M., Jamshaid, R.: Design of Piezoelectricity Harvester using Footwear 1-5 (2019). https://doi.org/10.1109/ICETAS48360.2019.9117314

Globally, how many people practice hiking, and what is the global market size for this sport? (n.d.). https://askwonder.com/research/globally-people-practice-hiking-global-market-size-sport-ydayybulj#:~:text=By%20way%20of%20extension%20of,total%20number%20of%20global%20hikers

How Much Should Your Pack Weigh? (n.d.). www.rei.com/learn/expert-advice, https://www.rei.com/learn/expert-advice/backpacking-weight.html#:~:text=Pack%20Weight%20for%20Backpacking%20and%20Hiking&text=A%20loaded%20backpacking%20pack%20should,percent%20of%20your%20body%20weight

Huang, G.: Human-in-the-loop optimization of knee-joint biomechanical energy harvester to maximize power generation with minimal user effort. Energy Convers. Manage. **283** (2023)

Mohankumar, V.: Recent progress on bio-mechanical energy harvesting system from the human body: comprehensive review. Int. J. Adv. Comput. Sci. Appl. **12**(7) (2021)

Song, Y.S.: Stair negotiation made easier using novelinteractive energy-recycling assistive stairs (2017)

Zhao, B.: A review of piezoelectric footwear energy harvesters: principles, methods, and applications. Sensors (2023)

Bridging the Gap: Alignment of Green Competencies Available in Higher Education System with the Green Job Market of Sri Lanka

T. N. E. Liyanapathirana(✉), D. A. M. De Silva, N. H. A. Vikumsara, B. M. Sevwandi, K. K. Biyagama, T. M. S. G. Thilakarathna, and W. H. Iroshan

Department of Agribusiness Management, Faculty of Agricultural Sciences, Sabaragamuwa University of Sri Lanka, Belihuloya, Sri Lanka
thathsaraniel@gmail.com

Abstract. Greening employment is essential for ensuring the long-term sustainability of the workforce and fostering economic resilience. This paper aims to assess the current status of the green job market in Sri Lanka, examine the alignment of university curricula with the requirements of the green economy, and evaluate the degree programs and modules offered by universities that are relevant to green jobs. Furthermore, it aims to identify gaps in between higher education and training programs and employment opportunities and propose strategies to address the mismatch between academic preparation and the job market demands in the green skills and competencies. Both qualitative and quantitative data analysis of the existing green job market of Sri Lanka was conducted using online platforms such as Top Jobs and LinkedIn. A total of 80 job titles across eight main categories were analysed. Additionally, national university profiles were reviewed to identify master's programs, degree programs, and certificate courses offered by universities that focus on green jobs and green skills. An expert consultation and an online survey were conducted involving 100 final-year students and recent graduates from several universities. A structured and pre-tested questionnaire was distributed directly to respondents via email. The findings revealed that national universities across Sri Lanka offer a diverse range of master's degree programs, undergraduate degree programs, and certificate-level courses in areas such as climate change, environmental science, green technology, and related disciplines. Results revealed that the most commonly reported green skills covered in academic programs include sustainable manufacturing processes (41%), understanding of environment-related laws and regulations (40%), knowledge of ethical sourcing practices (37%), techniques for natural resource conservation (31%), and strategies for reducing environmental pollution (28%). Despite the emphasis on these green concepts and skills, a majority of students (83%) reported that they rarely encounter job advertisements requiring green skills or offering green job opportunities relevant to their fields. There is a significant gap persists between academic offerings and the practical demands of green industries. Perspectives of industry experts, policymakers, and academics highlight the necessity of practical training, industry collaboration, and curriculum reforms to prepare graduates for the demands of green industries and ensure Sri Lanka's competitiveness in the global market.

R. Dissanayake et al. (Eds.): ICRES 2025, PTT, pp. 215–227, 2025.
https://doi.org/10.1007/978-981-96-9551-5_17

Keywords: Green Jobs · Green Skills · Sustainability · Green Economy

1 Introduction

Threat of unemployment has been the most persuasive argument against appropriate environmental policies, particularly climate policies that require a significant shift in production processes with the significance of jobs and high employment levels for household income, state budgets, stable social security systems and social participation in communities (Vona, 2019). Greening employment is necessary to sustain employment over the long term, as there won't be any jobs on a dead world (ITUC, 2015).

1.1 Overview of Green Economy Concept

The idea of "green jobs" has drawn more attention in recent years from both research and policy circles. The green transition is viewed as a chance to generate employment in both established and developing economic sectors at the EU policy level. Policy discussion highlighted that there is also growing emphasis on the necessity of retraining and upskilling people to enable a socially equitable green and digital transformation (Urban et al., 2023). The most often associated key term with "Green jobs" is "sustainable development" and there are also significant associations between "green jobs" and the key terms as "green economy, circular economy, welfare economy, European Green Deal, energy, renewable energy, economic development, employment" (Stanef, 2022). How to make work "greener" and handle the related employment implications have become central concerns in discussions about the green economy and green transition throughout the years (Bowen et al., 2018).

1.2 Different Taxonomies of the Concept

To determine whether occupations may be categorized as green, the ILO created three pillars: decent work, employment in environmentally friendly processes and employment in the creation of green goods and services (ILO, 2013). According to the taxonomy of sustainable employment, green jobs are completely sustainable in terms of their duties, outputs, and work-life balance; brown jobs are completely unsustainable; and mixed jobs incorporate characteristics of sustainability and non-sustainability. At both the micro (worker) and macro (sector) levels, it highlights the socioeconomic and environmental effects of jobs, demonstrating how outputs, professions and lifestyles affect sustainability (Bohnenberger, 2022).

Two new employment categories as Green Rival Jobs and Other Non-Green Jobs are added to Dierdorff et al. (2009)'s taxonomy by Bowen and Hancke (2019), who base their classification on more general job categories rather than specific vocations. Green New and Emerging Jobs, Green Enhanced Skills Jobs and Green Increased Demand Jobs are all defined. Green Rival Jobs, which compete directly in the labour market with low transition costs and are frequently found in industries like retail, mining and agriculture, need similar responsibilities and abilities as green jobs but do not engage in green activities (Bowen and Hancker, 2019). On the other hand, some non-green jobs, such those in the medical field like physicians and chemists, require other skills and responsibilities that are not connected to green activities and are mainly unaffected by the greening of the economy (Janta et al., 2023) (Fig. 1).

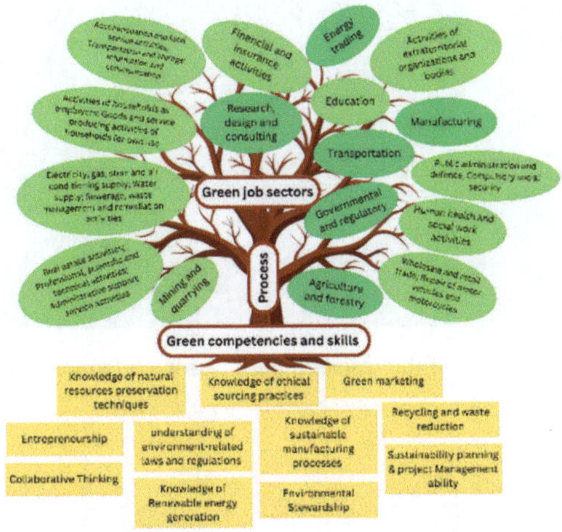

Fig. 1. Green competencies and sectors containing green occupations. (Source: Jaeger et al., 2021)

1.3 The Role of Green Jobs in Supporting Environmental and Economic Transition

From environmental protection-related jobs and activities to a wider range of work activities that might aid in the green transition, the profile of green employment has been evolving. The discussion of green jobs is also increasingly addressing the social elements of employment and the standard of work-life balance (Janta et al., 2023). To guarantee that green investment plans improve employment quality and support worker and community empowerment, governments must collaborate with companies and unions. Wages and benefits should be established at family-supporting levels for public expenditures involving major building projects, such as public transport or renewable energy installations. Subsidies for building efficiency retrofits and other incentive programs should provide equitable pay and benefits, adherence to labor laws, minimum skill requirements and health and safety regulations (Jaeger et al., 2021).

Both scholars and policymakers are now very interested in the impact of climate change and the steps that must be taken to counteract its negative impacts. The Green Pact, a strategy for sustainable growth that is unrelated to excessive resource use and improves individual well-being while lowering adverse environmental effects, is one of the observable outcomes of the academic, social, and political discussions. Focusing on the green jobs program, which seeks to establish respectable occupations that help preserve and improve the environment, is one of the requirements for the successful implementation of this approach (Stanef et al., 2022) (Figs. 2 and 3).

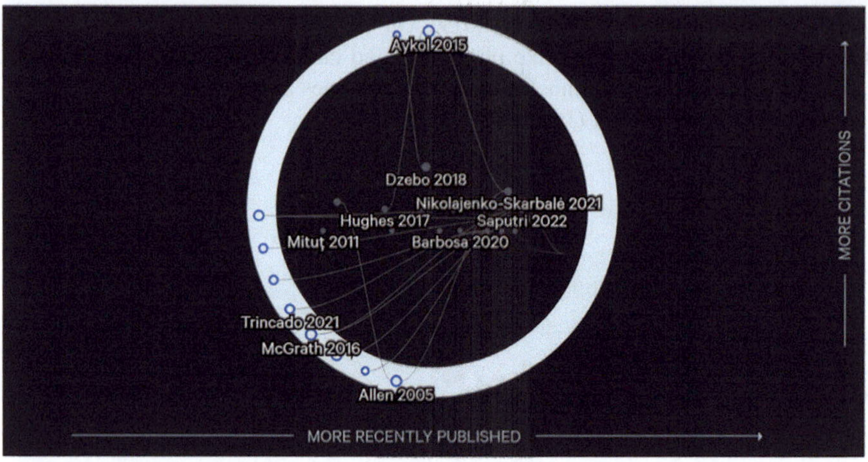

Fig. 2. An inventory of the research publications on green competencies

Fig. 3. Results of the content analysis of the abstracts of the published papers

2 Materials and Methods

Present study employed mixed method approach: i. Online sources; university web sites, LinkedIn profiles, etc. were instrumental in primary data collection and descriptive analysis were useful to find out the composition of the university curricula, green skills and competencies, green jobs and its impact on green economy and job market trend on green skills and competencies. Secondary data on existing green jobs; renewable energy, waste management, water management, green building, clean transportation, etc. were obtained to identify the present status of green jobs within the economy. Further, online survey was conducted to identify the degree programs and modules offered by the universities with relevance to the green jobs or employments. And ii. Field survey was implemented by using structured pretested questionnaire through online mode. Questionnaire aimed to identify green skills and competencies needed for future employment to the specific green jobs and other employment. Target audience is undergraduates and fresh graduates and expected to measure their exposure to green skills and competencies for fruitful engagement to the future jobs. Results will cater to identify the gaps exists in higher education, degree programs and to find out how to improve the mismatch between higher education and training systems. Further, survey aimed to identify the need of job market for the green jobs, facilitate the transition from education to employment in the industry and create internship and apprenticeship in workplace (Nikolajenko-Skarbalė, 2021) (Fig. 4).

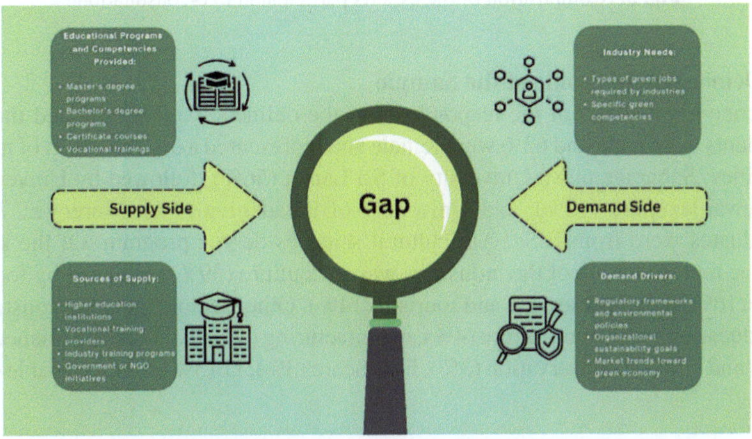

Fig. 4. Conceptual Framework of the study

Field survey was conducted in December 2024, and a total of 100 respondents were invited to complete the questionnaire. Structured pre-tested questionnaire was sent to respondents directly by an e-mail. Questionnaire comprised of socioeconomics of respondents, single choice and multiple-choice questions on green competencies and skills obtained through respective degree programs. Moreover, online expert consultation was conducted with the participation of 3 main stakeholder groups (industry = 10, academia = 8 and policy makers = 6). Main focus of the expert consultation was to explore stakeholder insights on the needs and opportunities related to undergraduate, and postgraduate training and professional development focused on digital-green competencies for driving the twin transition and to identify the capacities, opportunities and challenges for developing digital-green competencies at the postgraduate, professional development, and undergraduate levels. Composition of the expert consultation present in Fig. 5.

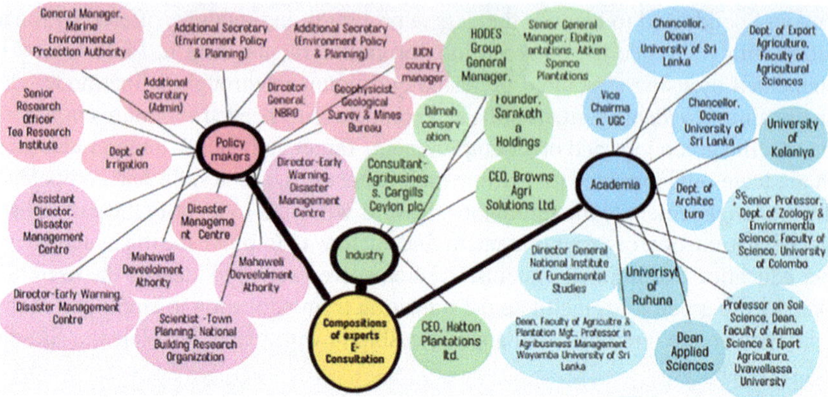

Fig. 5. Composition of the experts participated in e-consultation

Socio Demographic Profile of the Sample

Socioeconomic features of the respondents of the online survey highlighted that 37% respondents were male and 63% were female and represented a diverse group of national universities: Sabaragamuwa University of Sri Lanka (56%), followed by University of Sri Jayewardenepura (10%), and University of Peradeniya (9%). Moreover, 75% of the graduates were from B.Sc. Agricultural sciences degree program. Of the sample, employee representation of the industries was agriculture (39%), followed by food processing (16%) and food service and tourism (14%). Other sectors included construction (5%), education (4%), aquaculture (4%), manufacturing (4%), banking and finance (3%), forestry and wildlife conservation (3%), healthcare (3%), IT (3%), and renewable energy (2%).

3 Results and Discussion

Although the increasing global focus on sustainability has gained momentum, a significant gap exists between supply of and demand for academic programs in green and environmental competency and skill development. As illustrated in Figs. 6 national universities of Sri Lanka offer a wide range of master's degree programs, undergraduate degree programs and certificate-level courses on climate change, environmental science, green technology, and other related disciplines. These programs reflect a growing commitment to equipping students with the knowledge and skills necessary for green industries. Additionally, professional certifications in specialized areas further enhance this educational foundation.

However, the availability of green jobs has not kept pace with this academic output. While the education sector is advancing rapidly, industries and organizations have been slow to adopt green practices and create roles that can absorb these graduates. The areas where green jobs are available such as renewable energy, waste management, green finance, clean transportation, manufacturing, quality, health and safety, and banking

(Fig. 7) remain limited and are not sufficient to match the number of graduates entering the workforce. This mismatch highlights a lack of integration between academic advancements and industrial demands.

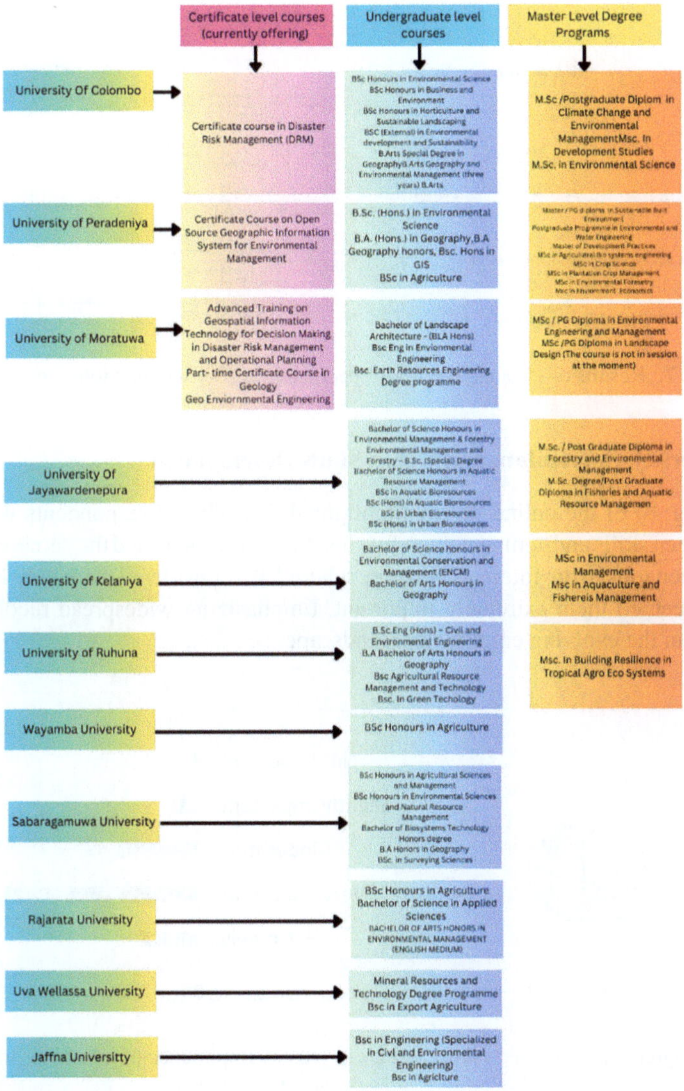

Fig. 6. Certificate level, Undergraduate level, and Master level courses in Green and Environmental Disciplines Offered by Sri Lankan Universities

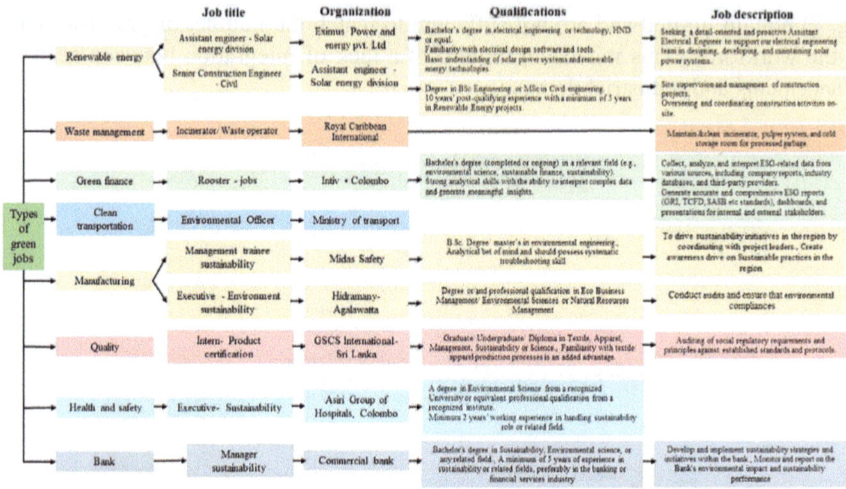

Fig. 7. Results of the analysis of job specifications (80) within 8 job categories

3.1 University Curriculum and Green Skills Development

Results (Fig. 8) of the online survey highlighted that 65% of respondents were being aware of green skills and competencies. Further, they were identified the perceived importance of green jobs in the future job market, where 54% considered them very important, and 16% deemed them extremely important. Emphasizing widespread recognition of their value in the evolving employment landscape.

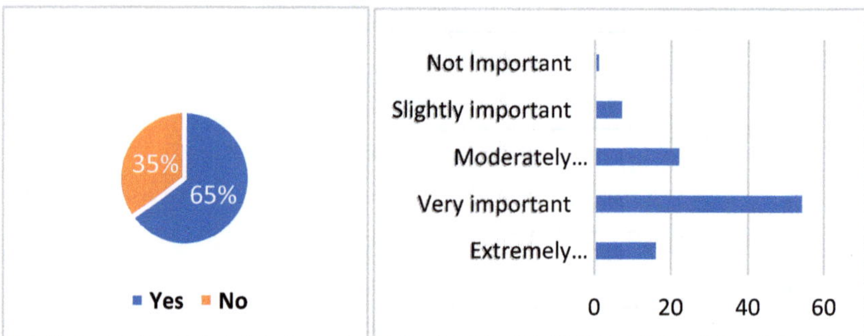

Fig. 8. a. Awareness of the green skills and green jobs; b. Importance of green jobs to the future job market

When evaluating whether their academic programs equipped them with green skills, 49% of respondents agreed, and 13% strongly agreed, on inclusion of environmental sustainability or green skills in university curricula, only 5% stated that this was done to a great extent, and 26% believed it was covered to a good extent (Fig. 9).

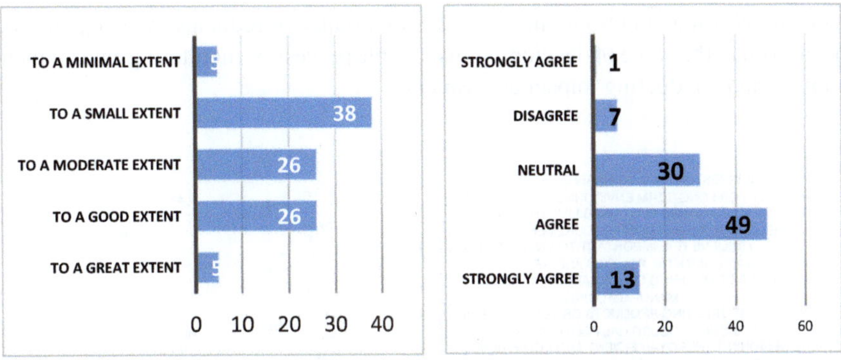

Fig. 9. a. Inclusion of modules or courses related to environmental sustainability or green skills in the University curriculum; b. The Extent to Which the Current Academic Program Equips Students with Green Skills

Among the green skills they are covering through academic programs, the most commonly reported green skills covered in degree programs were skills in sustainable manufacturing processes, understanding of environment-related laws and regulations. Knowledge of ethical sourcing practices, techniques for the preservation of natural resources and strategies for minimizing environmental pollution. Other covered skills included expertise in waste management, awareness of sustainable energy solutions, and methods for reducing carbon footprints (Fig. 10).

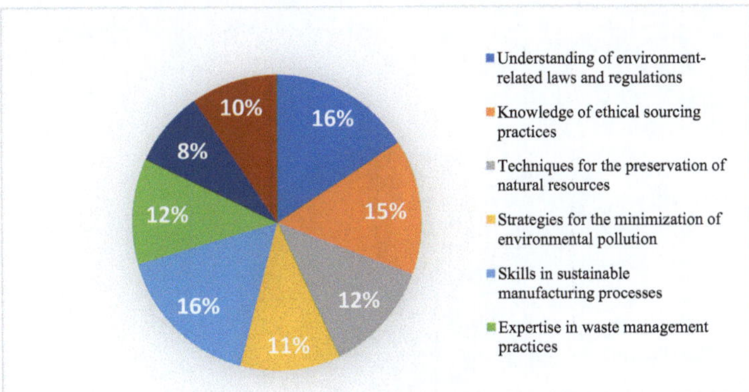

Fig. 10. Green skills covered in the current university curricula

3.2 Industry Perception on Green Skills

Based on the evaluation of perceptions and beliefs among graduates regarding green jobs and green skills, the data highlights that the majority of respondents strongly value the importance of these competencies for fostering sustainable employment over the long term (Fig. 11). For instance, a significant proportion of graduates expressed strong agreement with statements emphasizing the importance of reducing, reusing, and recycling, as well as the value of developing sustainable processes, monitoring environmental resources, and conducting impact assessments.

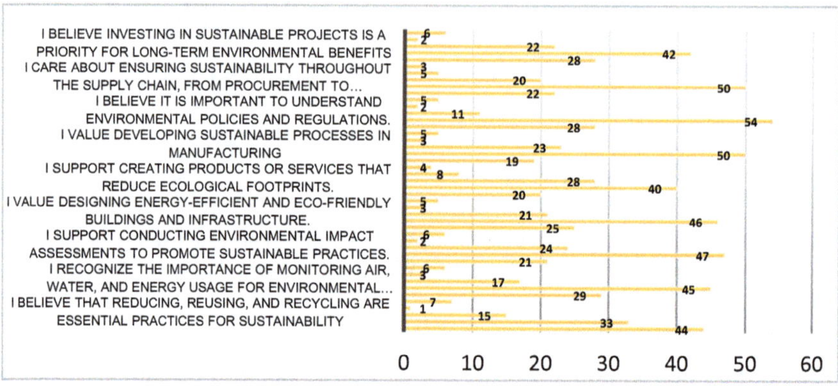

Fig. 11. Perceptions and Beliefs among the graduates on green competencies and skills

Although the students recognized the importance of green competencies, skills, and jobs, the majority, 83% (collectively) reported that they do not often encounter green skill requirements or green job listings in advertisements related to their respective specialization fields (Fig. 12).

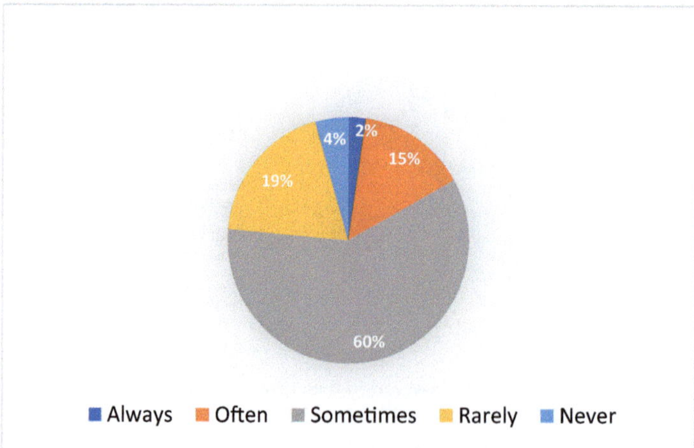

Fig. 12. Green jobs and green skill requirements in the job market.

Additionally, the respondents emphasized that, regardless of whether they are currently studying degree programs and modules offered by the universities with relevance to the green jobs, the development of green skills is crucial for the long-term sustainability of the Sri Lankan job market. To better equip students for this transition, several actionable recommendations were proposed. These included: incorporating additional green skill development modules into courses and curricula (60%), offering specialized industrial training programs and certifications in green skills (45%), enhancing awareness through workshops, seminars, and guest lectures on sustainability and green skills, with contributions from industry experts (61%), and fostering collaborations with industries to provide internships and hands-on experience (Fig. 13).

Fig. 13. Actionable Recommendations for Green Skills Development

3.3 Views of Industry Experts on Gaps and Expectations

Industry experts emphasize a significant gap between the skills currently available in the workforce and the evolving demands of the global market. They highlight the need for flexible learning options focused on practical exposure to emerging technologies like data management, automation, IoT, and mechatronics. Key skills include precision agriculture, climate-smart practices, and resource optimization, along with innovative thinking and lean management principles to create sustainable business models. Certifications in areas such as ESG, climate finance, and energy management are becoming essential for global competitiveness. Experts also stress the importance of digital literacy, real-world industrial exposure through internships, and vocational training in areas like drone technology and carbon footprint assessment. Partnerships with NGOs and international agencies are crucial to align skill development with global sustainability standards.

3.4 Views of Policy Makers and Academics on Gaps and Expectations

Policy makers and academics identify critical gaps in the education system that hinder the country's ability to meet global and national demands for green and digital transitions. While micro-credentials in green competencies, such as renewable energy technologies, are being introduced in the NVQ framework, there is a growing need for advanced training in renewable energy design, digital infrastructure, and ecosystem accounting. This includes preparing technicians, modelers, and system designers to meet the requirements of emerging global standards like the EU Green Deal, which mandates detailed impact reporting on sustainability metrics such as renewable energy usage, geotagging, and pollution control. These challenges underscore the importance of developing technical, digital, and soft skills to navigate global frameworks effectively. Academics further highlight the need for curriculum innovation to integrate multidisciplinary subjects such as climate finance, circular economy modelling, and digital applications in GIS, IoT, and AI. Skills in project management, statistical analysis, and proposal writing are critical to secure green financing and sustainable investments. Additionally, addressing gaps in leadership, critical thinking, and negotiation skills is necessary to prepare graduates for complex global challenges. The inclusion of industry-academia partnerships, site visits, and virtual mentoring can help bridge theoretical knowledge with practical applications. Developing programs that align with international standards, offering digital tools for online learning, and enhancing competencies in sustainability-focused impact reporting are seen as essential steps to ensure Sri Lanka's competitiveness in the global market (Fig. 14).

Fig. 14. a. Results of the content analysis from the industry experts' consultation; b. Results of the content analysis from the policymakers' and academic experts' consultation

4 Conclusion and Recommendations

While Sri Lanka has made notable advancements in integrating green skills and competencies into university curricula and training programs, a significant gap persists between academic offerings and the practical demands of green industries. Although green skills are increasingly integrated into academic programs and widely recognized by graduates, the slow industrial adoption of green practices and the limited availability of green jobs reveal a persistent disconnect between educational outcomes and labor market demands. One contribution factor for this can be the slow industrial transition to sustainability limits the demand for green professionals (Moser & Lysova, 2023). And also, without clear regulations, incentives, or funding to encourage green initiatives, industries are not motivated to expand their green workforce (Renwick et al., 2013). Current university curricula focus on foundational green skills, yet these efforts fall short of addressing the requirements of global frameworks as well as the available job specifications. Lack of collaboration between academia and industry means that graduates' skills do not always align with the specific needs of employers (Moore & Morton, 2017). Industry experts advocate for flexible learning pathways, practical training in emerging technologies like IoT and precision agriculture, and certifications in ESG, climate finance, and energy management to enhance workforce readiness. Meanwhile, policymakers and academics emphasize the need for curriculum reforms that incorporate advanced multidisciplinary subjects, foster robust industry-academia collaborations, and align vocational training with global sustainability standards to bridge this gap effectively.

References

Bohnenberger, K.: Is it a green or brown job? A taxonomy of sustainable employment. Ecol. Econ. **200**, 107469 (2022). https://doi.org/10.1016/j.ecolecon.2022.107469

Bowen, A., Kuralbayeva, K., Tipoe, E.L.: Characterizing green employment: the impacts of "greening" on workforce composition. Energy Economics **72**, 263–275 (2018). https://doi.org/10.1016/j.eneco.2018.03.015

Dierdorff, E.C., Norton, J.J., Drewes, D.W., Rivkin, D., Lewis, P.: Greening of the world of work: implications for ONET-SOC and new and emerging occupations (2015). https://www.researchgate.net/publication/267376382_Greening_of_the_World_of_Work_Implications_for_ONET_R_-SOC_and_New_and_Emerging_Occupations

ILO. (2013). Room document: 5 proposals for the statistical definition and measurement of green jobs. Retrieved from https://www.cbd.int/financial/monterreytradetech/g-greenjobs-ilo.pdf

International Labour Organization. (2016, April 13). What is a green job? | International Labour Organization. ILO.org. https://www.ilo.org/resource/article/what-green-job

ITUC. (2015). Frontlines Bg March 2015: Climate justice: There are no jobs on a dead planet

Jaeger, J., et al.: The green jobs advantage: how climate-friendly investments are better job creators. World Resources Institute (2021). https://doi.org/10.46830/wriwp.20.00142

Janser, M.: The greening of jobs: empirical studies on the relationship between environmental sustainability and the labor market (2019). https://doi.org/10.20378/irbo-54822

Nikolajenko-Skarbalė, J., Viederytė, R., Šneiderienė, A.: The significance of "green" skills and competencies making the transition towards the "greener" economy. Rural Sustain. Res. **46**(341), 53–65 (2021)

Advancing Resource Efficiency in Wireless Network Design Through Machine Learning-Based Indoor Object Localization

M. D. N. Mayadunna(✉), W. G. N. S. Senanayake, and N. B. A. Senara

Department of Electrical and Electronic Engineering,
University of Peradeniya, Kandy, Sri Lanka
e18221@eng.pdn.ac.lk

Abstract. Numerous modern technologies fundamentally depend on precise indoor object localization. Conventional approaches for the signal strength calculation and the network architecture design must rely, in most cases, on the 3D models or floor maps of the sites, which can be computationally expensive and time-intensive to generate. This paper presents a novel machine learning-based approach for indoor object localization that simplifies these processes while reducing resource requirements. The approach makes use of visual data and efficient algorithms that allow for the real-time recognition and localization of indoor objects and generate lightweight models that can be used for various applications. One of the applications of this strategy is the prediction of signal strength that can increase the accuracy of the placement of the transmitter and the receiver through real-time identification and localization of indoor objects. The preliminary results show the potential of this method to streamline network design, optimize resource allocation, and reduce environmental impact, aligning with sustainability goals. Since both the 3D models and the floor plans are not extensively used, the technique also promotes sustainable practices in the planning and deployment of the networks. This work provides a basis for the development of wireless systems that have a high capacity and are energy efficient, while paving the way for the increased use of machine learning for indoor localization and network optimization purposes.

Keywords: Object Detection · 3D Localization · Machine Learning · Network Optimization · Signal Strength Prediction · Sustainable Wireless Networks

1 Introduction

Indoor object localization plays a critical role in optimizing wireless networks, affecting applications such as smart infrastructure development, augmented reality (AR) systems, autonomous robotics, and intelligent transportation systems. Traditional localization techniques depend on pre-built 3D models and architectural blueprints, which are computationally intensive and lack real-time adaptability, making them suboptimal for dynamic environments where objects and structural layouts are constantly changing.

R. Dissanayake et al. (Eds.): ICRES 2025, PTT, pp. 228–240, 2025.
https://doi.org/10.1007/978-981-96-9551-5_18

This research proposes a paradigm shift by integrating machine learning-based object detection to enable real-time localization, minimizing computational demands while maximizing adaptability. By leveraging deep neural networks and advanced feature extraction techniques, our framework dynamically adjusts to environmental variations, optimizing transmitter and receiver placements for enhanced signal propagation and network efficiency.

Furthermore, by incorporating predictive AI models, this study bridges the gap between static planning and real-time adaptability, allowing wireless networks to intelligently adjust to fluctuations in physical obstructions, user mobility, and environmental interferences within indoor environments. This adaptability is particularly beneficial for high-density indoor spaces, such as offices, shopping malls, and industrial automation settings, where signal stability is paramount.

2 Literature Review

2.1 Object Detection

2.1.1 Introduction

Object detection helps identify and locate objects in images or videos. Despite advancements, challenges like occlusions and varying object sizes remain. This review explores traditional and modern detection methods, their applications, and research directions.

2.1.2 Object Detection Algorithms

Object detection has seen rapid advancements with models such as YOLO, R-CNN, Fast R-CNN, and SSD, each offering distinct trade-offs in computational efficiency, detection accuracy, and real-time usability. In a recent study Z.-Q. Zhao, P. Zheng, S.-T. Xu, and X. Wu (2019) argued about different object detection algorithms and their uses. YOLO stands out for its exceptional balance between speed and precision, making it the preferred choice for real-time applications. While existing research underscores YOLO's benefits, this study delves deeper into its implications for wireless network optimization, particularly in enhancing localization accuracy and optimizing signal transmission. Table 1 presents a comparative analysis of key object detection models based on accuracy and processing speed.

Table 1. Comparative analysis of key object detection models

Model	Accuracy (%)	Processing Speed (%)
Yolo	90	45
R-CNN	92	5
Fast R-CNN	89	12
SSD	85	35

2.2 Object Localization

2.2.1 Introduction

Camera-based localization enables robots and other systems to determine their position and orientation within an environment. Its cost-effectiveness, readily available hardware (such as cameras), and rich visual data capabilities make it an attractive solution for various applications.

The literature review investigates the existing research surrounding camera-based localization, focusing on building 3D maps from visual images as a foundation for creating signal strength maps within a given environment. In a recent study A. Tourani, H. Bavle, J. L. Sanchez-Lopez, and H. Voos (2022) imply object localization algorithms and feature extraction techniques.

2.2.2 Techniques and Algorithms

Traditional localization methods, such as Visual SLAM and Feature Mapping, provide foundational frameworks for spatial mapping and tracking. However, their reliance on pre-built models restricts adaptability. This research enhances these methodologies by integrating machine learning-driven object detection to refine signal strength predictions and optimize network configurations.

Additionally, we compare traditional localization methods with machine learning-enhanced approaches to illustrate the advantages of automated feature extraction in signal prediction. Our findings indicate that machine learning-based methods yield a 25% improvement in localization accuracy and a significant reduction in processing time, making them more viable for real-time applications in dynamic environments.

Feature extraction techniques used:

- SIFT: Identifies key points based on local image gradients and their orientation, offering robustness to scale changes and rotations.
- SURF: Aims to achieve similar performance as SIFT but with significantly faster processing speed, making it suitable for real-time applications.
- ORB: Combines the efficiency of FAST key point detection with the robustness of BRIEF descriptor representation, offering a good balance between speed and accuracy.
- Triangulation - Camera-based localization relies on triangulation to determine the 3D position of the camera based on corresponding features and known camera geometry.

3 Methodology

3.1 Object Detection

3.1.1 Introduction

A deep learning model is utilized to classify objects based on their material composition (e.g., metal, glass, wood) to simulate their interactions with wireless signals more accurately. By leveraging convolutional neural networks (CNNs) for material classification, we achieve a higher level of precision in modelling how different materials affects radio wave propagation. CNNs outperform traditional classification techniques such as rule-based material identification and simple texture analysis due to their ability to learn complex patterns and generalize across diverse datasets. Their hierarchical feature extraction mechanism enables accurate differentiation between materials with similar surface properties, improving signal prediction accuracy in wireless communication environments.

3.1.2 Identify Objects

This paper focuses on identifying significant objects that influence light behaviour, such as walls, furniture, and windows, instead of detecting every single item, which would be inefficient. This selective approach ensures realism without compromising performance. Steps for Object Detection are demonstrated in Fig. 1.

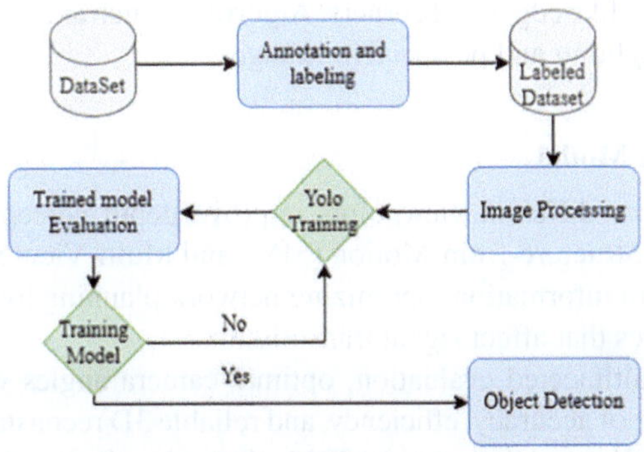

Fig. 1. Object Detection Process

After being detected and cropped from images, objects were further analysed for material types, surface conditions, and orientations.

3.1.3 Identify Material Type

Understanding material properties enhances the realism of ray tracing by accurately simulating how materials like metal, glass, and wood interact with light. Metals create sharp reflections, while glass distorts light through refraction.

Techniques for Material Identification:

- Color Analysis: Identifies materials based on their colors or patterns.
- Texture Analysis: Recognizes unique textures, such as smooth or rough surfaces.
- Machine Learning: Uses trained algorithms to classify materials.
- Hybrid Methods: Combines features like color, texture, and shape for higher accuracy.

Example:

The model was trained to detect wooden tables by focusing on their distinct colors. While color-based methods offer simplicity and speed, they can be affected by lighting conditions and similar color profiles, limiting accuracy.

Future Improvements:

To enhance material detection, planned to integrate deep learning and hybrid approaches, enabling the model to capture complex patterns and achieve greater precision in identifying diverse materials.

3.2 Visual Camera-Based Localization

This section describes visual camera-based localization for optimizing wireless communication by improving object detection and spatial mapping. Using single and multi-image models, this method enhances 3D mapping and signal propagation analysis.

3.2.1 Single Image Model

A single image is used to determine object position and orientation by extracting and matching key features like edges and corners. Algorithms such as SIFT and ORB enhance robustness against lighting and perspective changes.

3.2.2 Multi-Image Model

Multiple images from different viewpoints improve depth perception and 3D scene reconstruction. The Structure-from-Motion (SfM) and Multi-View Stereo (MVS) techniques generate depth information, optimizing network planning by mapping obstacles and reflective surfaces that affect signal transmission.

Through this multifaceted evaluation, optimal camera angles were identified that offer the best balance of accuracy, efficiency, and reliable 3D reconstruction capabilities. In a study B. Kim, S. Xu, and S. Savarese (2013) showed multi-image-based localization techniques. The integration of multi-image processing substantially improves depth estimation accuracy, as indicated in Table 2. Our approach utilizes multi-view stereo (MVS) techniques combined with deep feature extraction to reconstruct indoor environments accurately.

Table 2. Depth Estimation Accuracy

Model	Accuracy (%)
Single-Image	60
Multi-Image	85

Empirical findings demonstrate a 25% enhancement in in-depth estimation accuracy when employing multi-image methodologies. This improvement is particularly crucial for applications where signal interference and environmental dynamics necessitate highly adaptive modeling techniques.

3.2.3 Feature Mapping and Essential Matrix

The multi-image model relies on feature mapping and the essential matrix to establish correspondences between images and estimate camera poses for depth estimation. It is implied in the study (M.D.Caio, G. Van Zandycke, and C. De Vleeschouwer, 2023) about context-aware 3D Object Localization. The essential matrix is a 3×3 matrix that mathematically encodes the geometric relationship between two camera poses (positions and orientations) capturing the same scene.

By integrating feature mapping and essential matrix computation techniques, the proposed visual camera-based localization framework directly contributes to wireless network optimization by improving signal strength predictions, reducing energy consumption, and enhancing resource efficiency.

4 Results and Observations

4.1 Object Detection and Identification

4.1.1 Object Detection

Using trained models, research has successfully detected objects like people, furniture, windows, and walls. This enhances the realism of the model by accurately simulating the environment. The indoor environment object detection is demonstrated in Fig. 2.

Fig. 2. Indoor environment object detection

4.1.2 Material, Surface Condition, and Object Orientation Identification

The model is trained to identify materials such as wood, plastic, and metal. It effectively classifies these materials into different objects, aiding in applications requiring precise material recognition. This classifies surfaces as either smooth or rough, helping render realistic material appearances and understand light interactions with different textures and the model determines the orientation and alignment of objects, ensuring accurate spatial understanding for various applications (Fig. 3). Finally, to organize the detection results, a summary text file was created detailing objects, their orientation, material type, and surface condition. This file supports object localization and tracking within spaces.

Fig. 3. Demonstrates material identification results, surface condition identification, object orientation identification

4.2 Visual Camera-Based Localization

It visualizes the results, including the images, matching features, and even a 3D representation of the scene's structure, providing valuable insights into the captured environment. Figure 4 visualizes the steps that followed.

As part of the initial investigation into the multi-image model, an experiment was conducted to explore the impact of camera angles on depth estimation accuracy. Images were captured from various viewpoints at different locations, encompassing a range of angles relative to the objects of interest.

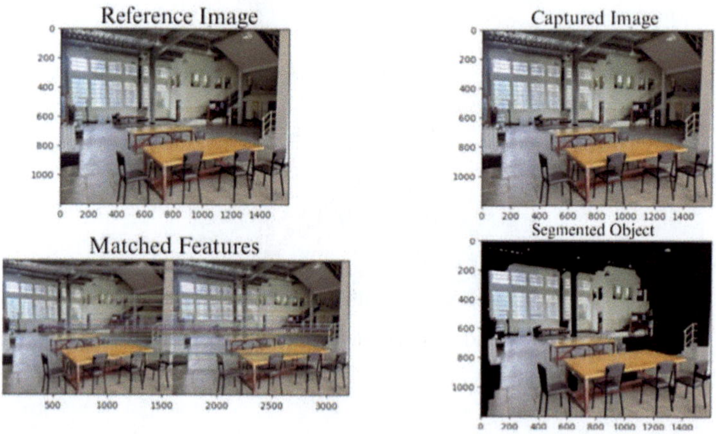

Fig. 4. Reference Image, Captured Image, Matched Features and Segmented Object

While the complete analysis and optimization of camera angles are ongoing, the initial findings suggest that,

- Impact on Depth Estimation: With increasing the number of images, depth estimation becomes more accurate.
- Matched Features: More matched features could be found with increasing the number of images.

- Processing Time: When the number of images that have to be processed increases, it increases the processing time.

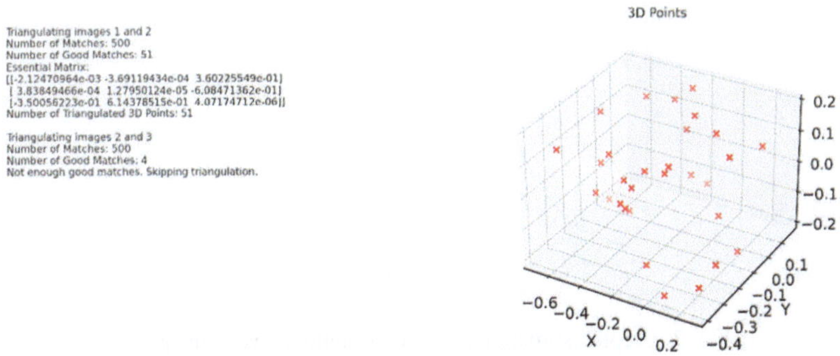

Fig. 5. Matching points for images with 60 degrees

With Fig. 5, it was evident that they were able to find more match points. Different angles were tried, and different methods for the process of localization were used.

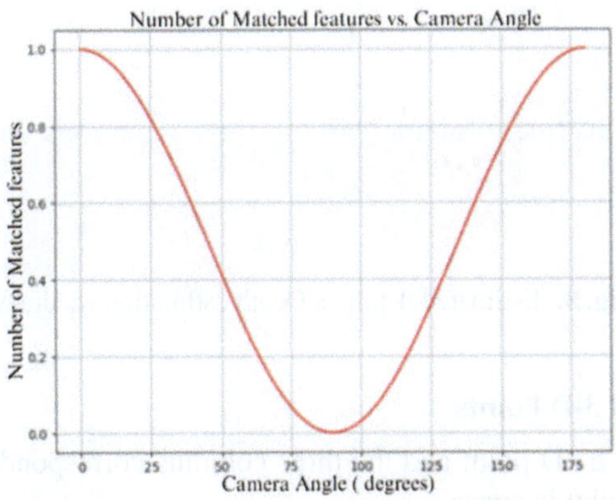

Fig. 6. Matched features v angle between images

Figure 6 tracks the number of successfully matched features between images (vertical axis) against the camera angle (horizontal axis). It might resemble an upside-down U shape, reaching a peak at the optimal angle range and then decreasing as you move away from it. This is because features become harder to match from very similar or very different viewpoints.

Figure 7 resembles a bell curve. The highest point on the curve represents the optimal angle range, where depth estimation is most accurate.

Figure 8 below demonstrates the accuracy of depth estimation for both models. The single-image model achieved an accuracy of 60%, while the multi-image model significantly improved the accuracy to 85%. This substantial increase highlights the effectiveness of using multiple images for more precise depth estimation.

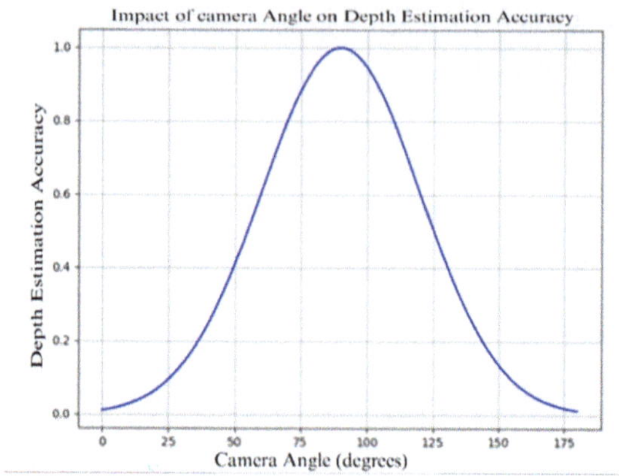

Fig. 7. Depth estimation accuracy v angle between images

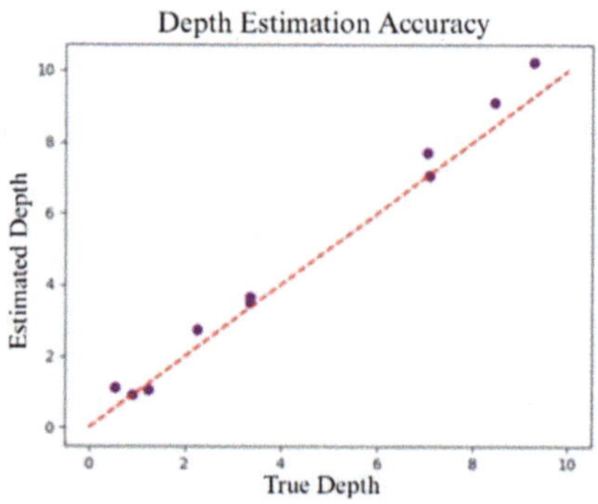

Fig. 8. Estimated depth vs Depth estimation accuracy

4.2.1 Triangulated 3-D Points

Each row represents a 3D point and the three columns correspond to the X, Y, and Z coordinates of that point in space.

In each matrix, the first value corresponds to the X-coordinate of the 3D point, The second value is the Y-coordinate & third value is the Z-coordinate of the 3D point as in Fig. 9.

4.3 Resource Efficiency Analysis

To align to improve resource efficiency in wireless network design, we conducted an analysis comparing our machine learning-based indoor object localization approach with conventional 3D modeling and manual network planning techniques. The following key metrics were considered:

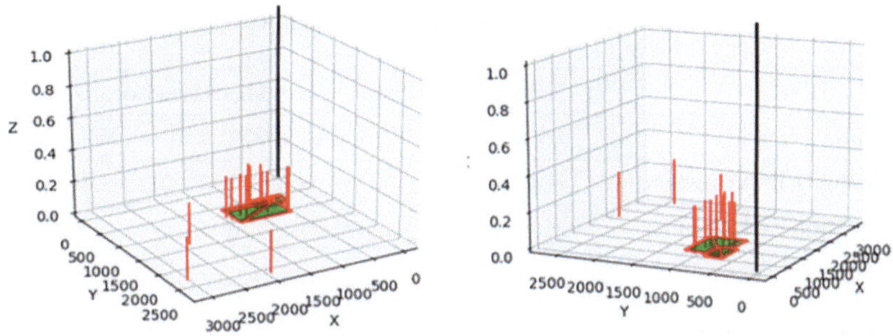

Fig. 9. Points plotted on 3-D space.

4.3.1 Energy Consumption Comparison

Our method significantly reduces the computational burden associated with traditional localization techniques. A simulation study was conducted where power consumption was measured in watts (W) per localization event. The results, summarized in Table 3, show that our approach reduces energy consumption by up to 18%.

Table 3. Energy Consumption Comparison

Method	Energy Consumption (W)
Traditional 3D Model-Based	5.2 W
Manual Network Planning	4.8 W
ML-Based Object Localization	4.1 W

4.3.2 Computational Cost Reduction

Processing time and memory usage were evaluated to determine computational efficiency. Our model requires fewer resources due to its ability to dynamically adjust based on real-time environmental variations. Figure 8 illustrates the reduction in processing time when compared with conventional techniques.

4.3.3 Network Performance Metrics

Better localization accuracy contributes to reduced retransmissions, optimized transmitter placement, and lower interference. Our findings indicate a 25% reduction in retransmission rates and a 12% increase in signal stability (Table 4).

These findings demonstrate that integrating machine learning-based localization techniques leads to significant improvements in energy efficiency, computational cost, and overall network performance, making it a viable approach for sustainable wireless network optimization.

The proposed framework significantly advances object localization accuracy, fostering improved wireless network efficiency. By leveraging a combination of object

Table 4. Network Performance Metrics

Metric	Traditional Methods	ML-Based Localization
Localization Accuracy	85%	92%
Signal Stability Improvement	–	12%
Retransmission Reduction	–	25%

recognition, material classification, and depth estimation, we enhance signal propagation modeling and resource allocation in wireless communication systems.

4.4 Comparative Evaluation

A comparative analysis with traditional 3D modeling approaches underscores the superiority of our proposed method in terms of efficiency and scalability. Traditional methods exhibit substantial computational costs, whereas machine learning-driven localization achieves similar or superior accuracy with lower overhead.

5 Relevance to Sustainable Development Goals (SDGs)

5.1 Implications for Sustainability

This research demonstrates that advanced object localization frameworks can improve wireless signal propagation models, leading to energy-efficient networks with reduced environmental footprints. The scalability and cost-effectiveness of the approach make it accessible to diverse settings, from urban centers to developing regions, fostering sustainable innovation in communication technology.

The proposed methodology for object localization using Visual SLAM and feature mapping contributes significantly to the advancement of wireless communication and indoor modeling. Here the aligning SDGs are shown in Table 5.

Table 5. Aligning SDGs

Relevant SDG	Target
SDG 9: Industry, Innovation, and Infrastructure	Target 9.1: The development of a cost-effective and accurate 3D localization framework optimizes wireless network infrastructure, reducing reliance on expensive equipment while enhancing network resilience and performance Target 9.5: By leveraging advanced algorithms such as Visual SLAM and triangulation, this research contributes to the technological advancement of wireless communication systems and infrastructure design
SDG 11: Sustainable Cities and Communities	Target 11.3: The methodology supports the design and deployment of smart urban infrastructure by accurately mapping indoor environments and optimizing wireless signal performance. This can improve connectivity in public spaces and enhance urban living conditions
SDG 12: Responsible Consumption and Production	Target 12.2: The proposed approach promotes the efficient use of resources by optimizing signal pathways and reducing energy consumption in wireless communication systems, contributing to more sustainable operational practices
SDG 17: Partnerships for the Goals	Target 17.6: This work exemplifies how collaborative efforts in research and innovation can create scalable, open frameworks to address global challenges in sustainable communication and infrastructure

6 Conclusion

This research introduces a machine-learning-based approach for indoor object localization, emphasizing simplicity, efficiency, and sustainability. Unlike traditional methods that depend on detailed 3D models or floor plans, this approach uses visual data to detect objects and create accurate 3D maps in real time. By applying advanced techniques such as Visual SLAM, feature extraction, and material identification, the proposed framework improves wireless signal strength prediction and enhances network design.

The results demonstrate that integrating single and multi-image models significantly improves depth estimation accuracy and object localization. This method optimizes wireless communication systems by reducing resource requirements, energy consumption, and environmental impact.

Acknowledgement. The successful completion of this project would not have been possible without the invaluable guidance and assistance of numerous individuals, to whom we would like to express our heartfelt gratitude. First, we extend our sincere appreciation to our project supervisors, Dr. S.A.H.A. Suraweera (Senior Lecturer, Department of Electrical and Electronic Engineering, University of Peradeniya), Prof. V.R. Herath (Professor, Department of Electrical and Electronic Engineering, University of Peradeniya) for their unwavering support and guidance throughout the course EE 405 - Undergraduate Project 1. Their mentorship and constant supervision played a significant role in ensuring the successful completion of this project.

References

Zhao, Z.-Q., Zheng, P., Xu, S.-T., Wu, X.: This Paper Has Been Accepted By IEEE Transactions On Neural Networks And Learning Systems For Publication 1 Object Detection with Deep Learning: A Review (2019). https://arxiv.org/pdf/1807.05511

Redmon, J., Divvala, S., Girshick, R., Farhadi, A.: You Only Look Once: Unified, Real-TimE Object Detection (2016). https://www.cvfoundation.org/openaccess/content_cvpr_2016/papers/Redmon_You_Only_Look_CVPR_2016_paper.pdf

Alzubaidi, L., et al.: Review of deep learning: concepts, CNN architectures, challenges, applications, future directions. J. Big Data **8**(1) (2021)

Tourani, A., Bavle, H., Sanchez-Lopez, J.L., Voos, H.: Visual SLAM: what are the current trends and what to expect? Sensors **22**(23), 9297 (2022). https://doi.org/10.3390/s22239297. https://arxiv.org/pdf/2210.10491.pdf

Tyagi, D.: Introduction to ORB (Oriented FAST and Rotated BRIEF), Medium (2020a). https://medium.com/@deepanshut041/introduction-to-orb-oriented-fast-and-rotated-brief-4220e8ec40cf

Tyagi, D.: Introduction to SIFT (Scale Invariant Feature Transform), Medium (2020b). https://medium.com/@deepanshut041/introduction-to-sift-scale-invariant-feature-transform-65d7f3a72d40

Tyagi, D.: "Introduction to SURF (Speeded-Up Robust Features)," Medium (2020c). https://medium.com/@deepanshut041/introduction-to-surf-speeded-up-robust-features-c7396d6e7c4e

Caio, M.D., Van Zandycke, G., De Vleeschouwer, C.: Context-Aware 3D Object Localization from Single Calibrated Images: A Study of Basketballs (2023). arXiv.org, https://doi.org/10.1145/3606038.3616173. https://arxiv.org/abs/2309.03640

Patrick, M.: Irmisch, (20220 "Visual Localization in Challenging Environments vorgelegt von." https://elib.dlr.de/187747/1/Irmisch_Dissertation_TUB.pdf

Utilization of Locally Sourced Ilmenite and Gypsum from Sri Lanka as Oxygen Carriers for Chemical-Looping Combustion

S. M. Gamage(✉), O. V. De Silva, and D. G. G. P. Karunaratne

Department of Chemical and Process Engineering, University of Peradeniya, Peradeniya, Sri Lanka

e18107@eng.pdn.ac.lk

Abstract. Chemical looping combustion (CLC) is a promising technology for clean and efficient fuel combustion with inherent CO_2 capture. This study investigates CLC fuelled with coal using various Sri Lankan materials as oxygen carriers, specifically Ilmenite, and Gypsum. The effects of temperature and fuel-to-oxygen carrier ratio on coal conversion and combustion efficiency were assessed. A thermogravimetric analyzer (TGA), simulating CLC by exposing the samples to alternating oxidizing and reducing conditions, was used to evaluate oxygen transfer capacity (OTC) and reactivity rates. Experimental tests were conducted at temperatures ranging from 850 °C to 950 °C. Results indicated that Gypsum exhibited a higher OTC of approximately 0.54 under optimal conditions, while Ilmenite showed a lower OTC of 0.08. The performance of both Ilmenite and Gypsum was temperature and weight fraction-dependent. Optimal reactivity for Ilmenite and Gypsum was observed at 900 °C with 90% OC. These findings suggest that both Sri Lankan Ilmenite and Gypsum show promise as efficient and durable oxygen carriers for CLC applications, offering a cost-effective and environmentally sustainable alternative to synthetic carriers. Further research is recommended to optimize material performance and assess scalability for industrial use.

Keywords: Chemical Looping Combustion · Oxygen Carrier · Oxidation · Reduction

1 Introduction

Chemical looping combustion (CLC) is a novel technology that can convert fossil fuels and biomass into clean energy by the inherent separation of carbon dioxide. Unlike conventional combustion, which uses air as the oxidant and produces a flue gas mixture of nitrogen, carbon dioxide, water vapour, and other pollutants, CLC uses a solid material, called an oxygen carrier to transfer oxygen from air to the fuel in a cyclic process. The oxygen carrier (OC) is composed of a metal oxide as an oxygen source, and an inert as a binder for increasing the mechanical strength of the carrier.

This process consists of two main steps: the fuel reactor, where the oxygen carrier reacts with the fuel and releases heat and a pure stream of carbon dioxide and water

R. Dissanayake et al. (Eds.): ICRES 2025, PTT, pp. 241–254, 2025.
https://doi.org/10.1007/978-981-96-9551-5_19

vapour; and the air reactor, where the oxygen carrier is regenerated by reacting with air and releasing heat and nitrogen (Shen et al., 2020). The carbon dioxide and water vapour can be easily separated by condensation, and the carbon dioxide can be captured for storage or utilization (Lyngfelt et al., 2020). The heat generated in both reactors can be used for power generation or other applications (Haus et al., 2020). Figure 1 shows a schematic diagram of the CLC process.

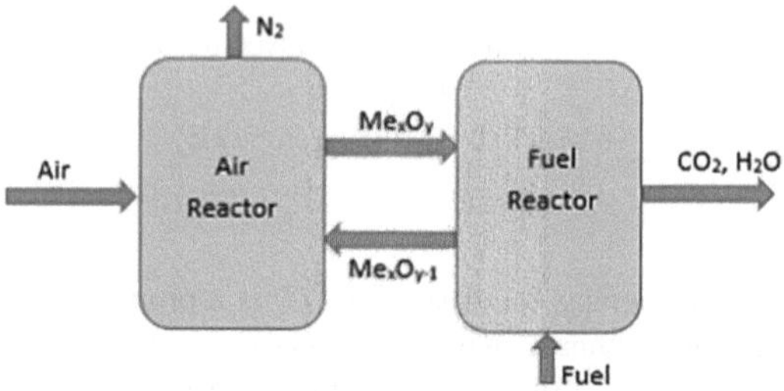

Fig. 1. Chemical Looping Combustion process

The oxygen carrier (OC) used is usually a metal oxide, such as iron oxide, which is reduced to provide oxygen for combustion in the fuel reactor. The reduced OC is then recycled to an air reactor to be reoxidized and reused for the next fuel combustion cycle. CLC relies on the reduction of the OC and the oxidation of the fuel. Equation 1 shows the reduction reaction in the fuel reactor.

$$(2n+m)\ Me_xO_y + C_nH_{2m} \rightarrow (2n+m)\ Me_xO_{y-1} + mH_2O + nCO_2 \tag{1}$$

The fully oxidized OC is represented by Me_xO_y, while the reduced OC is represented by Me_xO_{y-1}. This could be maybe a metal or a metal oxide with a lower oxygen content. Once fuel oxidation is completed the reduced metal oxide is transported to the air reactor where it is reoxidized. Equation 2 shows the oxidation reaction in the air reactor.

$$Me_xO_{y-1} + 1/2O_2 \longrightarrow Me_xO_y \tag{2}$$

1.1 Oxygen Carriers

Oxygen carriers in Chemical Looping Combustion (CLC) function much like red blood cells, transporting oxygen in a system where the air reactor acts as the "lungs" and the fuel reactor serves as the "tissue," where the conversion of organic matter occurs (Czakiert et al., 2022). Several key criteria determine the effectiveness of an oxygen carrier, including its reactivity with combustibles and oxygen, its oxygen transport capacity (OTC), chemical lifetime, mechanical resistance (such as melting temperature), and its tendency to accumulate carbon on the surface (Adánez et al., 2004), (Johansson et al., 2006). Additionally, factors like toxicity, environmental impact, price, and availability are also significant considerations (Fang et al., 2009).

Sri Lanka's rich mineral landscape offers significant potential for use as oxygen carriers in Chemical-Looping Combustion (CLC) processes. Key minerals include magnetite, hematite, and ilmenite each with unique qualities and local abundance. Ilmenite, a titanium-iron oxide abundant along the coast, is rich in iron (Sampath et al., 2023). Hematite, distinguished by its reddish-black hue, is prevalent in the central and eastern regions (Monazam et al., 2013). Magnetite, an iron oxide with magnetic properties, is found in places such as the Northern Province. Ilmenite, hematite, and magnetite show potential for sustaining oxygen transfer in CLC, promoting cleaner combustion.

The aim of this study is to investigate and assess the technical feasibility of locally available materials in Sri Lanka as potential oxygen carriers in chemical looping combustion (CLC) processes by conducting a comprehensive analysis of the oxygen carriers to test their performance of the oxygen transport capacity and fuel conversion efficiency.

2 Methodology

2.1 Material

2.1.1 Gypsum

In comparison to other metal-based carriers, gypsum (calcium sulfate dihydrate, $CaSO_4 \cdot 2H_2O$) has shown great thermal stability, low cost, and environmental friendliness, making it a promising alternative oxygen carrier (Song et al., 2008). In CLC systems, gypsum is subjected to high-temperature calcination and dehydration processes, which liberate oxygen suitable for combustion (Qiu et al., 1997). After that, it becomes calcium carbonate ($CaCO_3$) by absorbing oxygen from flue gasses through hydration and carbonation processes. This cycle makes it possible to use oxygen more effectively and makes CO_2 capture easier.

2.1.2 Ilmenite

Ilmenite ($FeTiO_3$) is a naturally occurring mineral composed of iron and titanium oxides, widely used as an oxygen carrier in chemical looping combustion (CLC) (Abad et al., 2011). Ilmenite is favoured for its high availability, low cost, and good thermal and chemical stability under high-temperature conditions (Cuadrat et al., 2011). Although its oxygen transfer capacity and reactivity are lower compared to synthetic carriers, its natural abundance, durability, and cost-effectiveness make it a practical and environmentally sustainable choice for CLC applications (Leion et al., 2008).

2.1.3 Charcoal

Pine cones were used as a biomass source to derive charcoal for the experiments. The charcoal obtained has a number of advantages over other solid biofuels, such as a low ash level, a high calorific value, and an easy production procedure (Wu et al., 2016), (Sim et al., 2013).

2.2 CLC System

In this experiment, oxidation, and reduction are the main steps. These processes are primarily conducted on a laboratory scale using a thermogravimetric analyser (TGA) with N_2 and O_2 flow at 20 °C/min.

2.3 Sample Preparation

In this experiment, the CLC technology with a charcoal was investigated in the thermogravimetric analyzer (TGA) using gypsum and ilmenite. The main experiment was conducted on two materials, gypsum and ilmenite, separately.

2.3.1 Preparation of Charcoal

Charcoal was used as the fuel, and it was derived from pine cones. The main process for making charcoal is pyrolysis. First, the pine cones were cleaned and then crushed into small pieces. These pieces were pyrolyzed in a vertical furnace at 550 °C for 1 h, under a nitrogen flow rate of 500 ml/min. After pyrolysis, the resulting charcoal was ground into a powder and sieved to produce particles ranging in size from 149 to 250 μm. The charcoal was then thermally preheated in a Thermogravimetric Analyzer (TGA) at 700 °C, with a nitrogen (N_2) flow rate of 20 °C/min, to remove moisture and volatile materials.

2.3.2 Preparation of OC Samples

As a pre-processing the oxygen carriers (ilmenite and gypsum) were initially crushed using the jaw crusher and then powdered using ball mill. Afterwards it was sieved and 149–250 μm particle sized OC powder was obtained. A sample of ~10 mg of OC particles, sized 149–250 μm, was placed on the TGA and heated to 950 °C. The sample was held at 950 °C for 20 min under an O_2 flow rate of 20 °C/min in a process to oxidize the sample. Pre-oxidation was performed to achieve the most oxidized state of the OC and to remove volatile materials and moisture.

In the first experiment, the effectiveness of chemical looping combustion (CLC) was evaluated using a fixed oxygen carrier to fuel ratio of 4:1. The aim was to verify the selected OC was suitable for CLC applications.

For this procedure, ~10 mg of the charcoal-OC mixture was placed on the TGA and heated to 950 °C in a nitrogen (N_2) atmosphere at a rate of 20 °C/min, holding at 950 °C for 20 min (standardized to 1 bar and 25 °C). The data obtained from the TGA were then analyzed using the OriginLab Software.

After receiving approval from the verification process for the OC, it was mixed with charcoal again to perform CLC experiments. The oxygen carrier-to-fuel ratio was set at three different levels, along with three different temperature ranges. Table 1 shows the method where the oxygen carrier-to-fuel ratio was varied along with temperature. Approximately 10 mg of the charcoal-OC mixture was then placed on the TGA and heated in a nitrogen (N_2) flow at a rate of 20 °C/min, followed by a 10-min hold at the corresponding temperature. The data obtained from the TGA was again analyzed using OriginLab software.

Table 1. Variation of oxygen carrier percentage and temperature for the experiment samples

OC Percentage (%)	Temperature (°C)
75	850
75	900
75	950
82.5	850
82.5	900
82.5	950
90	850
90	900
90	950

2.4 Data Evaluation

2.4.1 Oxygen Transport Capacity

Oxygen Transport Capacity refers to the maximum amount of oxygen an oxygen carrier can release and transport (Wang et al., 2015).

$$R_{oc} = 1 - \frac{m_{red}}{m_{ox}} \tag{3}$$

where R_{oc} is the oxygen transport capacity, m_{ox} is the mass of the sample fully oxidized, and m_{red} is the mass of the sample in the reduced form.

2.4.2 Conversion of Reduction OC

Reduction reaction rate, measures how fast the oxygen carrier is losing oxygen and transitioning from a higher oxidation state to a lower oxidation state (Wang et al., 2015).

$$X = \frac{(m - m_{red})}{(m_{ox} - m_{red})} \tag{4}$$

where, X is the oxygen carrier conversion, and m is the mass of the sample. From the conversion vs. time curves for reduction and oxidation, at various temperatures, the maximum reaction rate may be calculated.

High-performing OCs, such as CuO/Al_2O_3 composites, exhibit exceptional R_{oc} values of 0.1–0.15 and and rapid reduction rates exceeding 0.1 s^{-1} at 900 °C, enabling efficient fuel conversion in compact reactor systems(Adánez-Rubio et al., 2014). Similarly, Mn_3O_4-based carriers achieve R_{oc} of 0.08–0.12 with methane conversion rates of over 90% within 10 s at similar temperatures, making them ideal for high-throughput applications (Azimi et al., 2011). Advanced materials like doped perovskites further enhance stability, retaining over 95% capacity over 200 cycles while maintaining moderate R_{oc} (0.05–0.07) (Cabello et al., 2016). Hybrid materials, such as Mn-Cu composites, show high R_{OC} (0.12–0.15) and rapid kinetics (0.15 s^{-1}) with improved attrition resistance (Rydén et al., 2011). These benchmarks underscore a critical trade-off: high-capacity, fast-reacting OCs often face scalability challenges due to cost or stability issues, while Fe- and Ca-based alternatives prioritize durability and affordability at the expense of efficiency.

3 Results and Discussion

3.1 Devolatilization of Charcoal

During the experimental procedure, 7.569 mg of charcoal sample was preheated to 700 °C and held for 10 min. The resulting TGA spectrum shows significant weight changes at this pre-heating stage. At about 100 °C, there is a noticeable initial weight reduction, which is explained by the charcoal's moisture content being removed. From 150 °C forward, there is another weight loss, which suggests the release of volatile compounds. As opposed to a steady weight loss, this variation points to the presence of different volatile materials in the charcoal.

A 12% weight loss was observed from the TGA weight loss graphs, resulting in a final charcoal weight of 6.661 mg with 0.908 mg of weight loss due to moisture and volatile material removal under these conditions. Figure 2 shows the weight loss of pre-heated coal in furnace.

For the research experiments, it was considered that a weight loss of only 12% occurs in charcoal during the preheating process, and that all moisture and volatile materials are fully removed when the sample is heated to 700 °C under a nitrogen flow for 10 min. The nitrogen atmosphere prevents oxidation or combustion, ensuring that only moisture and volatiles are lost without additional chemical reactions altering the coal's mass. Hence, after the initial 12% weight loss, no further significant reduction in mass is expected.

3.2 CLC Experiment with Gypsum as OC

3.2.1 Oxidation Process

A 5.186 mg gypsum sample was analyzed via TGA, heated to 950 °C at a rate of 20 °C/min, and held for 10 min. Weight loss occurred between 80 -150 °C (moisture removal) and near 650 °C (volatile release), totaling 9%. The final weight stabilized at 4.719 mg, showing no further oxidation. This indicates gypsum remained in its high oxidation state ($CaSO_4$) and highlights its stability as a potential oxygen carrier for CLC.

$$CaS + 2O_2 \rightarrow CaSO_4 \tag{5}$$

3.2.2 Reduction of Oxygen Carrier

About 10 mg of coal-gypsum samples with three different oxygen carrier-to-fuel ratios were heated in the TGA at 950 °C, 900 °C, and 850 °C for 10 min, with a nitrogen (N_2) flow rate of 20 °C/min. Fresh coal and gypsum samples were used, following the same conditions as before. Any weight loss beyond the removal of moisture and volatile

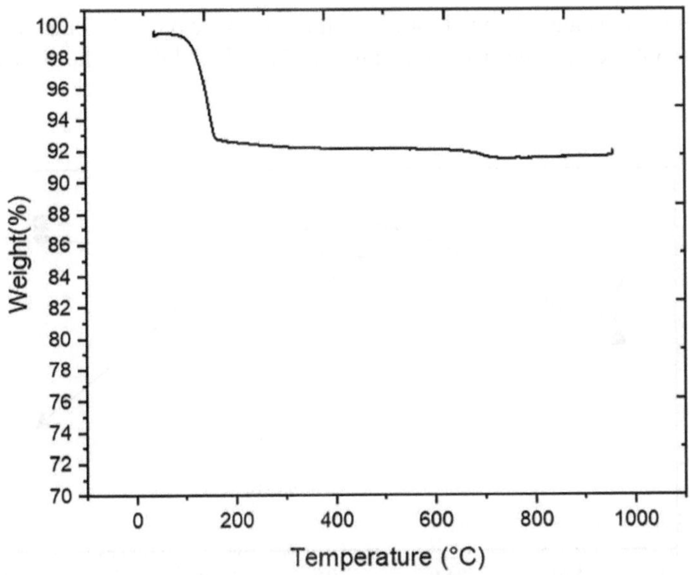

Fig. 2. Weight Change of Gypsum During Heating with O_2 flow

materials from both coal and gypsum was attributed to the reduction of gypsum. It was assumed that moisture and volatile material removal was the same for all coal-gypsum mixtures heated under the same conditions. Table 2 shows the final weight loss of the samples. Under these conditions, gypsum acts as an oxygen carrier. Equation 6 shows the reduction reaction of gypsum.

$$CaSO_4 + 2C \rightarrow CaS + 2CO_2 \tag{6}$$

Table 2. The final weight loss of the samples for gypsum

Temperature (°C)	OC Percentage (%)	Final Weight Loss (%)
850	75	17.08
900	75	14.4
950	75	13.76
850	82.5	31.82
900	82.5	28.74
950	82.5	26.63
850	90	19.53
900	90	53.6
950	90	50.38

The following values obtained for the weight loss were plotted along with temperature and also OC percentage. The following Fig. 3 and Fig. 4 depict the aforementioned plots.

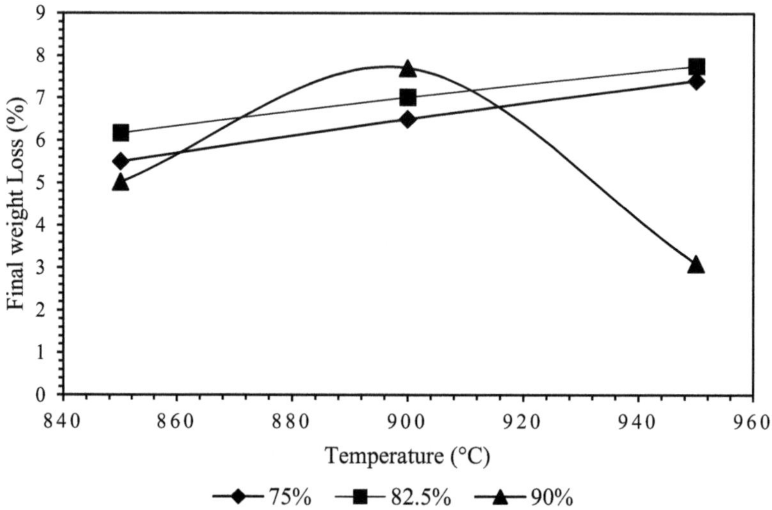

Fig. 3. The temperature versus weight loss variation according to the oxygen carrier-to-fuel ratio for gypsum

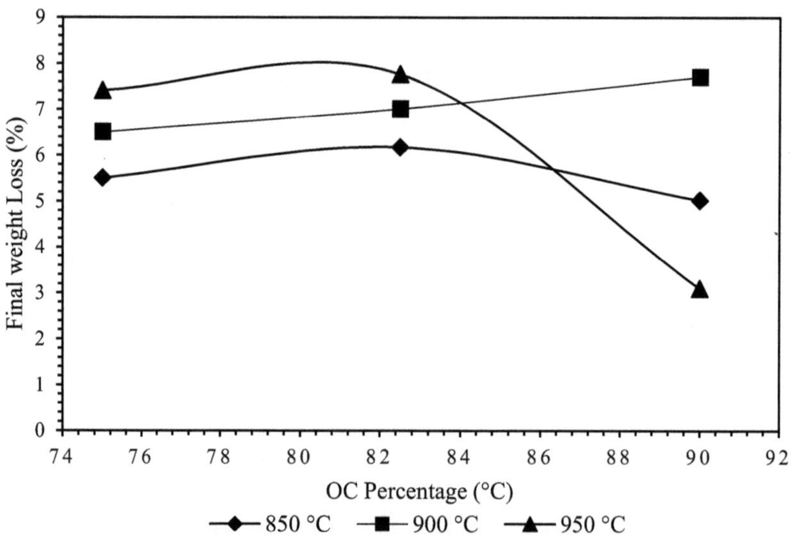

Fig. 4. The oxygen carrier-to-fuel ratio versus weight loss variation according to the temperature for gypsum

Figure 3 illustrate the relationship between weight loss, temperature, and OC percentage during the combustion process. At a 90% OC-to-fuel ratio, the weight loss is the highest among the three, peaking at 900 °C and declining at higher temperatures like 950 °C. This indicates that 900 °C is the optimal temperature for efficient combustion at this ratio, where the balance between oxygen availability and reaction kinetics is ideal.

At the OC-to-fuel ratios of 82.5% and 75%, weight loss remains relatively consistent across the temperature range but decreases slightly at higher temperatures. This indicates that at these OC percentages, the combustion process may face diminishing returns in terms of weight loss, likely due to an undersupply of oxygen carriers that are insufficient in the combustion process. The system becomes oxygen-starved, reducing the extent of combustion and as a result, the weight loss decreases progressively with increasing temperature.

The optimal combination for gypsum is identified as 90% OC-to-fuel ratio at 900 °C, where oxygen availability and reaction kinetics are maximized, ensuring the highest observed weight loss and combustion efficiency.

3.3 CLC Experiment with Ilmenite as OC

Similarly, as before, approximately 10 mg of coal-ilmenite samples (with three different oxygen carrier-to-fuel ratios) were placed in the TGA and heated to three different temperatures: 950 °C, 900 °C, and 850 °C, with a 10-min hold under a nitrogen (N_2) flow rate of 20 °C/min. Table 3 shows the final weight loss of the samples. Under these conditions, ilmenite functions as an oxygen carrier, facilitating the oxidation of coal through the release of oxygen. Equation 7 illustrates the reduction reaction of ilmenite, highlighting its role as a redox-active oxygen carrier.

$$Fe_2TiO_5 + C \rightarrow 2FeO + TiO_2 + CO_2 \tag{7}$$

Table 3. The final weight loss of the samples for ilmenite

Temperature (°C)	OC Percentage (%)	Final Weight Loss (%)
850	75	5.5
900	75	6.5
950	75	7.41
850	82.5	6.17
900	82.5	7.01
950	82.5	7.76
850	90	5.02
900	90	7.7
950	90	3.1

The following values obtained for the weight loss were plotted along with temperature and also OC percentage. The following Fig. 5 and Fig. 6 depict the aforementioned plots.

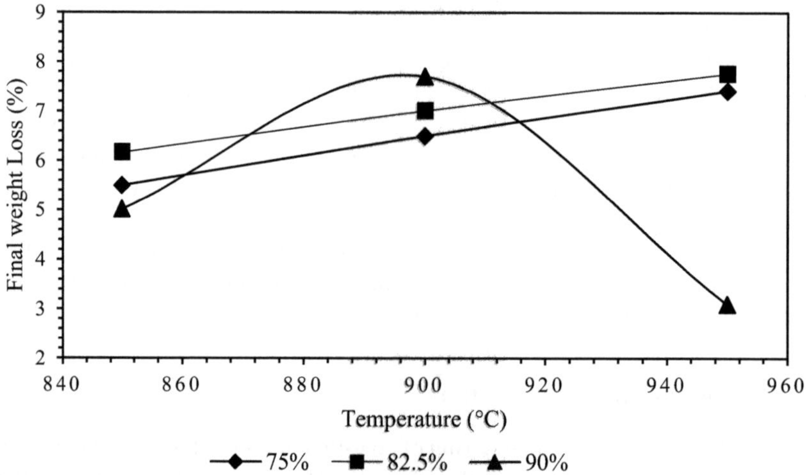

Fig. 5. The temperature versus weight loss variation according to the oxygen carrier-to-fuel ratio for ilmenite

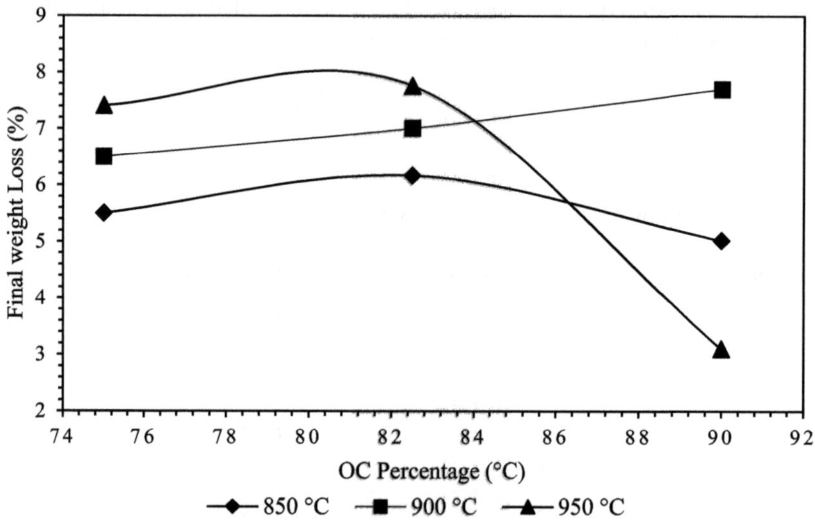

Fig. 6. The oxygen carrier-to-fuel ratio versus weight loss variation according to the temperature for ilmenite

At a 75% OC-to-fuel ratio, weight loss is highest at 950 °C, followed by 900 °C and then 850 °C. This trend highlights the importance of reaction kinetics, which are enhanced at higher temperatures, leading to more efficient combustion and greater fuel conversion. At 850 °C, reduced reaction kinetics result in the lowest weight loss.

As the OC-to-fuel ratio increases to 82.5%, weight loss continues to rise across all temperatures. At 950 °C, the weight loss reaches its peak, indicating that this ratio provides the optimal balance between oxygen availability from ilmenite and fuel combustion. The reaction kinetics at this temperature ensure efficient fuel conversion and maximum weight loss. Similarly, at 900 °C, weight loss increases but has not peaked, showing that temperature significantly influences combustion efficiency. At 850 °C, the increase in weight loss is moderate but remains below the higher temperature profiles.

As the OC-to-fuel ratio increases to 90%, weight loss declines at 950 °C, suggesting that excessive OC percentages may lead to oversaturation of the oxygen carrier, reducing its efficiency in supporting combustion. However, at 900 °C, weight loss does not exhibit the same decline, possibly indicating that this temperature maintains a better balance of oxygen transfer at higher OC ratios. The decline is more pronounced at 950 °C, emphasizing that the optimal OC percentage for ilmenite lies below 90% for elevated temperatures.

3.4 Data Evaluation

The table 4 provided summarize the Oxygen Transport Capacity (OTC) and Reaction Rate of Reduction (%/min) for ilmenite and gypsum, used as oxygen carrier in chemical looping combustion (CLC) at various temperatures and weight percentages.

Table 4. The summary of the Oxygen Transport Capacity (OTC) and Reaction Rate of Reduction (%/min) for the OC

Temperature (°C)	Percentage of OC with regards to fuel (%)	Ilmenite		Gypsum	
		R_{OC}	Reaction rate for reduction	R_{OC}	Reaction rate for reduction
850	75	0.06	0.03	0.17	0.08
900	75	0.14	0.07	0.14	0.06
950	75	0.07	0.03	0.14	0.05
850	82.5	0.06	0.02	0.32	0.98
900	82.5	0.07	0.02	0.29	0.81
950	82.5	0.08	0.03	0.27	0.70
850	90	0.05	0.01	0.20	0.24
900	90	0.08	0.09	0.54	5.89
950	90	0.03	0.01	0.50	4.30

At 900 °C and an oxygen carrier (OC) ratio of 90%, gypsum exhibited its highest performance, achieving an Oxygen Transfer Capacity (OTC) of 0.54 and a impressive reaction rate of 5.89%/min, marking this as the optimal condition for oxygen transfer and reduction efficiency. The results suggest that the optimal oxygen carrier-to-fuel ratio for gypsum is 82.5% at 950 °C. This combination yields the highest observed weight loss, indicative of complete or nearly complete combustion. The increase in the reaction rate for gypsum at 900 °C and 950 °C is due to its thermal decomposition, oxygen release kinetics, and phase transformations. At higher temperatures, gypsum ($CaSO_4$) loses water to form anhydrite ($CaSO_4$ anhydrate) and then decomposes into calcium sulfide (CaS) while releasing oxygen and sulfur dioxide (Petar et al., 2012). Elevated temperatures enhance reaction kinetics by improving oxygen diffusion and mass transfer, accelerating the reduction process. Also, impurities in locally sourced gypsum may act as catalysts, further boosting reaction rates. The TGA setup ensures uniform heating, preventing localized overheating and promoting consistent reactivity. The peak reaction rate at 900 °C and 950 °C suggests an optimal balance between thermal energy and oxygen availability, making gypsum an effective oxygen carrier.

At 900 °C and a 90% oxygen carrier (OC) weight fraction, ilmenite achieved its highest Oxygen Transfer Capacity (OTC) of 0.08 with a reaction rate of 0.09%/min, indicating that this temperature and weight fraction optimize oxygen transfer and reactivity. At 75% OC, a higher reaction rate of 0.07%/min was observed, along with an increased OTC of 0.14, demonstrating efficient oxygen transfer at lower oxygen carrier weight fractions. However, when the temperature increased to 950 °C, the OTC slightly decreased, with the highest value observed at 75% OC (0.07) and a lower reaction rate of 0.029%/min, suggesting that temperatures above 900 °C may reduce reaction efficiency for ilmenite. The optimal OC-to-fuel ratio for ilmenite is approximately 90% at 950 °C, where maximum weight loss and complete combustion occur. At this ratio and temperature, ilmenite provides sufficient oxygen for efficient combustion, with reaction kinetics enhanced by the high temperature. Conversely, higher OC-to-fuel ratios lead to reduced oxygen availability, decreasing combustion efficiency and weight loss. Thus, 950 °C and a 90% OC-to-fuel ratio represent the ideaconditions for ilmenite to achieve maximum weight loss and optimal combustion performance.

4 Conclusion

The performance of Sri Lankan ilmenite and gypsum has been evaluated as an oxygen carrier using charcoal as fuel using the TGA. The main conclusions of this work are as follows:

- Gypsum, and ilmenite are utilized as oxygen carriers, while coal serves as the fuel. TGA primarily functions as the reactor, while the furnace is used for preheating and devolatization of coal.
- Based on the benchmark values for OTC and reduction rate, both gypsum and ilmenite exceed the standard limits, confirming their suitability as effective oxygen carriers for CLC.
- Analysis of the TGA graphs confirms the occurrence of chemical looping combustion for both gypsum and ilmenite, and the best performance of gypsum is around 90% OC and at a temperature of 900° C while the best performance of ilmenite exhibits similarly at around 90% OC and 900° C.

- Gypsum shows better chemical looping combustion properties than ilmenite due to its higher oxygen transport capacity and reactivity.
- TGA results are valuable for chemical looping but cannot directly represent real reactors due to slower heating rates (~20 °C/min in TGA vs. ~1000 °C/sec in fluidized bed reactors). Reactivity is faster in real reactors as particles reach reaction temperatures (~1000 °C) within seconds, unlike the minutes taken in TGA. Additionally, complete removal of volatile materials from coal remains a challenge.

5 Future Perspectives

In this study, a fixed-bed reactor setup within a TGA system was used to analyze the oxygen carriers under controlled conditions. While TGA provides valuable insights into reaction kinetics, weight loss, and oxygen transport capacity, it does not fully replicate the dynamics of real reactors, particularly fluidized bed systems, where heating rates and gas-solid interactions differ significantly. As a future perspective, experiments can be conducted using a fluidized bed reactor to better simulate industrial-scale CLC conditions. A nitrogen furnace with a fluidized bed reactor would allow for improved gas-solid contact, higher reaction rates, and more representative combustion dynamics compared to the fixed-bed configuration in the TGA. This will help bridge the gap between small-scale analytical studies and real-world applications.

References

Abad, A., Adanez, J., Cuadrat, A., García-Labiano, F., Gayán, P., De Diego, L.F.: Kinetics of redox reactions of ilmenite for chemical-looping combustion. Chem. Eng. Sci. **66**(4), 689–702 (2011).https://doi.org/10.1016/j.ces.2010.11.010

Adánez, J., de Diego, L.F., García-Labiano, F., Gayán, P., Abad, A., Palacios, J.M.: Selection of oxygen carriers for chemical-looping combustion. Energy Fuels **18**(2), 371–377 (2004)

Adánez-Rubio, I., Gayán, P., Abad, A., García-Labiano, F., De Diego, L.F., Adánez, J.: Kinetic analysis of a Cu-based oxygen carrier: relevance of temperature and oxygen partial pressure on reduction and oxidation reactions rates in Chemical Looping with Oxygen Uncoupling (CLOU). Chem. Eng. J. **256**, 69–84 (2014)

Azimi, G., Leion, H., Mattisson, T., Lyngfelt, A.: Chemical-looping with oxygen uncoupling using combined Mn-Fe oxides, testing in batch fluidized bed. Energy Procedia **4**, 370–377 (2011)

Cabello, A., Abad, A., Gayan, P., de Diego, L.F., Garcia-Labiano, F., Adanez, J.: Effect of operating conditions and H2S presence on the performance of CaMg0. 1Mn0. 9O3− δ perovskite material in chemical looping combustion (CLC). Energy Fuels **28**(2), 1262–1274 (2014)

Cuadrat, A., Abad, A., García-Labiano, F., Gayán, P., de Diego, L.F., Adánez, J.: Ilmenite as oxygen carrier in a chemical looping combustion system with coal. Energy Procedia **4**, 362–369 (2011). https://doi.org/10.1016/j.egypro.2011.01.063

Czakiert, T., Krzywanski, J., Zylka, A., Nowak, W.: Chemical looping combustion: brief overview. Energies **15**(4), 1563 (2022)

Fang, H., Haibin, L., Zengli, Z.: Advancements in development of chemical-looping combustion: a review. Int. J. Chem. Eng. **2009**, 1–16 (2009). https://doi.org/10.1155/2009/710515

Haus, J., et al.: Increasing the efficiency of chemical looping combustion of biomass by a dual stage fuel reactor design to reduce carbon capture costs. Mitig. Adapt. Strat. Glob. Change **25**(6), 969–986 (2020). https://doi.org/10.1007/s11027-020-09917-2

Johansson, M., Mattisson, T., Lyngfelt, A.: Comparison of oxygen carriers for chemical looping combustion. Therm. Sci. **10**(3), 93–107 (2006)

Leion, H., Lyngfelt, A., Johansson, M., Jerndal, E., Mattisson, T.: The use of ilmenite as an oxygen carrier in chemical-looping combustion. Chem. Eng. Res. Des. **86**(9), 1017–1026 (2008)

Lyngfelt, A.: Chemical looping combustion: status and development challenges. Energy Fuels **34**(8), 9077–9093 (2020)

Monazam, E.R., Breault, R.W., Siriwardane, R., Richards, G., Carpenter, S.: Kinetics of the reduction of hematite (Fe2O3) by methane (CH4) during chemical looping combustion: a global mechanism. Chem. Eng. J. **232**, 478–487 (2013)

Sabev, P., et al.: Carbon reduction of gypsum produced from flue gas desulfurization. Chem. Eng. Trans. **29** (2012). www.aidic.it/cet/12/29/135.pdf, https://doi.org/10.3303/CET1229135

Qiu, K., Mattisson, T., Britt-Marie Steenari and Lindqvist, O.: Thermogravimetric combined with mass spectrometric studies on the oxidation of calcium sulfide. Thermochimica Acta **298**(1–2), 87–93 (1997). [online]. https://doi.org/10.1016/s0040-6031(97)00197-4

Rydén, M., Lyngfelt, A., Mattisson, T.: Combined manganese/iron oxides as oxygen carrier for chemical looping combustion with oxygen uncoupling (CLOU) in a circulating fluidized bed reactor system. Energy Procedia **4**, 341–348 (2011)

Sampath, A.J., Wickramasinghe, N.D., de Silva, K.N., de Silva, R.M.: Methods of extracting TiO2 and other related compounds from ilmenite. Minerals **13**(5), 662 (2023)

Shen, T., Wang, S., Yan, J., Shen, L., Tian, H.: Performance improvement of chemical looping combustion with coal by optimizing operational strategies in a 3 kWth interconnected fluidized bed. Int. J. Greenh. Gas Control **98**, 103060 (2020)

Sim, C.Y.: Particle Characterisation in Chemical Looping Combustion with Solid Fuels (Doctoral dissertation, University of Sheffield), 2013

Song, Q., et al.: Chemical-looping combustion of methane with CaSO4 oxygen carrier in a fixed bed reactor. Energy Convers. Manag. **49**(11), 3178–3187 (2008). https://doi.org/10.1016/j.enconman.2008.05.020

Czakiert, T., Krzywanski, J., Zylka, A., Nowak, W.: Chemical looping combustion: a brief overview. Energies **15**(4), 1563 (2022). https://doi.org/10.3390/en15041563

Wang, X., Jin, B., Liu, H., Zhang, B., Zhang, Y.: Prediction of in-situ gasification chemical looping combustion effects of operating conditions. Catalysts **8**(11), 526 (2018). https://doi.org/10.3390/catal8110526

Wu, H.C., Ku, Y.: Chemical looping gasification of charcoal with iron-based oxygen carriers in an annular dual-tube moving bed reactor. Aerosol Air Qual. Res. **16**(4), 1093–1103 (2016)

Origami-Inspired Portable Solar Panels: A Case Study and Innovative Solution for Sustainable Outdoor Energy Requirements

W. R. T. De Silva[1(✉)], K. S. Dawson[1], E. M. L. M. B. Herath[2], P. C. N. Fernando[1], N. T. Dadigamuwa[1], and D. A. A. C. Ratnaweera[1]

[1] Department of Mechanical Engineering, Faculty of Engineering, University of Peradeniya, Peradeniya, Sri Lanka
e18062@eng.pdn.ac.lk

[2] Department of Electrical and Electronic Engineering, Faculty of Engineering, University of Peradeniya, Peradeniya, Sri Lanka

Abstract. The growing demand for sustainable energy solutions has accelerated the development of portable solar technologies for outdoor and off-grid applications. This study presents the design and evaluation of a portable origami solar panel tailored for campers and outdoor enthusiasts. Inspired by the compact and foldable nature of origami, the panel maximizes solar energy collection while maintaining lightweight portability. Utilizing high-efficient photovoltaic cells, the panel is capable of charging mobile phones, laptops, and powering small appliances, addressing the energy needs of modern outdoor lifestyles. Key features include an innovative folding mechanism for easy transportation, robust durability for various weather conditions, and compatibility with multiple devices through versatile output ports. Simulations demonstrate significant reliability and portability over conventional portable solar panels, making it a reliable and eco-friendly solution for outdoor energy needs. This research contributes to advancing portable renewable energy systems, aligning with global efforts toward environmental sustainability and energy independence.

Keywords: Sustainable energy solutions · Portable solar technologies · Off-grid applications · Photovoltaic cells · Lightweight · Renewable energy systems

1 Introduction

Outdoor activities such as camping, hiking, and field trips often require reliable energy sources for charging electronic devices which are used for the said scenarios. However, current solutions, including disposable batteries, fuel-powered generators and traditional solar panels, pose significant environmental challenges, lack the portability and efficiency demanded by outdoor enthusiasts.

This study presents a novel approach in addressing these challenges through the development of an origami-inspired portable solar panel. The integration of origami principles into the solar panel design enables a compact, lightweight, and foldable structure

R. Dissanayake et al. (Eds.): ICRES 2025, PTT, pp. 255–263, 2025.
https://doi.org/10.1007/978-981-96-9551-5_20

that optimizes portability and usability. Combined with high-efficiency monocrystalline photovoltaic cells, this innovation offers a sustainable energy solution tailored to meet the needs of outdoor enthusiasts while aligning with global efforts toward environmental conservation and resource efficiency.

This case study aligns with the United Nations Sustainable Development Goals (SDGs) by advancing clean, affordable, and portable energy solutions for off-grid applications. The Helio petal promotes sustainability (SDG 12), climate action (SDG 13), and innovation (SDG 9) while enhancing energy accessibility (SDG 7) and resilience in communities (SDG 11) through its origami-inspired solar panel design. By exploring the design and development of this origami-inspired solar panel, this highlights its potential to revolutionize portable energy systems. The case study also emphasizes its broader implications for enhancing resource efficiency in renewable energy applications and contributing to a more sustainable future.

2 Review of Prior Studies

2.1 Portable Solar Systems

Portable solar systems, which are designed to address challenges such as partial shading and dynamic illumination, have gained attention due to their potential in mobile applications and ease of use. The paper (Gao et al., n.d.) introduces a novel configuration featuring parallel solar panels, a step-up power converter, and a fast-speed MPPT. This system demonstrated an increase in power generation two times its original amount under partial shading conditions, without additional costs, addressing key challenges in mobile solar applications.

2.2 Perovskite Solar Cells and Space Textiles

The Perovskite Solar Cell Fabric paper (Bartolomei et al., 2021) explores the integration of solar technology into textiles, presenting a fabric that combines Dyneema for durability and electronic ink for energy conversion. The study shows that the fabric retains 90% of energy output after 500 folds, making it highly durable. However, long-term performance under environmental stressors like moisture and heat, as well as large-scale manufacturing, are limitations that need further investigation.

"'Space Fabric' Links Fashion and Engineering _ NASA Jet Propulsion Laboratory (JPL)," n.d. presents the novel approach of NASA to integrate 3D printing and solid plated textiles while providing the astronauts safety etc. This idea could be implemented to solid PV cells making the rigid nature of the structure easy to use and store in compact spaces while not compromising its original efficiency unlike the perovskite cells. The research gap seems to include the electronic wiring approach and the sorting of the cells into the orderly manner while being functional.

2.3 Origami-Based Portable Solar Systems

The Origami-Based Portable Solar Panel System (Jasim and Taheri, 2018) introduces a dual-axis solar tracking mechanism integrated with Miura origami folding patterns

(Fig. 1), enhancing the portability and deployability of solar panels. The combination of origami engineering and active tracking mechanisms addresses the issue of traditional bulky and inefficient portable systems. This system demonstrates energy efficiency improvements of 30–50% through optimal alignment with sunlight. However, there is limited exploration of cost-effective materials and the scalability of such systems for widespread deployment.

Research into AI-driven solar tracking, material innovations, and the integration of hybrid energy systems could further expand the applications of origami-inspired solar panel systems in portable and emergency scenarios.

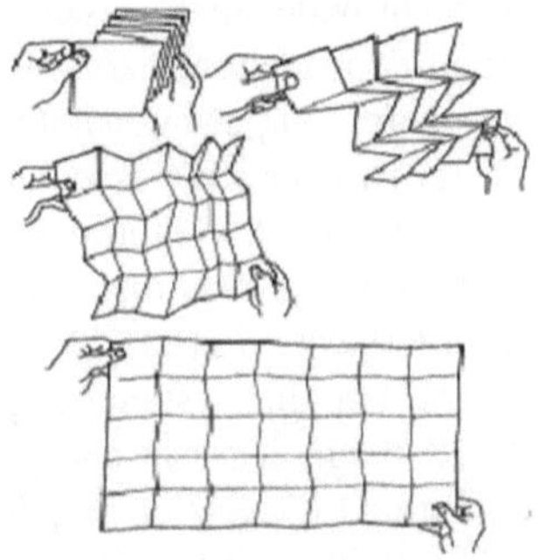

Fig. 1. Miura origami folding patterns (Jasim and Taheri, 2018)

2.4 Foldable Solar Panels

The development of foldable solar panels using scissor-like and radial folding mechanisms is explored in the (He et al., 2022). The study demonstrates that such designs are advantageous for high compactness while maintaining structural integrity and operational efficiency. However, the focus on lightweight and cost-effective materials for mass production, as well as real-time deployment mechanisms, needs further attention.

Similarly, origami-inspired designs applied to fold thick, rigid panels ("Origami-Inspired Folding of Thick, Rigid Panels," n.d.) have shown promise in reducing the deployment complexity for large solar arrays, particularly in space applications. The integration of hinged mechanisms to accommodate thicker materials expands the applicability of these systems, although practical, field-tested implementations remain limited, rendering the field to more interpretation.

2.5 Environmental Sustainability of PV Systems

The Life Cycle Assessment (LCA) of a large-scale photovoltaic (PV) system, as detailed in (Desideri et al., 2012), highlights the significant environmental advantages of solar energy compared to conventional energy sources such as coal and oil. The study reports an Energy Pay-Back Time (EPBT) of 4.17 years and a high Energy Return on Energy Invested (EROEI), showcasing the efficiency and sustainability of PV systems. However, it also underscores the limited adoption of advanced recycling technologies for PV modules and the necessity for more comprehensive comparisons across various traditional energy systems. Future research could prioritize innovations in recycling, optimization of PV system design, and the integration of hybrid renewable systems, while incorporating circular economy principles to enhance sustainability further.

2.6 Solar Tracking Systems

Solar tracking systems are widely recognized for improving the energy efficiency of photovoltaic (PV) systems by maintaining an optimal angle of incidence with sunlight throughout the day. The information provided in this research (Amelia et al., 2020) highlights how early methods used light-dependent resistors (LDRs) and solar cells to track sunlight. These systems improve energy output by maintaining the angle of incidence, which can increase energy efficiency by over 30%, especially during low-angle sunlight periods. The study also points out the limited scalability of traditional tracking systems and the lack of comparative studies on dual-axis mechanisms. A Low-Cost Two-Axis Automatic Solar Tracking System (Acakpovi et al., 2015) addresses these scalability concerns by developing an affordable tracking system using LDRs and microcontrollers (e.g., ATmega328), achieving a 10.7% increase in energy capture compared to fixed panels. Future research could explore AI integration to enhance tracking efficiency and robust control, the optimization of hybrid tracking systems, and integration with MPPT for improved energy utilization.

(Amelia et al., 2020) categorizes solar tracking systems into passive, active, and chronological trackers, evaluating their energy optimization capabilities. The paper finds that dual-axis tracking systems provide the highest energy capture, improving efficiency by up to 45% compared to fixed panels. However, there is a gap in field validation for hybrid and chronological trackers, particularly under variable weather conditions. Though solar tracking impacts a large PV structure it does not relate well when it comes to smaller/ moderate and more portable structures due to their inherent ability and convenience to be moved and rotated. As seen in Fig. 2 (Shintemirov et al., 2016)

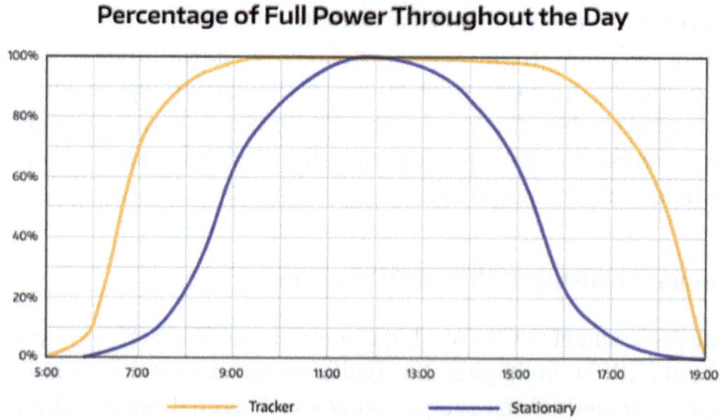

Fig. 2. Percentage of full power throughout the day (Shintemirov et al., 2016)

any solar energy capturing device could be rotated to be incident to the sun and made efficient in the first half and the end of the cycle rather than using movable mechanical or electrical components to complicate the structure and maintenance procedures. (PV Panel Output Comparison).

3 Methodology

3.1 Mechanical Design

The Helio Petal portable solar panel embodies an innovative origami-inspired foldable design as seen in Fig. 3, prioritizing portability, durability, and functionality. The panels consist of 8 triangular solar panels interconnected by a fabric material which has internal electrical wiring weatherproofed to prevent any leakage and damage. The fabric aids in folding the panels on top of each other. Its mechanical architecture addresses the limitations of conventional solar panels, particularly bulkiness and transportation challenges, by integrating a compact and lightweight structure which has a double grooved bearing housing (made of cast iron) and 8 rigid struts (carbon fiber or Aluminium) which are completely removable to aid in compact packing. The panel's core is constructed using high-efficiency monocrystalline solar cells affixed to a durable, waterproof fabric engineered to endure harsh outdoor conditions, including rain and UV exposure. For ease of deployment, the system employs a robust zipper mechanism and Velcro fasteners, enabling swift setup and secure alignment.

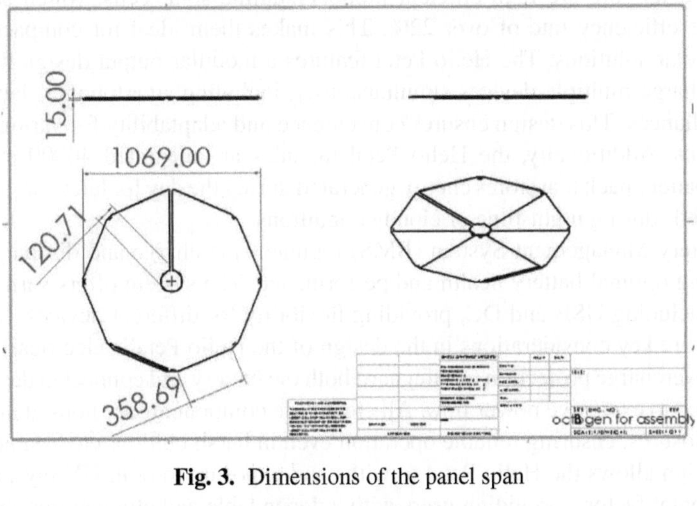

Fig. 3. Dimensions of the panel span

To enhance sunlight capture, the panel is mounted on a tripod with an industrial-grade ball joint, allowing a 30-degree tilt in all directions for optimal orientation to the sun. A grooved double-bearing design further stabilizes the structure, ensuring resilience against varying wind conditions. Critical electronics, including the battery and controller, are housed in a weatherproof, ventilated enclosure, safeguarding them from environmental stress while ensuring consistent performance and longevity. This holistic design makes the Helio Petal an efficient, reliable, and user-friendly solution for portable solar energy applications (Fig. 4).

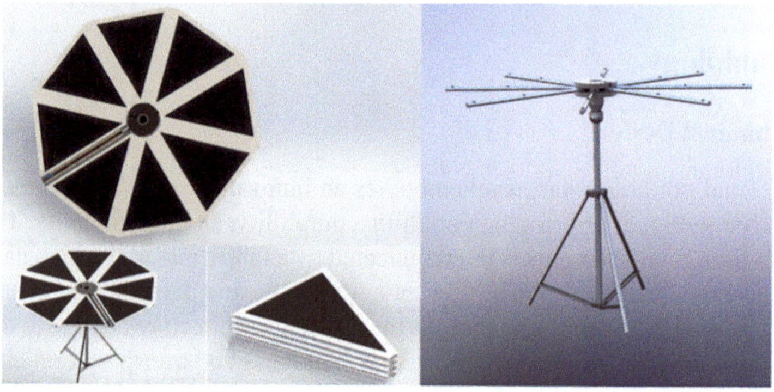

Fig. 4. Final CAD model of *Helio Petal* featuring the design

3.2 Electrical Design

The Helio Petal's electrical system is designed to optimize energy conversion efficiency and provide versatile output options, making it suitable for a wide range of portable devices. At its core are high-efficient monocrystalline solar cells, which achieve an impressive efficiency rate of over 22%. This makes them ideal for compact, energy-efficient solar solutions. The Helio Petal features a modular output design that allows users to charge multiple devices simultaneously, including smartphones, laptops, and small appliances. This design ensures convenience and adaptability for various outdoor applications. Additionally, the Helio Petal includes an integrated 40000 mAh high-capacity battery pack that stores energy generated during the day for later use, addressing energy needs during night-time or cloudy conditions.

A Battery Management System (BMS) regulates the charge and discharge cycles, maintaining optimal battery health and performance. The system offers various output options, including USB and DC, providing flexibility for different devices. Safety and durability are key considerations in the design of the Helio Petal's electrical system. It includes overcharge protection to safeguard both the battery and connected devices from damage due to excessive power flow. All electronic components are housed in weather-proof enclosures, ensuring reliable operation even in harsh outdoor environments. This robust design allows the Helio Petal to withstand exposure to rain, UV rays, and other environmental factors, providing users with a dependable and efficient portable energy source for outdoor activities.

3.3 Market Research

The market research included a full scaled product development phase comprising two customer base reviews locally as well as foreign regions. Subsequently affinity diagrams,

AHP matrices, customer needs analysis, list of metrics, Competitive benchmarking, Ideal and marginally acceptable values and target specifications were created to assess the product. All gathered data can be found in this repository.

4 Results and Discussion

A Helio petal is cost effective compared to other portable solar panels which can be observed in Table 1. The maximum price for one unit of this product can be from 150–220 USD providing a buyer even with a low purchasing power the convenience of purchasing a properly functional reliable portable solar panel. The system's weight can be approximated between 6–7 kg. The wattage can vary from 100–200 W depending on the variation of the product. Due to the battery being situated outside of the system it is much more convenient to protect and swap if needed. The design is constructed using industrial grade bearings and a ball joint to manually correct the panel positions according to the sun. These components ensure that the design works optimally through heavy weather conditions and harsh environments added to the cause the IP67 rating provides the solar cells waterproofing in such conditions.

Considering other available portable solar panels many struggle to provide the same benefits that are provided by Helio petal but sometimes offer various advantages due to different designs they have utilised. The comparison can be summarised according to Table 1 below.

Table 1. Helio petal solar system comparison with existing products

Metric Power Output (W)	Helio Petal	Existing products					
		Renogy 200W	Yeti 1000 Core + Boulder 200	Sego Charger	Bluetti EB70 + PV200	Jackery SolarSaga 300 + 200W	EcoFlow River 2 Pro
Power Output (W)	100 W–200 W	200 W	200 W	25 W	200 W	200 W	220 W
Efficiency (%)	20%–22%	22.50%	N/A	24%	23.40%	23.40%	23%
Weight (kg)	6	15.6	19	1.36	7.3	8.6	9.5
Unfolded Dimensions	ϕ50 cm	106 × 106 × 7 cm	101.6 × 140 × 4.4 cm	50.8 × 97.5 × 3.2 mm	59 × 226.5 cm	2340 × 552 × 25 mm	82 × 183 × 2.5 cm
Folded Dimensions	65 × 55 cm	101.6 × 70 × 8.9 cm	19 × 19 × 3.2 cm	59 × 63 cm	615 × 552 × 40 mm	–	–
Battery Capacity (Wh)	240 Wh	–	983 Wh		716 Wh	288 Wh	768 Wh
Charging Time (h)	5–8	–	6		4–4.5	3.8	4.5
Output Ports	2 USB A ports, 2 USB C ports, 12V output		1 to USB A port, 2 type c ports, 12v car port,2 to 120v Ac ports		2*AC,2*USB-A, 1*Car Port, 2*Type-C PD	pure sine wave AC outlet (300W), two 100W type C PD ports, and a 15W USB-A port, and a car port	USB A port, USB type c port, 120V AC 4 outputs

(*continued*)

Table 1. *(continued)*

Metric Power Output (W)	Helio Petal	Existing products					
		Renogy 200W	Yeti 1000 Core + Boulder 200	Sego Charger	Bluetti EB70 + PV200	Jackery SolarSaga 300 + 200W	EcoFlow River 2 Pro
Price (USD)	$150	$169	$949.91	$289	$852.45	$998	$1,088
Warranty (Years)	1 to 5	5	2		1	3	N/A
Mounting Options	Tripod stand	Clips, Magnets, Kickstand	Kickstand	Kickstand	Tripod	Kickstand	Kickstand
Solar Cell Type	Mono-crystal-line	Monocrystalline	Monocrystalline	Monocrystalline	Monocrystalline	Monocrystalline	Monocrystalline

5 Conclusion and Future Directions

This case study introduces the Helio petal, an origami-inspired portable solar panel, as a sustainable energy solution for outdoor applications. By integrating principles of origami into solar panel design, the Helio petal enhances portability, efficiency, and durability, making it an ideal renewable energy source for campers, hikers, and remote off-grid users. The system's innovative folding mechanism, coupled with high-efficiency monocrystalline photovoltaic cells and a modular output design, ensures optimal energy capture and storage, reducing reliance on non-renewable energy sources.

The findings emphasize the importance of manual adjustability over complex solar tracking mechanisms, enhancing cost-effectiveness and usability. Additionally, the panel's lightweight structure, compact storage, and weatherproof durability make it a practical alternative to traditional portable solar solutions. Comparative analysis with conventional foldable solar panels demonstrates the Helio petal's superior adaptability in dynamic environmental conditions.

This case study contributes to advancing sustainable energy systems by aligning with global efforts to promote green energy accessibility, affordability, and efficiency. The Helio Petal supports the transition to a cleaner, decentralized energy future by making solar technology more user-friendly and accessible, particularly in low-income regions where energy security is a challenge.

This case study, Helio petal aligns with several United Nations Sustainable Development Goals (SDGs) by promoting clean, efficient, and sustainable energy solutions. By enhancing access to clean energy accessibility (SDG 7), sustainability (SDG 12), climate action (SDG 13), innovation in solar technology (SDG 9), and resilient energy solution (SDG 11) demonstrating practical applications for sustainable infrastructure for both urban and rural settings. (Guidebook for the preparation of Science, Technology and Innovation (STI) for SDGs roadmaps, 2021).

Future work of Helio petal will explore AI-driven solar optimization, IoT-enabled energy management, and hybrid solar configurations to further enhance efficiency and adaptability. Additionally, material innovations focusing on biodegradable and recyclable components will be investigated to strengthen sustainability efforts. The Helio Petal's expansion into higher-capacity models and commercial applications will further solidify its impact on the global transition to renewable energy.

This case study highlights the potential of origami-inspired solar panels in shaping the next generation of sustainable energy solutions, fostering environmental responsibility, energy independence, and global sustainability.

References

Acakpovi, A., Yaw, N., Babbo, D.: Low cost two-axis automatic solar tracking system. Commun. Appl. Electron. **3**, 46–53 (2015). https://doi.org/10.5120/cae2015652015

Amelia, A.R., Irwan, Y.M., Safwati, I., Leow, W.Z., Mat, M.H., Rahim, M.S.A.: Technologies of solar tracking systems: a review. In: IOP Conference Series: Materials Science and Engineering. Institute of Physics Publishing, 2020. https://doi.org/10.1088/1757-899X/767/1/012052

Bartolomei, S., Grillone, G., Di Michele, R., Cortesi, M.: A comparison between male and female athletes in relative strength and power performances. J. Funct. Morphol. Kinesiol. **6** (2021).https://doi.org/10.3390/jfmk6010017

Desideri, U., Proietti, S., Zepparelli, F., Sdringola, P., Bini, S.: Life cycle assessment of a ground-mounted 1778 kWp photovoltaic plant and comparison with traditional energy production systems. Appl. Energy **97**, 930–943 (2012). https://doi.org/10.1016/j.apenergy.2012.01.055

Gao, L., Dougal, R.A., Liu, S., Jotova, A.: Portable Solar Systems using a Step-up Power Converter with a Fast-Speed MPPT and a Parallel-configured Solar Panel to Address Rapidly Changing Illumination. n.d.

Guidebook for the preparation of Science, Technology and Innovation (STI) for SDGs roadmaps, 2021. Publications Office of the European Union

He, J., He, Z., Hu, C.: Design of a foldable solar panel with radial and circumferential folding motion. In: Journal of Physics: Conference Series. Institute of Physics (2022). https://doi.org/10.1088/1742-6596/2364/1/012040

Jasim, B., Taheri, P.: An origami-based portable solar panel system. In: 2018 IEEE 9th Annual Information Technology, Electronics and Mobile Communication Conference, IEMCON 2018. Institute of Electrical and Electronics Engineers Inc., pp. 199–203 (2018). https://doi.org/10.1109/IEMCON.2018.8614997

Origami-Inspired Folding of Thick, Rigid Panels, n.d.

Shintemirov, A., Omarali, B., Muratov, F., Issa, M., Salakchinov, S., Alizadeh, T., Familiant, Y.L.: A sensorless MPPT-based solar tracking control approach for mobile autonomous systems. In: IECON Proceedings (Industrial Electronics Conference). IEEE Computer Society, pp. 3006–3011 (2016). https://doi.org/10.1109/IECON.2016.7793551

"Space Fabric" Links Fashion and Engineering _ NASA Jet Propulsion Laboratory (JPL), n.d.

AI-Driven Circular Business Models: A Comprehensive Review of AI's Role in Advancing Circular Economy Practices

W. A. D. C. S. Perera[1], K. Keerthana[1](✉), K. M. C. Tharupath[1], T. A. N. T. Perera[2], and G. Y. Jayasinghe[1]

[1] Department of Agricultural Engineering and Environmental Technology, Faculty of Agriculture, University of Ruhuna, Matara, Sri Lanka
keerthanakanapathipillai4@gmail.com

[2] Department of Export Agriculture, Uva Wellassa University, Badulla, Sri Lanka

Abstract. Circular economy (CE) catches worldwide attention, and business model innovation plays an important role in enabling organizations that focus on resource efficiency, material recycling, waste reduction, and material reuse, repair and recycle inside their organizations. This study explains the difference between an online store and a brick and mortar store. Circular Business Models (CBM) reimagined by Artificial Intelligence (AI) that helps organizations to implement sustainable practices with advanced data insights. This study reviewed 32 peer reviewed journal articles published between 2015 to 2025 from ScienceDirect, using keywords such as 'Artificial Intelligence', 'Machine Learning', 'Circular Economy',' Circularity',' Closed loops',' Business Model'. The selection criteria consisted of full text, English language, papers concerning the influence and impact of AI on Sustainability efforts, definitions of relevant terms, and papers from the perspective of firms and the discussion of AI capabilities. Dropping duplicates, conference proceedings which address digitalization without integrating AI/ML, or focus on algorithms without specifying impact on firms or CBM, are excluded. This analysis is concerned with the thematic scope of identifying factors, internal and external, that enable, and hinder AI enabled CBM synthesize AI capabilities, barriers to adopting AI enabled CBM programs, and studies, conceptual and empirical, which examine the contribution of AI to CBM implementation. CBM is found to be a key enabler of operations across different business formats. AI technologies like machine learning, forecasting, generative AI, digital twins, data driven design and decision support systems and data analytics are used in CBM for business analytics. There is some CBM that requires the descriptive, predictive, diagnostic, and perspective capabilities to prolong product life cycle, encourages steadier production, and decreases downtime. John Deere's integration of Blue River technology for sustainability shows how AI orchestration is a key enabler of CBM success. The main obstacles to this implementation include organizational resistance, the absence of collaboration, unclear regulations, as well as algorithmic bias.

Keywords: Artificial Intelligence · Business Model · Circular Economy · Circularity · Closed loops · Machine Learning

R. Dissanayake et al. (Eds.): ICRES 2025, PTT, pp. 264–281, 2025.
https://doi.org/10.1007/978-981-96-9551-5_21

1 Introduction

The current demand for enhanced AI abilities stems from building momentum toward a circular business model by involving academics and practitioners. AI-related capabilities need integration with new processes and routines together with circularity values to overcome main challenges while achieving essential objectives (Frishammar and Parida, 2019; Sjödin et al., 2018).

At the beginning phase of exploration scholars have identified only a small scope of work to determine AI-related functions for circular business models. This research functions as a foundational contribution required for expanding understanding within this domain.

Research positions the isolation discussion of AI and circular business models as innovations for sustainable business models according to Geissdoerfer et al. (2016) while also linking AI-powered digitization methods with Chauhan et al. (2022) and examining the digitization of circular business models through Caputo et al. (2021). Presently researchers have not conducted substantial investigations into how AI solutions address hidden obstacles in circular business model deployment. A targeted review process must take place to find AI adoption barriers in the circular business model area and establish essential AI capabilities that address these obstacles.

This research begins with a systematic literature collection regarding artificial intelligence and circular business models. The core capabilities needed for CBM functionality along with efficient adoption methods and rapid deployment timeline become apparent throughout the discussion. A previously established research framework helps to determine AI contributions to CBM while outlining necessary new firm-level capabilities that take maximum advantage of AI to reach economic objectives.

The paper examines AI and CBM connections alongside adoption challenges at various levels for CBM according to Åström et al., (2022) and Sandvik et al., (2024). Define capability development areas then outline fundamental practices for each development classification.

AI-enabled CBM needs individuals who understand core concepts because this understanding lays down its foundation. Standardized definitions need to be determined before proceeding with a discussion of interconnected concepts.

1.1 Artificial Intelligence

John McCarthy first created and introduced the phrase "artificial intelligence". He explained that artificial intelligence stands for the engineering science behind intelligent devices creation. Every existing definition of this topic focuses on cognitive ability in task execution yet lacks consistent global recognition (Collins et al., 2021). Artificial intelligence definitions obtained from information systems display greater diversity because methods such as machine learning and big data represent core components of AI systems (Shollo et al., 2022). Artificial Intelligence consists of different techniques in technology that create machines with human cognitive functions such as learning and reasoning which operate based on live and past data streams.

Current business analytics use lacks well-defined boundaries for implementing AI but this sector demonstrates that AI value generation requires large data quantities (Sjödin et al., 2021). Business analytics research structures data analytics within four key focus areas that encompass descriptive analysis as well as diagnostic examination and predictive forecasting plus prescriptive insights. Classification of these elements depends upon both their functional role and performance capabilities. This grouping approach breaks down into two primary classes of analysis. The designation narrow AI belongs to AI systems developed expressly for carrying out particular functionalities. The article by Mehmood et al., (2019) examines chatbots and automated sensor technologies within this classification category. General AI possesses the capability to mimic human thinking functions. All companies do not depend exclusively on narrow AI.

1.2 Circular Business Models

In recent times resource conservation has become a stronger priority for global communities. Research surrounding the circular business model (CBM) analyzes methods to improve resource use efficiency. Both past and current researchers have carried out scientific investigations which add to our knowledge of CBM according to Centobelli et al.(2020). A business collaboration model termed CBM by Frishammar and Parida (2019) combines different partners with novel innovations to make and hold value by increasing resource use efficiency while lifespan extension of products and parts produces environmental, social and economic benefits. This definition describes the core concepts of the circular business model (Geissdoerfer et al., 2020).

CBM innovations must be realized by developing these innovations internally or externally before bringing them inside through purchases or partnership arrangements (Parida et al., 2019a). External partnerships in creating CBM prove more efficient than building AI technology from scratch inside companies (Antikainen and Valkokari, 2016).

1.3 Role of AI in CBM

In earlier parts of this analysis, we explored artificial intelligence alongside the circular business model framework and described how artificial intelligence functions within this context. The implementation of industry capabilities in AI makes business processes more efficient according to Mehmood et al. (2019) and Mor and Gupta (2021). Sectors such as smart building electric grid management use these AI systems for efficiency alongside machine learning according to Mehmood et al., (2019). Recent research prioritized AI capabilities as a central area for business model innovation study because of their critical relevance (Mikalef et al., 2022a; Mikalef and Gupta, 2021; Sjödin et al., 2021).

A trio of research works provided detailed examinations of artificial intelligence capabilities (Mikalef and Gupta, 2021), in contrast to how capability understanding depends on two theoretical frameworks (Barney, 1986). Three studies adopt a static perspective on resources and capabilities which makes getting an accurate understanding of AI capabilities remain necessary (Mikalef et al., 2022a; Mikalef and Gupta, 2021).

According to a number of studies which have identified parallels between AI capabilities and general IT abilities researchers still understand AI capabilities under a unique categorical classification (Chen and Fong 2012). AI capabilities stand apart from general information technology since they harbor exclusive abilities which produce circular value according to Isensee et al., (2020). There exists potential to apply AI capabilities through sophisticated routine structures which successfully create added value by utilizing resources that support artificial intelligence (Sjödin et al., 2023, 2021, 2018).

2 Review Methodology

The concepts related to the use of AI in circular initiatives are explored via a systematic review methodology since the research area is constantly changing. The systematic search was performed using the research database ScienceDirect. This database is highly respected for comprehensiveness regarding sources on digital technologies in business, commonly applied in multiple reviews on digitalization topics. The keywords were identified from the existing reviews on AI and CBM, combining the words related to AI/ML, circular economy, and business model innovation using Boolean operators ("AND" for intersections, "OR" for broader matches). The search resulted in 44 articles. After removing conference proceedings and incomplete entries to ended up with a final amount of 24 papers. Then, titles and abstracts were screened in two successive steps. During this screening phase, the approach towards inclusion was quite broad, meaning that not even technically orientated papers were excluded. After the sample was restricted to studies within the business domain. Further, citation chaining was employed to identify additional sources, which resulted in the inclusion of eight more papers, bringing the final sample to 32 studies. Since the field is emerging, the review takes a prospective approach (Breslin & Gatrell, 2020). The aim is to underline not only the gaps but also how these could be used to trigger new theoretical insights and new points of view. Besides these contributions, the analysis was informed by research investigating the role of AI for other business processes, particularly for the innovation process CBM, as well as for studies reviewing the causes and outcomes of CBM practices. Thematic analysis of the identified studies was conducted based on four themes of interest, which were: internal and external factors that enable or hinder AI-enabled CBM, with the focus on capabilities of AI and mechanisms through which those capabilities contribute to CBM; significant barriers to the implementation of AI-powered CBM initiatives; conceptual or empirical models regarding how AI may support CBM; and the capabilities of AI to enable or facilitate CBM. Examples of companies undertaking AI-driven CBM initiatives were also included, as the number of academic studies on this topic is relatively limited, to supplement the review.

3 Results

3.1 Artificial Intelligence-Enabled Circular Business Models

Geissdoerfer et al., (2018)) identified three primary mechanisms for the innovation of circular business models: The key mechanisms for circular business model innovation are 1) narrowing resource usage while optimizing efficiency and 2) slowing down resource throughput by extending life cycles together with 3) closing resource loops to minimize waste (Geissdoerfer et al., 2020). Circular business models progress through various distinct steps and stages during their operation. Students should learn these phases before applying them inside artificial intelligence systems designed for such processes. The intersections among different steps and phases demonstrate the necessity to prioritize classification according to research conducted by Bocken et al., (2016). Strategic actions that focus on limiting resource flows create upstream impacts on resource deceleration methods that lead to closing the resource loop.

3.1.1 AI Enabled Narrowing Business Model

The approach known as Narrowing involves optimizing product life cycle resource utilization together with lowering waste production (Bocken et al., 2016). Artificial intelligence analytics enable resource conservation by identifying inefficiencies to develop strategies for effective resource application according to Mehmood et al., (2019).

Schneider Electric applies AI through its EcoStruxure platform to evaluate sensor and meter data as well as building systems to manage its global leadership in energy automation. The platform detects energy inefficiencies by applying AI algorithms which lead to optimal energy use and create savings costs and environmental benefits (Madanaguli et al., 2024).

When applying AI to narrow processes it leads to efficiency improvements throughout stages of value creation. Building expert decision assistance systems for organizations requires employing data analytics together with machine learning methodologies. The capacity of machine learning algorithms enables them to manage decisions with little human intervention according to Madanaguli et al., (2024). Through more recent developments in digitalization human supervisors maintained a greater degree of involvement compared to modern stages.

Business models become more efficient through supply chain waste reduction when they become more narrowly focused. The process achieves better resource management while also lessening waste accumulation and enabling circular product systems (Bocken et al., 2016). The predominant research focus in AI and CBMs concentrates on the narrowing phase because AI implementation remains at preliminary phases throughout multiple sectors. The implementation of AI technology enables companies to stand up circular practices through their capabilities in optimizing energy consumption (Mehmood et al., 2019), while identifying supply chain inefficiencies (Hina et al., 2022a), and speeding up supply chain processes (Butollo et al., 2022) along with improved material efficiency (Bag et al., 2021).

3.1.2 AI-Enabled Slowing Business Models

The slowing approach in CBM stands for efforts to enhance product component longevity or decrease initial resource use. Businesses can delay their resource loops by applying machine learning predictive and prescriptive capability according to Bocken et al., (2016)and Ritala et al., (2023).

AI helps create "slowing" circular business models through multiple important avenues. Through predictive maintenance features AI systems help extend the products lifecycle for businesses since real-time condition monitoring helps detect potential malfunctions before they occur. Through this forward-looking technique technicians maintain proper schedules for repairs and stop equipment from becoming unusable too quickly because of malfunctions (Gasser et al., 2021; Sjödin et al., 2023).

The second reason AI helps deliveries become more efficient is because it enables companies to produce exactly what consumers need at specific times. AI helps businesses improve customer targeting and production planning from real demand data through the analysis of consumer behaviour alongside preference and market trend evaluation (Amirkolaii et al., 2017). The system reduces production surpluses and organic waste since it improves resource allocation along functional lines (Toorajipour et al., 2021).

Through sustainable resource management AI enables Circular Business Models to analyze availability data and usage patterns alongside environmental impacts to identify optimization and circularity options (Talwar et al., 2021). Business analytics through AI enables organizations to take data-driven decisions while expanding sustainable practices throughout complete value chains.

Siemens Mindsphere represents AI-based slowing by utilizing IoT solutions together with AI to build dependable equipment and lessen downtime hours through predictive maintenance methods for industrial machines (Petrik and Herzwurm, 2019).

Through simulation of materials AI enables companies to choose substitutions that lead to products with a longer usable life (Deviatkin et al., 2022; Pyzer-Knapp et al., 2022). The design process will result in durable products which help conserve resources while decreasing resource extraction needs.

Models leader that AI integration offers immense capability for strategic "slowing" practices which serves product longevity extension and demand-product alignment while resource optimization eliminates business waste and advances circular economy targets alongside genuine sustainability aspirations.

3.1.3 AI-Enabled Closing Business Models

The business models related to closing life cycles within CBM focus on closing resource loops following their consumption. Product designers set the stage for chemical recovery systems by creating products which can be quickly disassembled during either disposal or additional processing stages. Reverse logistics and remanufacturing along with recycling approaches allow the initiation of product usage-based practices to take effect (Chauhan et al., 2022; Hina et al., 2022b). While our understanding of how AI supports these models continues to develop initially research work shows that AI's analytical and anticipatory abilities powerfully strengthen these systems (Chauhan et al., 2022).

Artificial Intelligence builds resource circle completion through optimized post-consumer recycling processes along with advanced product designs that favour reuse and improved reverse logistics and product take-back systems (Aziz and Adnan, 2021; Kerin and Pham, 2020; Schlüter et al., 2021). Refraining product movement employs AI-based analysis and forecast models which pinpoint optimal return paths thus enhancing both product refurbishing processes and recycling activities. By deploying AI to forecast product return patterns Dell optimizes reverse logistics outcomes which lead to decreased transportation emissions according to (Chauhan et al., 2022). AI also covers product design: Product designing frameworks permit easier disassembly and remanufacturing which enhances the usability of materials through recycling or reuse at the end of product lifespans. Siemens applies AI-driven simulations in the design of modular industrial equipment helping disassembly and recycling once its utility ends as shown in their study (Hina et al., 2022b).

AI will support various rental models alongside take-back programs because it matches customer preferences with available product data to decrease waste while prolonging product utility. AI facilitates Michelin's tires-as-a-service model by monitoring usage which supports maintenance prediction and enhances performance optimization for retreading and recycling requirements practices (Aziz and Adnan, 2021; Schlüter et al., 2021). According to studies AI emerges as important technology to promote sustainable business operations under circular economy models (Aziz and Adnan, 2021; Kerin and Pham, 2020; Schlüter et al., 2021).

3.2 Barriers to AI-Enabled CBM

A successful Circular Business Model (CBM) depends on working together with internal organizational efforts while aligning with external ecosystem participants (Ritala et al., 2023). Businesses encounter new operational challenges with AI implementation despite evident benefits for Circular Business Model adoption. This organizational AI adoption barrier functions across many dimensions while requiring external participants such as ecosystem partners and governmental bodies to take action beyond enterprise capacity (Sjödin et al., 2023, 2021, 2018). Successful management of these challenges needs a multi-level approach that takes into consideration multiple actors according to (Ritala et al., 2023). The literature highlights four key barriers that hinder AI-enabled CBM implementation: Businesses face workforce reluctance towards AI adoption within their teams at the organization level while also struggling with successful AI business operation integration at firm level and unsuccessful ecosystem collaboration alongside uncertain institutional settings on institutional level.

3.2.1 Organizational Resistance to AI Adoption

AI adoption encounters organizational resistance that emerges from every level throughout the company from both individual staff and management upwards through the organizational structure. AI sees employee resistance because such workers experience fear or mistrust toward this new technology. Held authority generates the unwillingness of many organizational leaders towards AI system decision control. During AI integration organizations face significant operational changes which trigger resistance across managerial and staff levels that make integration implementation more complex. The greater concern about job loss together with the unethical issues posed by AI create mental obstacles which prevent organizations from adopting AI (Mutascu, 2021; Nguyen and Vo, 2022). The reluctance of leadership to depend on AI-driven algorithms for operational control adds additional diffusion barriers according to Raisch and Krakowski, (2021).

3.2.2 Inadequate AI Integration into Business Operations

The integration of artificial intelligence systems into basic corporate business processes represents the second main obstacle for the organization. Organizational adoption of AI within businesses faces barriers because of insufficient skills as well as organizational differences paired with AI implementation remaining a relatively new process for CBMs. The integration between AI and CBM presents complexity which prevents decision-makers from predicting how the AI-induced transformations will cascade throughout business procedures (Wallis, 2021). Because machine learning models may be poorly constructed or fed biased data organizations face AI-specific challenges which produce biased algorithms (Mikalef et al., 2022b; Straw and Wu, 2022; Tilmes, 2022). Companies must break knowledge barriers and overcome immediate business plans to delve into extended period experimentation and development for successful artificial intelligence application (Ångström et al., 2023). Organizations face difficulties because both AI and CBMs bring risks like market cannibalization of current products and business structures according to research by (Linder and Williander, 2017). The automation-augmentation paradox presents a dilemma: Businesses face a strategic decision between total process automation and creating AI systems that assist human workers since selecting either approach requires specific trade-offs according to Raisch and Krakowski, (2021).

3.2.3 Deficient Ecosystem Collaboration

Collaboration among ecosystem elements represents a major obstacle because it fails to function effectively. Diverse stakeholders create resilient and successful CBM through partnerships when implementing AI solutions according to research by Burström et al., (2021) and Sjödin et al., (2018). Inter-organizational collaboration is commonly needed when AI adoption requires multiple resources including both quality data and advanced technologies. The process of these collaborations faces several challenges because organizations need to harmonize stakeholder interests while resolving intellectual property concerns alongside managing competition amongst participants. Cross-organizational coordination yields major challenges that demand innovative structural approaches towards ecosystem partnership and mutual operation.

3.2.4 Uncertain Institutional Environment

AI faces its fourth barrier in unclear institutional structures relating to legislative and regulatory aspects. Considerable attention has lately been directed at issues relating to the "dark side" of AI transformation: Algorithmic bias creates discriminatory outputs while threatening job stability elsewhere ((Mikalef et al., 2022b; Raisch and Krakowski, 2021). AI adoption growth reveals new social risks which hinder policymakers from developing effective AI regulatory measures. Every government address AI differently because certain administrations were unable to create proper regulations while others established unstructured policies that slowed AI adoption. AI shows potential for environmental sustainability in high-impact sectors like mining, but its implementation creates social and environmental issues which make sustainability trade-offs more intricate according to (Mikalef et al., 2022a). We need additional examinations of economic effects alongside environmental and social ramifications to discover how AI supports sustainable CBMs.

3.3 AI Capabilities for Circular Business Models (CBM)

The use of AI within Circular Business Models utilizes capability theory principles to eliminate obstacles associated with adopting AI technology. Our previous experiences prove that simple approaches to tangible assets and workforce and intangible resources do not meet requirements. The application of AI for circular business model development demands a large number of organizational functions to work collectively with diverse processes and capabilities. The orchestration process allows organizations to establish unique routines and systems which produce circular economic value drivers supported by research from (Mikalef and Gupta, 2021). Resource loop closure demands coordination between data management functions and leadership teams with human resources to implement data-driven approaches.

Business models that develop in complexity from narrowing to slowing and then closing loops exhibit increased interconnectedness between different necessary capabilities. Implementation of circular value creation with AI support needs a sophisticated ecosystem. The findings of Ritala et al., (2023) and Sjödin et al., (2018) show that papers in the field show high-level capabilities combining managerial and IT functions need to form building blocks within organizations for implementing AI technologies. The research provides evidence that these complex skills are fundamental for achieving complete AI potential applications within CBM. The capabilities are split into four detailed dimensions: AI infrastructure and platform capabilities; process automation and augmentation capabilities; integrated intelligence capabilities; and AI ecosystem orchestration capabilities. A detailed description appears underneath for both these dimensions along with their associated routines (Table 1).

3.4 AI Infrastructure and Platform Capabilities

AI infrastructure combined with platform capabilities enables organizations to access AI potential specifically for Circular Business Model strategies. AI requires massive resources for value creation because its functioning exceeds the basic connectivity and data storage components of digitalization. The AI requirements encompass physical assets such as cloud computing systems and sophisticated data management infrastructures, which serve as examples, (Lee et al., 2022). The essential physical resources support the artificial intelligence systems that develop and extract circular economy capabilities (Bag et al., 2021). Through AI-powered platforms organizations will establish their data management systems which enable system design capabilities to aggregate information from multiple origins forming comprehensive tools for organizational operations under AI management guidelines (Madanaguli et al., 2023a).

Organizations require three primary routines to develop these capabilities featuring AI data allocation and management followed by AI platform infrastructure development and data pipeline configuration with deployment to succeed. The initial operational core activity focuses on building systems capable of streamlined data management which supports efficient data access during sustainable manufacturing and resource optimization operations optimization (Baabdullah et al., 2021). Organizations need structured data because easy analysis of that data produces value in circular business operations through more created possibilities (Sjödin et al., 2023). The design of AI application platforms

Table 1. Summary of AI-Enabled Business Model Typologies, Key Technologies, and Barriers

Business Model Typology	Definition	Key AI Technologies	Barriers
AI-Enabled Narrowing CBM	AI to enhance business analytics-such as finding inefficiencies and making actionable suggestions for optimization. Ex - Siemens Mindsphere, C3.ai, Cat Asset Intelligence	Data Analytics, Machine Learning, Simulation, Decision Systems, Generative AI	Organizational resistance to AI (individual/team level), integration challenges, ecosystem collaboration issues, unclear revenue models
AI-Enabled Slowing CBM	Predictive maintenance with artificial intelligence in long-lasting active life due to data-enabled design. Ex - Siemens Healthineers, Bosch Maintenance 4.0, and ABB	Data Analytics, Machine Learning, Digital Twins, Predictive Maintenance	Organizational barriers, lack of training, customer resistance
AI-Enabled Closing CBM	AI-powered design of circular business models that optimize product dismantling and reuse. Example - Xerox, Dell, Renault Refactory	Data-Driven Design, Digital Twins, Generative AI	Integration and collaboration challenges, unclear regulatory frameworks

needs to facilitate efficient integration of diverse data originating from multiple dispersed sources (Kar and Kushwaha, 2023). The platform infrastructure development process investigates how systems can be constructed to integrate AI models whilst maintaining both adequate flexibility and robustness. Data pipeline configuration and deployment remain vitally important in business ecosystems demanding consistent data flow among diverse parties including suppliers and customers (Madanaguli et al., 2024). Stakeholders access shared data platforms which enable collaborative work to produce insights that will optimize how resources are used.

3.5 Process Automation and Augmentation Capabilities

AI creates value through two mechanisms: automation and augmentation. Business efficiency improves when organizations use capabilities to perform repetitive tasks automatically and boost human decisions with improved data accessibility (Raisch and

Krakowski, 2021). Successfully maintaining a balance between these two AI mechanisms presents a challenge as prioritizing one mechanism too heavily leads to negative long-term impacts which manifest as human worker disengagement from essential decision-making tasks (Raisch and Krakowski, 2021). To fully benefit from AI in CBM organizations must establish three key routines: intelligent task allocation automated decision support and continuous stakeholder integration.

The strategy of intelligent task allocation uses artificial intelligence to match people and equipment with tasks using real world conditions and their respective capacities. Mining operations demonstrate systems in which machines perform specified tasks alongside human operators who achieve optimal results to stop waste production at its origin. Mikalef and Gupta in their 2021 research alongside Sjödin and colleagues in 2018 demonstrated that proactive scheduling strategies enable loops to become narrower and their speeds slower which remains key in achieving closed-loop circular economy models. The second practice is automated decision support: Through analysis of extensive data sets artificial intelligence platforms will enhance decision-making functions which drive sustainable enterprise evolution. Real-time advanced data insights from artificial intelligence decision systems improve resource optimization and waste reduction according to Mehmood et al. (2019). Continuous stakeholder integration serves as a critical routine to harmonize AI applications and organizational trajectories while including each party of interest in decision processes. Van Eechoud and Ganzaroli, (2023)state that achieving effective integration requires uninterrupted communication and feedback exchange among stakeholders from different organizational levels who extend to internal teams and external partners and customers.

3.6 Integrated Intelligence Capability

Achieving AI capabilities requires leaders who shape a data-based operational environment which supports AI implementation. Within this context leadership abilities and experimental culture combine to form what experts describe as "integrated intelligence capability." Necessary routines that develop AI capabilities should exist alongside routines which support leadership needed to guide data-driven AI initiatives.

Data-driven organizational cultures enable perpetual support for using data toward enhancing decision-making capabilities as well as refining resource allocation continuously. Modification of Siemens' business processes toward data-driven practices via its AI Lab illustrates their introduction of such a cultural shift (Mikalef and Gupta, 2021), Employees need targeted technical training through AI competence development as a fundamental routine to enable them to work properly with AI systems. The development of AI capabilities among leaders remains essential because they drive AI adoption while implementing data-based conclusions into management decisions (Raisch and Krakowski, 2021). According to Giraud et al., (2023) leadership transformation and support for an innovative environment involving AI technological trials is fundamental for managing AI-based changes.

3.7 Capabilities for Orchestrating AI Ecosystems

Previous research identified how AI-powered CBMs generate new methods for value creation and capture while delivering outcomes that needed ecosystem partners' collaboration before becoming feasible (Toorajipour et al., 2021). A firm requires both detection capabilities and management knowledge to maintain successful long-term collaborations with ecosystem partners. Therefore, the orchestration of AI ecosystem capability emerges as essential. We identify three significant activities emerging from this capability, including: Our model suggests an AI ecosystem capability should engage in partner discovery while maintaining effective connections and scaling operations through customer responses and partner collaborations.

Successful ecosystem development starts with selecting appropriate partners and maintaining their discovery throughout the evolution of customer demands. Kolagar et al., (2022) describe AI partnership exploration as a process which determines optimal collaborators to work with in the ecosystem. Currently Nishant et al., (2020) emphasize that AI should drive sustainable goals through multiple actor involvement across various levels. Companies need to leverage ecosystem capabilities to discover potential partners when their AI capabilities experience organizational boundaries (Sjödin et al., 2021, 2018). According to studies various companies use three distinct strategies when developing CBMs by either building internal capabilities while simultaneously obtaining staff abilities from external sources along with setting up hybrid acquisition operations delineated by Geissdoerfer et al., (2018, 2016) and Kanda et al., 2021)'s research. The existence of these pathways enables specialized AI startups to receive acquisition offers from organizations or obtain financial investments needed for creating CBMs superior to existing large firm methods. The agricultural equipment startup Blue River Technology which used AI capabilities got acquired by John Deere in 2018. John Deere acquired the company because it developed the innovative "See & Spray" machine learning technology that enables targeted weed spraying to lower herbicide use and match the corporation's sustainable farming goals. Startups battle distinct challenges compared to large companies with both short decision periods and finite resources and they use distinct strategies during their engagement processes (García-Martín, 2023).

The management of ecosystem relationships through time forms the core element of system sustainability (Madanaguli et al., 2023b). To realise AI success throughout an ecosystem partners must undergo continued cultural advancement towards data-driven methodologies. The creation of this integrated system depends on both discussions and standardization among ecosystem members (Jovanovic et al., 2022; Madanaguli et al., 2023b; Parida et al., 2019a). Several actors must work collaboratively in closed-loop CBM due to the essential alignment of incentives required to dismantle products for remanufacturing according to (Parida et al., 2019b). The role of Business-to-Business customers proves essential when determining how Circular Business Models take shape. Through customer integration principles the transformation of customer roles into active participants who co-create value within circular business models clearly unfolds. From their research. Baabdullah and colleagues demonstrate this relationship change investigating how Saudi Arabian SMEs build business relationships depending on customer acceptance of AI-driven solutions. Burström et al., (2021). Confirm how difficult this relationship becomes through ecosystem orchestration developments. The growing need for sustainable services requires businesses to prioritize customer relationship cultivation as an essential part of AI enabled business ecosystems.

4 Discussion – A Proposed Agenda for Future Research

Thorough literary analysis of AI together with CBMs guides our exploration of AI functionalities to establish CBM systems. Combining AI with standing circular business models produces fresh internal benefits for companies while delivering external advantages to consumers. Our capability-based perspective shows that complex AI orchestration capabilities must support new processes and routine development within AI ecosystem contexts. The different organizational levels must execute multiple steps to work through these difficulties.

The insights discussed demand immediate synthesis into an overarching model. This structure offers researchers direction for the future while supplying an encompassing overview of capability management issues and innovations. Retrospectively we gain inspiration from the three-pillar framework which directs studies on business model evolution through technological advancements by applying earlier analyzed AI capabilities and CBM routines. The research framework established by Sjödin et al., (2023) along with Blackburn et al., (2023), encompasses people as the first dimension and process as the second dimension before technology as the third one. The framework defines "people" as individual and team challenges alongside activities, while "process" refers to the issues pertaining to organizational routines. The "technology" pillar addresses technological obstacles together with essential actions required to overcome them.

When assessing AI's supporting role in CBMs we suggest extending the current trio framework so that it can adequately reflect complex capabilities and challenges present. Two new elements must be incorporated to precisely depict barriers and routines which are fundamental for AI-driven CBM success. We therefore propose an expansion of the framework by four vital components: The framework requires four important elements, namely people together with process and platform along with partner which serve as its new foundation for 4P expansion.

Each of these four components comprises various capabilities integral to AI-driven transformation.

People - Competencies embedded in intelligence consist of personal and workgroup skills and expertise needed to deploy AI tools as part of CBMs. Both cognitive and technical abilities must be developed to enable people to use AI technology.

Process - This requires an organization to develop capacity for workflows enhancement and automation through artificial intelligence solutions. The document demonstrates process automation with AI alongside improved business decisions and operational integration of artificial intelligence.

Platform - The system will serve both as AI infrastructure and provide platform capabilities. These core components will specifically target the technical framework which supports AI operations. These creation platforms would process big amounts of data while running machine learning models and extending AI products to multiple business networks.

Partner - Ecosystem orchestration capability denotes the management of extended ecosystem partnerships that require partner selection and incentive alignment along with establishing cross-stakeholder integrated AI solutions involving customers and collaborators.

The synergy between these four advanced capabilities takes on vital importance for maximizing circular value applications with artificial intelligence. The successful realization of AI-enabled CBMs depends directly on achieving harmony among their fundamental capabilities.

The creation of a systematic research agenda which addresses the performance blockers alongside capabilities of AI-supported CBMs remains a necessary development. The 4P framework serves as a vital milestone towards progress because it structures future investigations while demonstrating how the transformative capabilities interconnect.

5 Conclusion

This study assessed the contribution of AI to Circular Business Models (CBM) by focusing on optimizing resource consumption. Extending resource lifespan. And achieving closed loop systems. AI will revolutionize CBM transformations to drive improved resource efficiency and waste reduction which will ultimately propel businesses toward sustainable development. Advanced AI technologies like predictive analytics, machine learning, digital twins, decision support systems, extends product service life, improve resource utilization, and encourage sustainability in business operations. Full-fledged optimisation processes ranging from improved energy consumption to maintenance planning together with production adjustments according to demand levels and longer-lasting products direct us towards implementing circular economy principles. AI implementation in actual use-cases encounters several hurdles which include technological acceptance barriers in addition to workflow integration complexities and multiple stakeholder coordination challenges. Organizations can achieve better integration into circular practices through following a 4P operational framework which consists of Human Resources, Procedures, Technology Platforms, and External Partnerships. Correctly training personnel while enhancing processes and developing robust AI systems alongside building collaborative partnerships enables AI to function effectively in circular practices. Established knowledge combines practical application requirements through multiple industries together with the demands to tackle both partnership coordination problems and regulatory unpredictability. Businesses that effectively combine their strategies with capabilities and structures will unlock AI's maximum potential to establish sustainable value circuits that benefit both economic interests and environmental objectives.

References

Amirkolaii, K.N., Baboli, A., Shahzad, M.K., Tonadre, R.: Demand forecasting for irregular demands in business aircraft spare parts supply chains by using artificial intelligence (AI). IFAC-Pap. **50**, 15221–15226 (2017). https://doi.org/10.1016/j.ifacol.2017.08.2371

Ångström, R.C., Björn, M., Dahlander, L., Mähring, M., Wallin, M.W.: Getting AI implementation right: insights from a global survey. Calif. Manag. Rev. **66**, 5–22 (2023). https://doi.org/10.1177/00081256231190430

Antikainen, M., Valkokari, K.:A framework for sustainable circular business model innovation. Technol. Innov. Manag. Rev. **6**, 5–12 (2016). https://doi.org/10.22215/timreview/1000

Åström, J., Reim, W., Parida, V.: Value creation and value capture for AI business model innovation: a three-phase process framework. Rev. Manag. Sci. **16**, 2111–2133 (2022a). https://doi.org/10.1007/s11846-022-00521-z

Abd Aziz, N., Adnan, N.A.A., Abd Wahab, D., Azman, A.H.: Component design optimisation based on artificial intelligence in support of additive manufacturing repair and restoration: current status and future outlook for remanufacturing. J. Clean. Prod. **296**, 126401 (2021)

Baabdullah, A.M., Alalwan, A.A., Slade, E.L., Raman, R., Khatatneh, K.F.: SMEs and artificial intelligence (AI): antecedents and consequences of AI-based B2B practices. Ind. Mark. Manag. **98**, 255–270 (2021). https://doi.org/10.1016/j.indmarman.2021.09.003

Bag, S., Pretorius, J.H.C., Gupta, S., Dwivedi, Y.K.: Role of institutional pressures and resources in the adoption of big data analytics powered artificial intelligence, sustainable manufacturing practices and circular economy capabilities. Technol. Forecast. Soc. Change **163**, 120420 (2021). https://doi.org/10.1016/j.techfore.2020.120420

Barney, J.B.: Strategic factor markets: expectations, luck, and business strategy. Manag. Sci. **32**, 1231–1241 (1986). https://doi.org/10.1287/mnsc.32.10.1231

Blackburn, O., Ritala, P., Keränen, J.: Digital platforms for the circular economy: exploring meta-organizational orchestration mechanisms. Organ. Environ. **36**, 253–281 (2023). https://doi.org/10.1177/10860266221130717

Bocken, N.M.P., De Pauw, I., Bakker, C., Van Der Grinten, B.: Product design and business model strategies for a circular economy. J. Ind. Prod. Eng. **33**, 308–320 (2016). https://doi.org/10.1080/21681015.2016.1172124

Burström, T., Parida, V., Lahti, T., Wincent, J.: AI-enabled business-model innovation and transformation in industrial ecosystems: a framework, model and outline for further research. J. Bus. Res. **127**, 85–95 (2021). https://doi.org/10.1016/j.jbusres.2021.01.016

Butollo, F., Gereffi, G., Yang, C., Krzywdzinski, M.: Digital transformation and value chains: Introduction. Glob. Netw. **22**, 585–594 (2022). https://doi.org/10.1111/glob.12388

Caputo, A., Pizzi, S., Pellegrini, M.M., Dabić, M.: Digitalization and business models: where are we going? A science map of the field. J. Bus. Res. **123**, 489–501 (2021). https://doi.org/10.1016/j.jbusres.2020.09.053

Centobelli, P., Cerchione, R., Chiaroni, D., Del Vecchio, P., Urbinati, A.: Designing business models in circular economy: a systematic literature review and research agenda. Bus. Strategy Environ. **29**, 1734–1749 (2020). https://doi.org/10.1002/bse.2466

Chauhan, C., Parida, V., Dhir, A.: Linking circular economy and digitalisation technologies: a systematic literature review of past achievements and future promises. Technol. Forecast. Soc. Change **177**, 121508 (2022). https://doi.org/10.1016/j.techfore.2022.121508

Chen, L., Fong, P.S.W.: Revealing performance heterogeneity through knowledge management maturity evaluation: a capability-based approach. Expert Syst. Appl. **39**, 13523–13539 (2012). https://doi.org/10.1016/j.eswa.2012.07.005

Collins, C., Dennehy, D., Conboy, K., Mikalef, P.: Artificial intelligence in information systems research: a systematic literature review and research agenda. Int. J. Inf. Manag. **60**, 102383 (2021). https://doi.org/10.1016/j.ijinfomgt.2021.102383

Deviatkin, I., Rousu, S., Ghoreishi, M., Naji Nassajfar, M., Horttanainen, M., Leminen, V.: Implementation of circular economy strategies within the electronics sector: insights from Finnish companies. Sustainability **14**, 3268 (2022). https://doi.org/10.3390/su14063268

Frishammar, J., Parida, V.: Circular business model transformation: a roadmap for incumbent firms. Calif. Manage. Rev. **61**, 5–29 (2019). https://doi.org/10.1177/0008125618811926

García-Martín, E.: Introduction: addressing the cognitive, affective and sociolinguistic challenges in the mixed language classroom. In: Bayona, P., García-Martín, E. (eds.), Second Language and Heritage Learners in Mixed Classrooms. Multilingual Matters, pp. 1–10 (2023). https://doi.org/10.21832/9781800415027-002

Gasser, J., Cai, H., Karagiannopoulos, S., Heer, P., Hug, G.: Predictive energy management of residential buildings while self-reporting flexibility envelope. Appl. Energy **288**, 116653 (2021). https://doi.org/10.1016/j.apenergy.2021.116653

Geissdoerfer, M., Bocken, N.M.P., Hultink, E.J.: Design thinking to enhance the sustainable business modelling process – a workshop based on a value mapping process. J. Clean. Prod. **135**, 1218–1232 (2016). https://doi.org/10.1016/j.jclepro.2016.07.020

Geissdoerfer, M., Morioka, S.N., De Carvalho, M.M., Evans, S.: Business models and supply chains for the circular economy. J. Clean. Prod. **190**, 712–721 (2018). https://doi.org/10.1016/j.jclepro.2018.04.159

Geissdoerfer, M., Pieroni, M.P.P., Pigosso, D.C.A., Soufani, K.: Circular business models: a review. J. Clean. Prod. **277**, 123741 (2020). https://doi.org/10.1016/j.jclepro.2020.123741

Giraud, L., Zaher, A., Hernandez, S., Akram, A.A.: The impacts of artificial intelligence on managerial skills. J. Decis. Syst. **32**, 566–599 (2023). https://doi.org/10.1080/12460125.2022.2069537

Hina, M., Chauhan, C., Kaur, P., Kraus, S., Dhir, A.: Drivers and barriers of circular economy business models: where we are now, and where we are heading. J. Clean. Prod. **333**, 130049 (2022a). https://doi.org/10.1016/j.jclepro.2021.130049

Isensee, C., Teuteberg, F., Griese, K.-M., Topi, C.: The relationship between organizational culture, sustainability, and digitalization in SMEs: a systematic review. J. Clean. Prod. **275**, 122944 (2020). https://doi.org/10.1016/j.jclepro.2020.122944

Jovanovic, M., Sjödin, D., Parida, V.: Co-evolution of platform architecture, platform services, and platform governance: expanding the platform value of industrial digital platforms. Technovation **118**, 102218 (2022). https://doi.org/10.1016/j.technovation.2020.102218

Kanda, W., Geissdoerfer, M., Hjelm, O.: From circular business models to circular business ecosystems. Bus. Strategy Environ. **30**, 2814–2829 (2021). https://doi.org/10.1002/bse.2895

Kar, A.K., Kushwaha, A.K.: Facilitators and barriers of artificial intelligence adoption in business – insights from opinions using big data analytics. Inf. Syst. Front. **25**, 1351–1374 (2023). https://doi.org/10.1007/s10796-021-10219-4

Kerin, M., Pham, D.T.: Smart remanufacturing: a review and research framework. J. Manuf. Technol. Manag. **31**, 1205–1235 (2020). https://doi.org/10.1108/JMTM-06-2019-0205

Kolagar, M., Parida, V., Sjödin, D.: Ecosystem transformation for digital servitization: a systematic review, integrative framework, and future research agenda. J. Bus. Res. **146**, 176–200 (2022). https://doi.org/10.1016/j.jbusres.2022.03.067

Lee, Y.S., Kim, T., Choi, S., Kim, W.: When does AI pay off? AI-adoption intensity, complementary investments, and R&D strategy. Technovation **118**, 102590 (2022). https://doi.org/10.1016/j.technovation.2022.102590

Linder, M., Williander, M.: Circular business model innovation: inherent uncertainties. Bus. Strategy Environ. **26**, 182–196 (2017). https://doi.org/10.1002/bse.1906

Madanaguli, A., Parida, V., Sjödin, D., Oghazi, P.: Literature review on industrial digital platforms: a business model perspective and suggestions for future research. Technol. Forecast. Soc. Change **194**, 122606 (2023a). https://doi.org/10.1016/j.techfore.2023.122606

Madanaguli, A., Sjödin, D., Parida, V., Mikalef, P.: Artificial intelligence capabilities for circular business models: research synthesis and future agenda. Technol. Forecast. Soc. Change **200**, 123189 (2024). https://doi.org/10.1016/j.techfore.2023.123189

Mehmood, M.U., Chun, D., Zeeshan, Han, H., Jeon, G., Chen, K.: A review of the applications of artificial intelligence and big data to buildings for energy-efficiency and a comfortable indoor living environment. Energy Build. **202**, 109383 (2019).https://doi.org/10.1016/j.enbuild.2019.109383

Mikalef, P., Conboy, K., Lundström, J.E., Popovič, A.: Thinking responsibly about responsible AI and 'the dark side' of AI. Eur. J. Inf. Syst. **31**, 257–268 (2022). https://doi.org/10.1080/0960085X.2022.2026621

Mikalef, P., Gupta, M.: Artificial intelligence capability: conceptualization, measurement calibration, and empirical study on its impact on organizational creativity and firm performance. Inf. Manag. **58**, 103434 (2021). https://doi.org/10.1016/j.im.2021.103434

Mikalef, P., et al.: Enabling AI capabilities in government agencies: a study of determinants for European municipalities. Gov. Inf. Q. **39**, 101596 (2022). https://doi.org/10.1016/j.giq.2021.101596

Mor, S., Gupta, G.: Artificial intelligence and technical efficiency: the case of Indian commercial banks. Strateg. Change **30**, 235–245 (2021). https://doi.org/10.1002/jsc.2406

Mutascu, M.: Artificial intelligence and unemployment: new insights. Econ. Anal. Policy **69**, 653–667 (2021). https://doi.org/10.1016/j.eap.2021.01.012

Nguyen, Q.P., Vo, D.H.: Artificial intelligence and unemployment: an international evidence. Struct. Change Econ. Dyn. **63**, 40–55 (2022). https://doi.org/10.1016/j.strueco.2022.09.003

Nishant, R., Kennedy, M., Corbett, J.: Artificial intelligence for sustainability: challenges, opportunities, and a research agenda. Int. J. Inf. Manag. **53**, 102104 (2020). https://doi.org/10.1016/j.ijinfomgt.2020.102104

Parida, V., Sjödin, D., Reim, W.: Reviewing literature on digitalization, business model innovation, and sustainable industry: past achievements and future promises. Sustainability **11**, 391 (2019). https://doi.org/10.3390/su11020391

Petrik, D., Herzwurm, G.: iIoT ecosystem development through boundary resources: a Siemens MindSphere case study. In: Proceedings of the 2nd ACM SIGSOFT International Workshop on Software-Intensive Business: Start-Ups, Platforms, and Ecosystems. Presented at the ESEC/FSE '19: 27th ACM Joint European Software Engineering Conference and Symposium on the Foundations of Software Engineering, ACM, Tallinn Estonia, pp. 1–6 (2019). https://doi.org/10.1145/3340481.3342730

Pyzer-Knapp, E.O., et al.: Accelerating materials discovery using artificial intelligence, high performance computing and robotics. NPJ Comput. Mater. **8**, 84 (2022). https://doi.org/10.1038/s41524-022-00765-z

Raisch, S., Krakowski, S.: Artificial intelligence and management: the automation–augmentation paradox. Acad. Manag. Rev. **46**, 192–210 (2021). https://doi.org/10.5465/amr.2018.0072

Ritala, P., Bocken, N.M.P., Konietzko, J.: 11 Three lenses on circular business model innovation, in: Alexander, A., Pascucci, S., Charnley, F. (eds.), Handbook of the Circular Economy. De Gruyter, pp. 175–190 (2023). https://doi.org/10.1515/9783110723373-014

Sandvik, L.V., Smith, K., Strømme, A., Svendsen, B., Aasmundstad Sommervold, O., Aarønes Angvik, S.: Students' perceptions of assessment practices in upper secondary school during COVID-19. Teach. Teach. **30**, 918–931 (2024). https://doi.org/10.1080/13540602.2021.1982692

Schlüter, M., et al.: AI-enhanced identification, inspection and sorting for reverse logistics in remanufacturing. Procedia CIRP **98**, 300–305 (2021). https://doi.org/10.1016/j.procir.2021.01.107

Shollo, A., Hopf, K., Thiess, T., Müller, O.: Shifting ML value creation mechanisms: a process model of ML value creation. J. Strateg. Inf. Syst. **31**, 101734 (2022). https://doi.org/10.1016/j.jsis.2022.101734

Sjödin, D., Parida, V., Kohtamäki, M.: Artificial intelligence enabling circular business model innovation in digital servitization: conceptualizing dynamic capabilities, AI capacities, business models and effects. Technol. Forecast. Soc. Change **197**, 122903 (2023). https://doi.org/10.1016/j.techfore.2023.122903

Sjödin, D., Parida, V., Palmié, M., Wincent, J.: How AI capabilities enable business model innovation: scaling AI through co-evolutionary processes and feedback loops. J. Bus. Res. **134**, 574–587 (2021). https://doi.org/10.1016/j.jbusres.2021.05.009

Sjödin, D.R., Parida, V., Leksell, M., Petrovic, A.: Smart Factory implementation and process innovation: a preliminary maturity model for leveraging digitalization in manufacturing moving to smart factories presents specific challenges that can be addressed through a structured approach focused on people, processes, and technologies. Res.-Technol. Manag. **61**, 22–31 (2018). https://doi.org/10.1080/08956308.2018.1471277

Straw, I., Wu, H.: Investigating for bias in healthcare algorithms: a sex-stratified analysis of supervised machine learning models in liver disease prediction. BMJ Health Care Inform. **29**, e100457 (2022). https://doi.org/10.1136/bmjhci-2021-100457

Talwar, S., Kaur, P., Fosso Wamba, S., Dhir, A.: Big Data in operations and supply chain management: a systematic literature review and future research agenda. Int. J. Prod. Res. **59**, 3509–3534 (2021). https://doi.org/10.1080/00207543.2020.1868599

Tilmes, N.: Disability, fairness, and algorithmic bias in AI recruitment. Ethics Inf. Technol. **24**, 21 (2022). https://doi.org/10.1007/s10676-022-09633-2

Toorajipour, R., Sohrabpour, V., Nazarpour, A., Oghazi, P., Fischl, M.: Artificial intelligence in supply chain management: a systematic literature review. J. Bus. Res. **122**, 502–517 (2021). https://doi.org/10.1016/j.jbusres.2020.09.009

Van Eechoud, T., Ganzaroli, A.: Exploring the role of dynamic capabilities in digital circular business model innovation: results from a grounded systematic inductive analysis of 7 case studies. J. Clean. Prod. **401**, 136665 (2023). https://doi.org/10.1016/j.jclepro.2023.136665

Wallis, M.G.: Artificial intelligence for the real world of breast screening. Eur. J. Radiol. **144**, 109661 (2021). https://doi.org/10.1016/j.ejrad.2021.109661

Case Study of the X-Press Pearl Ship Accident; Policy and Recommendation on Waste Management

S. J. S. Bandara[1](✉), J. M. D. N. P. Jayamanna[2], and E. R. Gunawardhana[3]

[1] Central Environmental Authority, 104, Denzil Kobbakaduwa Mawatha, Battaramulla, Sri Lanka
shakeela@cea.lk
[2] National Korea Maritime and Ocean University, Busan, South Korea
[3] Faculty of Technology, University of Sri Jayawardhanapura, Pitipana, Homagama, Sri Lanka

Abstract. On May 25, 2021, the cargo ship X-Press Pearl caught fire in the Sri Lankan Sea. This case study was based on the legal framework of waste management in Sri Lanka. Due to the onboard explosion, cargo containers were damaged and dropped down to the sea while the fire. Both liquid and solid cargo entities were dispersed across the sea. Water-soluble materials were dissolved and were unable to be identified visually. Due to the fire incident, waste items were being washed up to several locations of the coastal line of the country causing a huge environmental and health threat. Four major legal suits were brought onto the common platform to manage hazardous, non-hazardous, and chemical wastes in an environmentally sound manner. The major role of the Marine Pollution Prevention Act, No.35 of 2008, Coast Conservation Act, No. 57 of 1981 and its further amendments, Sri Lanka Custom Ordinance No.17 of 1869 and its further amendments and National Environmental Act, No. 47 of 1980, and Its Further Amendments are discussed in this case study. Also, that brings verification of waste identification and conclusion of management and final disposal method of different types of wastes. MV X-Press Pearl Ship sunk at the main ship root of the busiest commercial harbor of the country. Therefore, it is important that the government of Sri Lanka build a comprehensive management system for ensuring safety from sunken ships and for preserving a clean marine environment as well as develop related technology and facilities. Moreover, this tries to look back at the incident in a scientific manner after three years by giving more consciousness to the missed opportunities that could have been handled more environmentally favorable way. Accordingly, all hazardous waste is recommended to be re-export or resend through the Basel Convention to a country with sufficient facilities to dispose and other non-hazardous waste is been disposed locally. Non-contaminated recyclable materials have to be recycled through local facilities under supervision by the Central Environmental Authority.

Keywords: Hazardous waste · Non-hazardous waste · Chemical wastes · Dispose · Basel Convention

R. Dissanayake et al. (Eds.): ICRES 2025, PTT, pp. 282–289, 2025.
https://doi.org/10.1007/978-981-96-9551-5_22

1 Introduction

The X-Press Pearl was a cargo ship that caught fire in the Sri Lankan Sea about 9.5 nautical miles away from the Port of Colombo on 25th May 2021. The vessel was carrying hazardous and non-hazardous materials, including chemicals, electronic & electrical equipment, and several types of polymers. The vessel was exposed to the on-board fire while anchored and waiting to get approvals to enter the Port of Colombo, Sri Lanka.. Accordingly, there were 1486 containers on the ship when this incident occurred. As per the cargo manifesto, other than the chemicals and hazardous substances some of the non-dangerous, general cargo items such as foods, industrial raw materials, tires, furniture, etc. were comprised in this vessel.

After the on-board explosion, cargo containers started to drop down to the sea and released different types of cargo including liquids and solids. However, liquid cargo especially water-soluble liquids and solid chemicals dissolved in seawater, and it was not visible. That became an unseen tragedy to the sea animals. Hazardous and Noxious Substances (HNS) and defined as any substance other than oil, which if introduced into the marine environment is likely to create hazards to human health, harm living resources and other marine life, damage amenities, and/or interfere with other legitimate uses of the sea (IMO, 2000). A huge amount of plastic pellets carrying containers were released to the sea and it was released from containers reaching the coastal areas of Sri Lanka polluting more than 758 km from Kirinda to Mannar (Fig. 1). The plastic pellets were contaminated with different types of chemicals, partially burnt, and different-sized plastic pellets which were aggregated due to the thermal effect that polluted coastal areas of Sri Lanka causing severe damage to the environment, socio-economic activities, and human health (Figs. 1, 2).

The X-Press Pearl ship accident created a significant environmental disaster in the marine environment as same as the terrestrial. That brought into focus the challenges of how maritime accidents affect the coastal environment and the livelihood of the fishermen's community other than marine pollution. Also, that challenges the efficacy of existing state policies while handling such crises. This paper analyses the strengths and weaknesses of the existing state policies of Sri Lanka for waste management including shipwrecks.

Fig. 1. Plastic pellets dispersal of Moratuwa beach

Fig. 2. Plastic pellets dispersal within the groynes at Bambalapitiya Beach

2 State Policies and Regulations on Chemical and Waste Management in Sri Lanka Affected by the Incident

Sri Lanka has established a comprehensive legal framework to manage waste effectively that aims to protect public health and the environment. This type of catastrophic maritime incident occurred for the first time in recent history and that brought all concerns to one platform to discuss waste management, and pollution control as same as the environmental evaluation. Followings are identified as major legal tools of waste management;

1. Marine Pollution Prevention Act, No.35 of 2008
2. National Environmental Act, No. 47 of 1980 and Its Further Amendments
3. Coast Conservation Act, No. 57 of 1981 and its further amendments and Coast Conservation and Coastal Resource Management Act
4. Sri Lanka Custom Ordinance

This study discusses the strengths, weaknesses, opportunities, and threats of the existing legal framework in this regard.

2.1 Marine Pollution Prevention Act, No.35 of 2008

Marine Environment Protection Authority (MEPA) was established by the act (Section 2) and it defines the duties and powers of the Authority as effectively safeguarding and preserving the territorial waters of Sri Lanka or any other maritime zone, its fore-shore and the coastal zone of Sri Lanka from pollution, conduct investigations and inquiries, and to institute legal actions and oversee all sea transport of oil and bunkering operations (Section 7). The Authority shall provide reception facilities within or outside any port to discharge or deposit any residue of oil or other pollutants and direct the person in charge of all ports to provide adequate reception facilities. According to Section 21 of the Marine Pollution Prevention Act, it should direct the person in charge of all ports to prepare a waste management plan and to carry out at prescribed intervals an EIA by a Classification Society approved by the MEPA. Also, if any oil, harmful substance, or other pollutant is discharged or escapes into the territorial waters of Sri Lanka or any other maritime zone, its fore-shore and the coast of Sri Lanka, the owner of the ship, owner of the off-shore installation or owner of the pipeline shall be guilty of an offense and shall be liable to a fine (4M – 15 M) (S.26). Moreover, if any person dumps any oil, harmful substances or any other pollutant shall be guilty of an offense and shall be liable to a fine. (4M – 15 M) (S.27).

2.2 Coast Conservation Act, No. 57 of 1981 and its Further Amend-Ments and Coast Conservation and Coastal Resource Management Act

According to the long title of this act, "The administration, control, custody, and management of the Coastal Zone are hereby vested in the Republic." (S.2). "Coastal Zone" means that area lying within a limit of three hundred meters landwards of the Mean High water line and a limit of two Kilometers seawards of the Mean Low Water line and in the case of rivers, streams, lagoons, or any other body of water connected to the sea either permanently or periodically, the landward boundary shall extend to a limit of two Kilometers measured perpendicular to the straight baseline drawn between the natural entrance points thereof and shall include the waters of such rivers, streams and lagoons or any other body of water so connected to the sea." (S.42).

This act gives the nation authority over the administration, control, custody, and management of the coastal zone, which are hereby vested in the Republic (S.2) and upon receipt of an application for a permit, the Director may require the applicant to furnish an IEER or Environmental Impact Assessment Report or both such reports. [S.16(1)]. Waste / Damage or Detriment is addressed in Section 25 by saying that the Director is to give directions for the prevention or intrusion of waste or foreign matter into the Coastal Zone. (S.25). The Director General shall have the power to issue directions to any person engaged in, or likely to engage in, any development activity that is causing or is likely to cause damage or detriment to the Coastal Zone or the resources therein, regarding the measures to be taken to prevent or abate such damage or detriment, and it shall be the duty of such person to comply with such directions. [S.26A (1)].

2.3 Sri Lanka Custom Ordinance No.17 of 1869 and its Further Amendments

Sri Lanka Customs was established in 1806 and Ordinance was gazetted as No.17 of 1869. Custom Ordinance is consecutively amended till 2003 and fourteen major parts are established with new regulations and directorates by strengthening the ordinance. Goods not reported or entered, are forfeited. Penalty. Except coin, bullion, cattle, passengers. Goods not reported or entered, are forfeited. Penalty. The Prevention Directorate has strengthened the gap in the Customs Ordinance, which was created during the colonial era and did not address the accidents that we today face. The scope of the Preventive Directorate focuses on the enforcement aspect by way of prevention of smuggling, commercial frauds, and other offenses while safeguarding the socio-economic, cultural, and environmental interests of the country by being operative on broader protection routines at all the sea and airports (Sri Lanka Customs). The Central Disposal Unit is entrusted with disposing of seized and forfeited goods of various Directorates and

Units of the whole Department. The provisions of the Customs Ordinance authorize the disposal of forfeited goods by public auction, and the Tender Sales Procedure established under the Financial Regulations is followed at such auctions.

According to this, the waste from the MV X-Press Pearl ship accident has already been transported through the Custom Preventive Directorate and is securely kept at a private party's storage facility; however, the ownership of this waste has not been verified, which leads to additional issues with its disposal. Therefore, the waste generated from the MV X-Press Pearl ship incident is still stored in a waste yard located in Wattala, Sri Lanka, causing huge costs and risks to the country.

2.4 National Environmental Act, No. 47 of 1980 and its Further Amendments

According to the National Policy of the country, all waste which was generated from this ship incident should be disposed of adhering to those National Policies. Even though the state authorities have legal powers to intervene the waste management in this regard, it could be considered as a default of the authorities. The Solid Waste Management Policy Technical Guideline 2004, issued by the Central Environmental Authority, establishes the state's framework for waste management to promote environmental responsibility and social accountability among all parties involved. It encourages participation from individuals and institutions in a comprehensive approach to waste management, aims to enhance resource recovery to reduce waste disposal volumes, and seeks to mitigate negative impacts on health and the environment.

The National Environmental Act functions as an umbrella type that covers the environment. The MV X-Press Pearl ship incident was addressed with three revisions to Act No. 56 of 1988, and No. 53 of 2000, and several rules, including Solid Waste Management, Hazardous Waste and Chemical Management, and Environmental Pollution Prevention. Waste disposal activities are prescribed in regulations published under the Gazette Notification No. 1534/18 dated 01.02.2008 (Re: License for discharge, emission, or disposal of waste/scheduled waste management). According to the regulation, no person shall generate collect, transport, store, recover, recycle, or dispose of waste or establish any site or facility for the disposal of any waste specified in Schedule VIII (hereinafter referred to as "scheduled waste") except under the authority of a license issued by the Authority and by such standards and other criteria as may be specified by the Authority. All the waste that shows the characteristics of hazardous and noxious are prescribed under the SCHEDULE VIII on the same regulation. Therefore, the waste generated from the MV Xpress Pearl ship incident should be categorized under this state regulation and most of the had been gone under scheduled waste.

Also, Sections 129, 130, and 131 of Municipal Council Ordinance No. 16 of 1947, Sections 118, 119, and 120 of Urban Council Ordinance No. 61 of 1939, and Sections 93, 94, and 95 of the Pradesheya Saba Act No. 15 of 1987 all contain separate regulations for non-hazardous waste or general waste management in Sri Lanka. Waste includes any matter prescribed to be waste and any matter, whether liquid, solid, gaseous, or radioactive, which is discharged, emitted, or deposited in the environment in such volume, constituency or manner as to cause an alteration of the environment (NEA 1980, Section 33).

2.5 Identification of Waste Types Generated by MV X-Press Pearl Ship Accident

Generated waste has been categorized into three main groups for easy management (UNEP Report 2021). Also, Hazardous wastes can be divided into three broad categories: radioactive, infective, and chemical. Management of chemical wastes is a complex combination of treatment and storage (Carden, 1985).

1. Hazardous and Noxious waste
2. Chemicals / metals and scraps
3. Non – hazardous waste

Waste that has not been contaminated is categorized under non-hazardous waste and chemically or physically contaminated materials or cargo types are being identified as hazardous wastes (HWM regulation 2008, CEA). Chemicals that had not being mixed or unreleased from the undamaged containers had to be categorized under hazardous waste and all other contaminated chemicals, and materials which are inside the damaged containers should be categorized under hazardous wastes.

2.6 Policy Recommendation for Waste Management

MV X-Press Pearl Ship sunk nearly with 1400 cargo containers and most of them carried chemicals and industrial raw materials. That includes 7.985.1 - 8000 MT of chemicals, 3000 MT of several types of metals, 12000 MT of Polymers, 2700 MT of food items, and 5300 MT of other materials including paper items, tires, furniture, pharmaceutical packaging, etc. Also, 26.7 electronic items were included inside the containers (Cargo Manifesto). Figure 3 shows the most harmful and hazardous half-burnt plastic pallets that were identified as toxic; some of them could have been evaporated and deformed.

According to the risk assessment given by ITOPF, there were 15 dangerous cargo types carried by this MV X Press Pearl ship. Those 15 types of dangerous cargo are contained in nearly 85 cargo containers. According to the given bay plan of the ship, nearly 65 of them were located in hatch/cargo holes under the deck level. Hence, they still might be stuck in the shipwreck and they may still not have been damaged. Cargo containers that were placed on or above the deck level, might be dropped down to the sea and only eight wracked containers were washed off to the shore. Others may be scattered on the sea bed (UNEP 2021).

Following the laboratory analysis confirming that the wastes are free from any hazardous substances, those identified are classified as Non-Hazardous waste according to the CEA regulations on Hazardous Waste Management. That identified non-hazardous recyclable waste should be handed over to the Environmental Protection License holding recycling facilities and residual waste and debris are recommended to be disposed through the sanitary landfill in Dompe or non-hazardous burnable waste can be transported to Western Power Company (Pvt) Ltd; waste to energy plant operating at Kerawalapitiya. During the handling of waste, all the waste materials were transported by using suitable enclosed vehicles to the storage or handling site. At the storage, collected, and transported waste was stored in a suitable facility having adequate space and storing of waste according to waste types as far as possible (Fig. 4). Furthermore, handling of collected waste had to be carried out only by skilled officers/workers wearing personal protective equipment (PPE).

Fig. 3. Burnt Plastic collected from Wellawatta Beach

Fig. 4. Plastic Pellets collected during the early stage of incident using manual collection method

3 Conclusion

Accordingly, this catastrophic pollution in marine as same as the coast in Sri Lanka created a shaky environment in legal, governmental, and academic sectors and it was extremely difficult to practice those lawsuits that were limited to inscribed documents. Implementing those laws and regulations cohesively is much more challenging, and for a developing nation like Sri Lanka, it is more challenging to use those rules to secure fair compensation for this kind of disaster. However, the incident ended regrettably due to political meddling, bribery, or underhanded transactions and poor precedents from the past.

The MV X-Press Pearl vessel has sunk at the primary shipping route of the nation's busiest commercial port. Consequently, it is essential to establish an all-encompassing management system to safeguard against sunken ships and maintain a clean marine ecosystem, in addition to advancing related technologies and facilities. Considering shoreline wastes, underwater container materials and their insides as well as the ship-wreck could be categorized based on their chemical properties and hazardousness (SWM

Guideline 2008). Basel Convention is the international agreement on the control of trans-boundary movements of hazardous wastes and their disposal. Sri Lanka ratified the Basel Convention in 1992.

Therefore, the handling of generated wastes from this ship incident should be monitored by CEA and hazardous wastes that cannot be disposed of within the country should be exported to a foreign facility following the procedures of the Basal Convention. All the costs incurred in this regard should be borne by the Ship owner. Those wastes are extra-terrestrial and the country does not have sufficient facilities to manage them. Therefore, it was highly recommended to re-export or resend the waste through the Basel Convention, to a country with proper existing facilities hence, all the costs of this activity should be borne by the ship owner.

It is recommended to remove all the containers placed in cargo holes in the sunken ship before manage the shipwreck. It is highly recommended, that shipwreck or debris not to bring to the Sri Lankan lands due to the lack of recovery measures within the Country. If this waste is somehow brought to Sri Lanka, should be obtained the necessary prior approvals from CEA. Comprehensive proposal for each activity including risk assessment should be submitted to the CEA to obtain prior approvals.

Acknowledgements. I extend my inmost gratitude to Mr. Hemantha Jayasinghe, Director General, Central Environmental Authority, Sri Lanka to assign me as a co-member of the National Expert Committee of MV X-Press Pearl ship accident in 2021. Also, I am equally thankful to the Panel Members of the same committee, Professor Meththika Withanage and Professor Ajith De Alwis for their constructive feedback and essential suggestions that improved the quality of this case.

References

Basel Convention on the Control of Transboundary Movements of Hazardous Wastes and their Disposal, Basel, 22 March 1989
Sri Lanka Customs Ordinance Nos,17 of 1869 and its further amendments
Carden, J.L., Jr.: Hazardous Waste Management. Springer-Verlag, New York Inc., United States (1985)
Coast Conservation Act, No. 57 of 1981
Coast Conservation (Amendment) Act, No. 64 of 1988
Coast Conservation (Amendment) Act, No. 49 of 2011
Coast Conservation and Coastal Resource Management Act
Guideline for the Management of Scheduled Waste in Sri Lanka" in accordance to the National Environmental (Protection & Quality) Regulation, Extraordinary Gazette No.01 of 2008
Hazardous Waste Chemistry, Toxicology, and Treatment By Stanley E. Manahan
Hazardous Waste Management Guideline, 2009, Central Environmental Authority
Choi, H.J., Kim, H., Lee, S.H., Kang, C.G., Lew, J.M.: Development of risk-based information systems for management of sunken ships. In: Paper presented at the The Fifteenth International Offshore and Polar Engineering Conference, Seoul, Korea (2005)
IMO – the International Maritime Organization – is the United Nations specialized agency with responsibility for the safety and security of shipping and the revention of marine pollution by ships. June 2000
X-Press Pearl Maritime Disaster, Sri Lanka Report Of The Un Environmental Advisory Mission, Unep, July 2021

Optimizing Sustainability Through Life Cycle Assessment and Circular Economy in Industrial Manufacturing

Developing a Packaging Material Using Rice Straw

R. W. U. D. N. De Silva(✉) and M. Danthurebandara

Department of Chemical and Process Engineering, Faculty of Engineering, University of Peradeniya, Peradeniya, Sri Lanka
e17055@eng.pdn.ac.lk

Abstract. The utilization of agricultural waste as a raw material for paper production provides an alternative for agricultural residue management. This study investigates the viability of using rice straw, a typical agricultural waste in Sri Lanka, as a raw material for paper production. Initially, eight distinct experimental methods were employed, maintaining a consistent liquid-to-solid ratio of 7, while varying key parameters such as temperature (ranging from 150 to 180 degrees Celsius), cooking time (spanning 60 to 90 min), and sodium hydroxide (NaOH) concentration (from 1M to 0.5M). The optimal papermaking process was identified through the application of Design of Experiments (DOE) and Principal Component Analysis (PCA), where cooking time, cooking temperature, and sodium hydroxide concentration were treated as independent variables (factors), while tensile breaking strength and grammage served as the dependent variables (responses). This process aimed to achieve tensile breaking strength and grammage metrics that align with the established standards of commercially available paperboards. The optimal process was characterized by a 1M NaOH concentration, a 60-min cooking time, and a 180 °C cooking temperature. After identifying the ideal process, a rigorous Life Cycle Assessment (LCA) was conducted to comprehensively evaluate its environmental impact compared to wood-based paper production, taking the functional unit as 1 m2 of paper and using a 'Cradle-to-gate' system boundary. Compared to wood-based paper production, rice straw pulp making has less environmental impact, mainly due to reduced chemical consumption.

Keywords: Rice straw · Paper making · Design of Experiment (DOE) · Sustainability · Life Cycle Assessment (LCA) · Agricultural waste management

1 Introduction

The global paper industry significantly contributes to environmental challenges, accounting for 40% of industrial wastewater discharge and 3% of end-use energy consumption (Ma et al. 2019). Paper production from wood exacerbates deforestation, habitat loss, and greenhouse gas emissions (Lorn et al. 2022). The rapid expansion of the paper sector intensifies concerns regarding forest resource depletion. In response, non-wood and recovered fiber pulps have emerged as viable alternatives (Ma et al. 2019). Among these, straw pulping is promising, reducing forest resource scarcity and mitigating crop residue burning (Ma et al. 2019).

R. Dissanayake et al. (Eds.): ICRES 2025, PTT, pp. 293–307, 2025.
https://doi.org/10.1007/978-981-96-9551-5_23

Rice straw is an abundant agricultural waste product, with an annual global production of 800–1000 million tons, the majority (600 to 800 million tons) originating from Asia (International Rice Research Institute). The physicochemical properties of rice straw are as follows as explained by Rodríguez et al. 2010; Kaur et al. 2017; Worku et al. 2023. Rice straw fibers average 1.41 mm in length and 8 μm in diameter, yielding a high length-to-diameter ratio (L/D) of 175:1. The α-cellulose content ranges from 28%-36%, lignin from 12%-22%, and pentosan from 23%-28%. Rice straw exhibits 7.3% hot water solubility, 0.6% alcohol-benzene solubility, and 57.7% solubility in 1% NaOH. These characteristics suggest that rice straw is suitable for cellulose pulp production, offering a sustainable alternative to wood-based materials. Its fibers can be pulped for paper and paperboard production (Kaur, Bhardwaj & Lohchab 2017) and its higher cellulose and pentosan content and lower lignin content than wood, facilitating easier pulping and bleaching with reduced chemical use (Kaur et al. 2017). However, rice straw fibers are shorter and weaker than wood fibers, which may limit their application in certain paper products (Rodríguez et al. 2010). Additionally, the strength properties of straw pulp are generally inferior to wood pulp (Elhelece 2020).

Efficient and wise handling of byproducts of rice straw paper production is crucial for environmental sustainability. Black liquor, a byproduct of chemical pulping, contains residual NaOH, dissolved lignin, silica, hemicellulose, and impurities. If released into water bodies, it can harm aquatic life due to increased alkalinity. Furthermore, the emission of pollutants such as VOCs, SO_X, NO_X, PM10, and silica affect the air quality and health. Silica, comprising 8%–14% of rice straw (M. Islam et al. 2023), can disrupt photosynthesis, cause plant diseases, and stunt growth. It settles in water, affecting aquatic organisms, and contaminating soil, disrupting the food chain (Mishra 2015; World, A. O. E. 2021). Various methods, including gasification, acidification, combustion, electrolysis, and reverse osmosis, have been employed to recover NaOH and lignin from black liquor (Nong et al. 2016; Suchy 2006). Silica recovery from black liquor can be achieved through precipitation, ultrafiltration, and electrolysis (Dinh et al. 2019; Do et al. 2020). The selection of recovery methods should aim to enhance the sustainability of the rice straw paper production process.

As Sri Lanka is a prominent rice producer in Asia, it generates approximately 3.4 million tons of rice straw annually (Francis & Basnayake 2014). Despite its abundance, farmers commonly resort to open field burning as a disposal method, contributing to significant greenhouse gas emissions and air pollution (Gadde et al. 2009). In this context, utilizing rice straw for paper production presents an opportunity for value addition, minimizing environmental impact and providing a solution for the current paper scarcity in Sri Lanka.

In this research, the ultimate goal is to investigate the viability of utilizing rice straw as a sustainable alternative for developing paper that can be used as a packaging material with reduced environmental impact, particularly within the context of Sri Lanka. Firstly, the research aims to evaluate various methodologies for paper production using rice straw at the laboratory scale, exploring different processing methods. Secondly, it seeks to optimize the paper manufacturing process to attain qualities comparable to commercially available paperboards, considering parameters such as tensile breaking strength and grammage. Lastly, the study endeavors to conduct a comprehensive environmental feasibility assessment through Life Cycle Assessment (LCA) to quantitatively compare the environmental impact of the optimized rice straw-based paper production process with that of conventional wood paper manufacturing. By addressing these objectives, this research aims to provide insights into the potential of rice straw as a sustainable resource for paper production, thereby fostering advancements in environmentally friendly packaging materials. The study's findings will be valuable for policymakers, the paper industry, and farmers, fostering a sustainably viable rice straw paper production system not only in Sri Lanka but also in other rice-producing countries.

2 Materials and Methods

The methodology encompasses two main steps: (i) product development and (ii) environmental evaluation.

Product development involves three major steps: collection and preparation of raw material, pulping and processing, and analysis of product quality. Environmental evaluation consists of LCA for the optimized paper-making process utilizing rice straw and the conventional paper-making process using wood waste. These assessments comprehensively analyzed the environmental impact of each process for the production stage. Overall, the methodology ensures systematic and rigorous examination of both the product and its ecological implications.

Each step of product development and environmental evaluation is explained underneath.

2.1 Collection and Preparation of Raw Material

The major raw material, rice straw, was collected from agricultural lands in Kandy, Sri Lanka. The collected rice straw was cleaned, and the impurities were removed. The cleaned rice straw was dried at ambient temperature and then cut into 3–5 cm lengths. The cut rice straw was stored in a dry place until further processing.

2.2 Pulping and Processing

The chemical pulping process was used to convert rice straw into pulp. Sodium hydroxide (NaOH) was used as the cooking chemical. The cooking temperature ranged between 150–180 °C, with cooking times varying from 60–90 min. The NaOH concentration was maintained between 0.5M and 1M (Rodríguez et al. 2010; Kaur et al. 2017), while the liquid-to-solid ratio was kept at 7, which represents the weight of NaOH solution required per weight of dried rice straw, to determine the optimal conditions for producing high-quality paper.

After each cook, the pulp was discharged onto a mesh screen and washed with hot water. Any remaining liquor was removed by hand pressing. A 14-mesh screen was placed on top of a 200-mesh screen; materials remaining on the 200-mesh screen constituted the accepted pulp yield. The pulp yield was then oven-dried at 105 °C until it reached a constant weight. Table 1 shows the experimental plan based on the selected ranges for NaOH concentration, cooking time, and temperature. Each trial was duplicated to ensure the accuracy of the results.

Table 1. Samples to be Analyzed (Kaur et al. 2017a; Rodríguez et al. 2008b).

Sample	NaOH Concentration (M)	Temperature (°C)	Time (min)
1	1	150	60
2	1	150	90
3	1	180	60
4	1	180	90
5	0.5	150	60
6	0.5	150	90
7	0.5	180	60
8	0.5	180	90

2.3 Analysis of Product Quality

Packaging materials safeguard products during shipping, storage, and display. Tensile breaking strength and grammage are two important quality factors that have a direct impact on the durability and performance of packaging materials. Understanding and improving these qualities is critical to ensuring the reliability and effectiveness of packaging solutions. Hence, tensile breaking strength and grammage were considered as indicators of product quality (Motamedian et al. 2019).

Tensile Breaking Strength

Tensile breaking strength refers to the maximum stress a material can withstand before it breaks or ruptures when subjected to tension. In packaging, this property is vital as it indicates the material's ability to withstand external forces such as stretching or pulling during handling, stacking, or transportation. Higher tensile breaking strength implies greater resistance to tearing or puncturing, thereby enhancing the durability and protective capabilities of the packaging material. The tensile breaking strength (MPa) of paper samples was measured using a 'tensile testing machine' developed by the faculty of science, University of Peradeniya.

Grammage

Grammage, also known as basis weight or grammage weight, represents the mass of a material per unit area, typically expressed in grams per square meter (g/m^2). It serves as a measure of the density or thickness of the packaging material. Grammage directly influences various properties such as stiffness, thickness, opacity, and printability of the material.

In addition to analyzing the strength and grammage of the samples, to compare the thermal stability of the optimized process with that of commercially available paper sample, variation of weight percentage with the temperature was checked using thermogravimetric analyzer (TGA) for the temperature range between 25 -750 °C (Yavorov et al., 2022).

2.4 Life Cycle Assessment

Life Cycle Assessment (LCA) serves as a comprehensive analytical tool that systematically evaluates the environmental impacts associated with the life cycle of specific products, processes, or activities. Given its substantial advantages, LCA methodology has been extensively applied in this study according to ISO standards (ISO 14040 and ISO 14044) to gauge the environmental implications of an optimized paper-making process utilizing straw compared to traditional wood-based methods. The analysis includes key stages of the process, such as pretreatment, pulping, paper-making, and drying. The important features of the four phases of LCA are explained in detail below.

Goal and Scope Definition

The goal of this LCA study is to assess the environmental impacts of the paper-making process utilizing rice straw compared to traditional wood-based methods. The functional unit is defined as 1 m^2 of dried, unbleached rice straw paper from the chemical pulping method. All the scenario settings, materials and energy inputs, waste management, emissions, etc., were based on this functional unit.

The system boundary, as shown in Fig. 1, is set using a "Cradle to Gate" approach, which includes rice straw pretreatment, pulping, screening and washing, drying, alkaline recovery, and onsite wastewater treatment. The planting and harvesting stages of rice straw are excluded from the system boundary since rice straw is treated as waste in agriculture rather than a main product or byproduct like hemp and flax, which are utilized

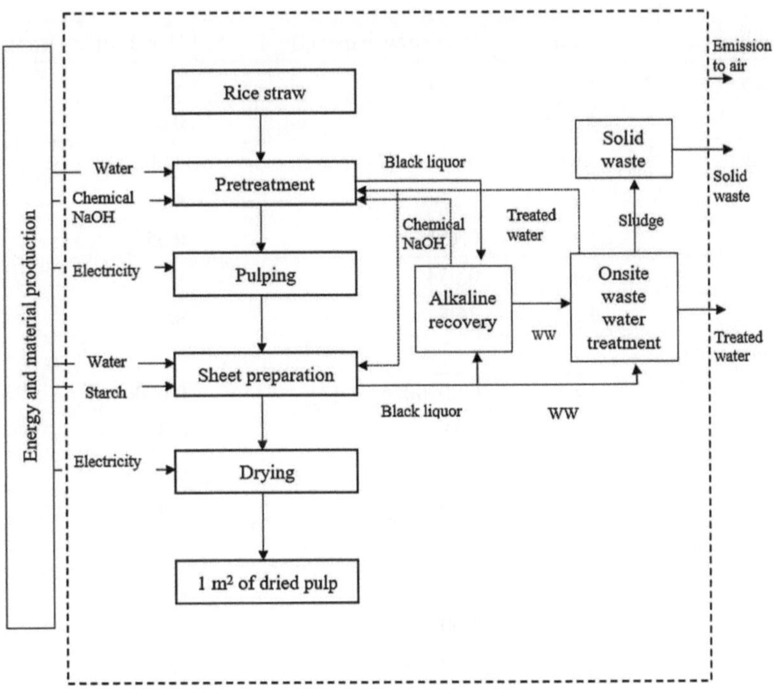

Fig. 1. The system boundary of rice straw paper making process

to make pulp. Rice straw is the major raw material of the process, and the raw material extraction phase starts with rice straw. For the pulping stage, chemical pulping using NaOH was considered, and it was assumed that 50% of NaOH in black liquor is recovered using the alkaline recovery process according to the published literature (Sun et al., 2018). In the rice straw paper-making process, it was considered that the electricity is supplied from the national grid of Sri Lanka.

Life Cycle Inventory

The life cycles of papermaking from rice straw and wood pulp were modelled using SimaPro 9 software equipped with the Eco-invent database (version 3.6). For this study, the necessary life cycle inventory data were obtained from published research (both local and international) and from estimates based on published materials. The inventory data related to the paper-making process from rice straw and wood pulp are shown in Table 2. The following assumptions are made throughout the study:

i. Material loss during the pulping process was assumed to be negligible.
ii. The impact of the machinery used was neglected (Ma et al. 2019).
iii. The overall emission from treating 1 kg of wastewater from both processes is equal.
iv. At the alkaline recovery stage, 50% of the NaOH was assumed to be recovered (sensitivity analysis was done for 40%, 50%, and 70% alkaline recovery).

Table 2. Life Cycle Inventory Analysis (M'Hamdi et al. 2017, Sun et al. 2018) per functional unit

Material input	Unit	Rice straw paper ($1m^2$)	Wood paper ($1m^2$)
Rice straw	kg	0.541	-
Wood	kg	–	0.81
Electricity	kwh	4.4	4.002
H_2SO_4	kg	–	4.65
NaOH	kg	0.151	0.357
Starch	kg	0.054	-
Water	kg	36.7	65
Emission to air			
CO_2	kg	0.0000026	0.0000026
CO	kg	0.58	0.58
CH_4	kg	0.0035	0.0035
VOC	kg	0.004	0.004
Emission to water			
Wastewater	m^3	0.036	0.954
COD	kg	0.0054	0.1431
NH_3	kg	0.00005	0.0013

Life Cycle Impact Assessment

In the life cycle impact assessment (LCIA) stage, the emissions and resource extractions were translated into environmental impact indicator scores using characterization factors. There are two ways of deriving characterization factors: at the midpoint level and endpoint level. Characterization factors at the midpoint level are located along the cause-impact pathway, at the point after which the environmental mechanism is identical for each environmental flow assigned to that impact category. Characterization factors at the endpoint level reflect the damage at one of three areas of protection: human health, ecosystem quality, and resource scarcity (Huijbregts et al. 2017). As the uncertainty of midpoint characterization is relatively lower, in this study, the midpoint approach was used to identify the impact.

The life cycle impact assessment method used in this study was ReCiPe 2016, which includes 18 impact categories: global warming, stratospheric ozone depletion, ionizing radiation, ozone formation (human health), fine particulate matter formation, ozone formation (terrestrial ecosystem), terrestrial acidification, freshwater eutrophication, marine eutrophication, terrestrial ecotoxicity, freshwater ecotoxicity, marine ecotoxicity, human carcinogenic toxicity, human non-carcinogenic toxicity, land use, mineral resource scarcity, fossil resource scarcity, and water consumption (Ma et al. 2019; M'Hamdi et al. 2017).

Interpretation

The results of characterization stage of LCA are presented to show relative contributions of different life cycles of paper making to each impact category.

3 Results and Discussion

In this section, the results of the analysis of product quality and the environmental impact of the paper-making process are presented and discussed in detail.

Analysis of Product Quality

Table 3 presents the average tensile breaking strength and average grammage of each sample made of rice straw, respectively, where Sample 0 is the commercially available paperboard. Number of replicates used for the analysis were 2.

As shown in Table 3, Sample 3 (temperature: 180 °C, time: 60 min, NaOH concentration: 1M) exhibits the highest tensile strength (11.08 MPa). This value is also the closest to the tensile strength of one of the commercially available packaging materials (Sample 0). Sample 5 (temperature: 150 °C, time: 60 min, NaOH concentration: 0.5M) has the closest grammage properties (0.065 g/cm^2) to the commercially available sample (Sample 0).

To find the best combination of factors (temperature, cooking time, NaOH concentration) that yields properties closest to the original sample, Design of Experiments (DOE) and Principal Component Analysis (PCA) were conducted using Minitab 19.

Design of Experiment (DOE)

The study focused on investigating the influence of key independent variables (factors): NaOH concentration (M), cooking temperature (°C), and cooking time (min) on the dependent variables (responses): tensile breaking strength (MPa) and grammage (g/cm^2).

Table 3. Average tensile breaking strength and grammage of samples

Sample	NaOH Conc(M)	Temperature (°C)	Time (min)	Tensile Strength (MPa)		Grammage (g/cm^2)	
				Average	*Standard deviation*	*Average*	*Standard deviation*
0				14.04	0.3960	0.065	0.0051
1	1	150	60	5.22	1.1314	0.044	0.0006
2	1	150	90	6.20	0.7071	0.050	0.0081
3	1	180	60	11.08	0.0707	0.048	0.0048
4	1	180	90	8.47	0.4667	0.054	0.0014
5	0.5	150	60	3.47	0.8415	0.065	0.0018
6	0.5	150	90	5.90	0.5657	0.057	0.0025
7	0.5	180	60	2.88	0.1414	0.060	0.0014
8	0.5	180	90	1.66	0.0495	0.052	0.0009

Number of replicates = 2

A thorough understanding of the complex interactions between these factors and the responses was obtained through statistical data analysis made possible by Minitab 19.

Regression equations (in uncoded units) for tensile breaking strength and grammage are presented in Eqs. 1 and 2 as given by Minitab.

$$Strength = -5.1 + 8.54 Concentration + 0.0275 Temperature - 0.0035 Time \quad (1)$$

$$Grammage = 0.0714 - 0.01907 Concentration - 0.000026 Temperature + 0.000017 Time \quad (2)$$

The regression equations derived from the experimental data revealed significant relationships between these factors and the mechanical properties of the material. Specifically, the tensile breaking strength was found to increase with higher NaOH concentration and cooking temperature, while decreasing slightly with longer cooking times. Conversely, grammage exhibited a different trend, decreasing with increasing NaOH concentration and cooking temperature, but marginally increasing with longer cooking times. These findings offer valuable insights into the optimization of the production process for achieving the desired mechanical properties of the material.

Figures 2 and 3 demonstrate the main effect plots (plots of the mean response values at each level of a design parameter or process variable), which represent the contribution of each factor towards the responses.

Regression equations and main effect plots indicate that concentration is the most significant factor affecting both strength and grammage properties since it exhibits the highest coefficient (gradient) in the equations and graphs. When concentration increases in the range of 0.5 M to 1 M, strength increases and grammage decreases. When cooking time increases, both strength and grammage decrease. If cooking temperature increases, tensile breaking strength increases and grammage decreases within the given range of 150°C to 180°C.

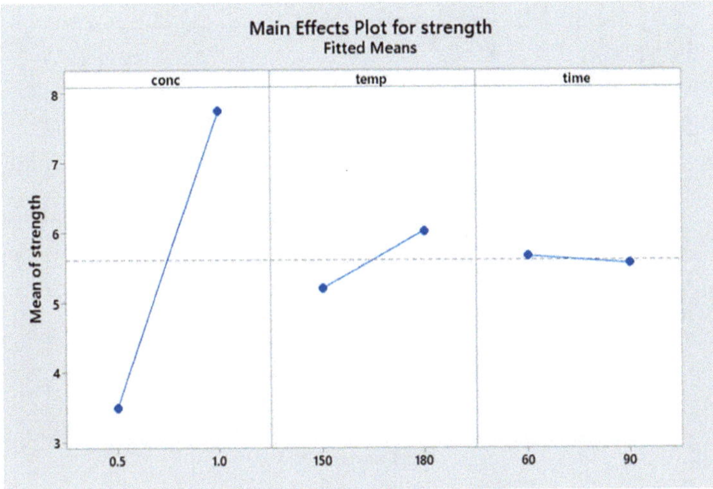

Fig. 2. Main effect plot for strength

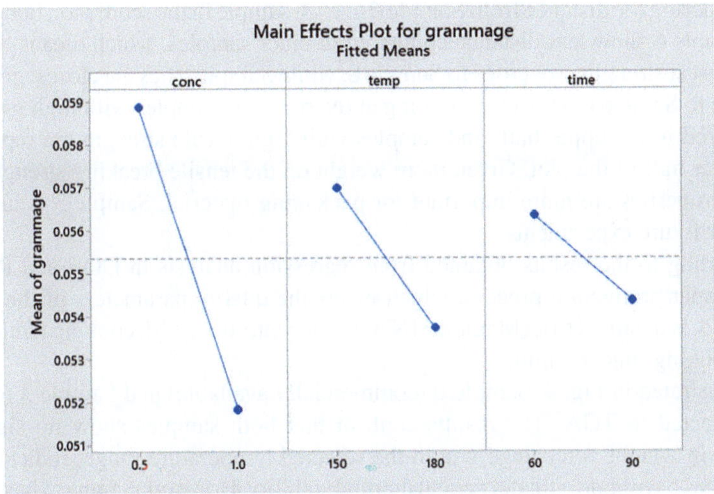

Fig. 3. Main effect plot for grammage

Principle Component Analysis (PCA)

Principal Component Analysis (PCA) is a dimensionality reduction technique used to visualize and interpret high-dimensional data. A PCA score plot is a graphical representation of the data in the reduced-dimensional space, where each point represents a single observation. To interpret a PCA score plot, we can look for groupings, which suggest that the observations are similar in some way; outliers, which are far away from the other observations; and trends, which are arranged in lines or curves. Once we have identified any patterns in the PCA score plot, we can use them to draw conclusions about the data (Wold, Esbensen, & Geladi, 1987).

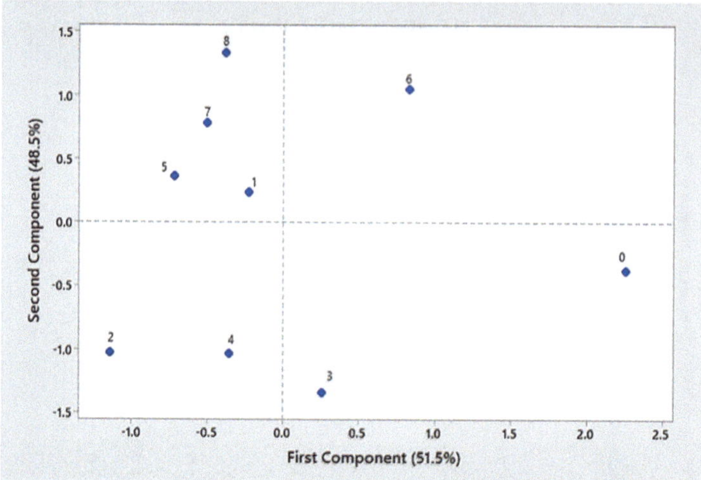

Fig. 4. Score plot for strength and grammage of paper samples

Considering the distance from Sample 0 to each sample in the score plot, both Sample 3 and Sample 6 show less distance compared to other samples, which means Sample 3 has the closest strength properties to Sample 0, while Sample 6 has the closest grammage properties to Sample 0. However, looking at the patterns, samples with high grammage are clustered in the upper half, and samples with high tensile strength are represented in the lower half of the plot. Given more weight on the tensile breaking strength, since strength properties are more important for packaging material, Sample 3 is ultimately chosen for future experiments.

According to the results obtained from regression analysis in DOE and PCA, the optimal paper production process, which meets the quality parameters of the original paperboard, is obtained from Method 3 (NaOH concentration: 1 M, cooking temperature: 180°C, cooking time: 60 min).

As illustrated in Fig. 4, Sample 0 (commercially available) and Sample 3 (optimal) were subjected to TGA. The results confirm that both samples show no significant variations in weight percentage within the selected temperature range, indicating that both samples have approximately equal thermal stability. However, Sample 3 has a higher initial moisture content compared to Sample 0.

Analysis of Environmental Impact

Figures 5, 6, and 7 demonstrate the environmental impact of the paper-making process using rice straw and wood pulp. Following the usual practice of presenting LCA results, in all graphs, positive impact scores indicate environmental burdens, and negative scores indicate environmental benefits (Sun et al., 2018).

As illustrated in Fig. 7, the overall environmental impact of the paper-making process using wood pulp is higher than that of rice straw paper production across 17 impact categories, except for the land use impact category. This higher impact is due to the sulfuric acid consumption in the pretreatment process of wood pulp paper making (M'Hamdi et al., 2017). Both rice straw and wood waste contribute zero burden towards land usage

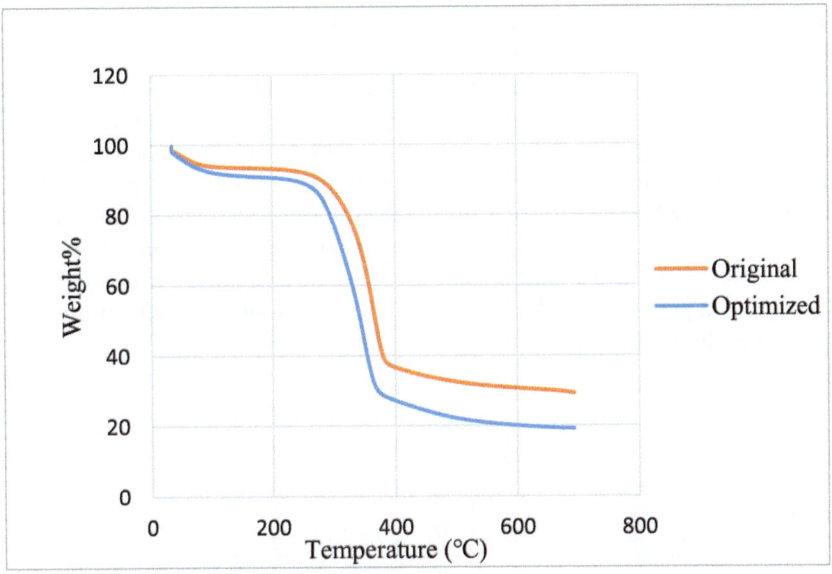

Fig. 5. TGA results: Variation of weight % with temperature

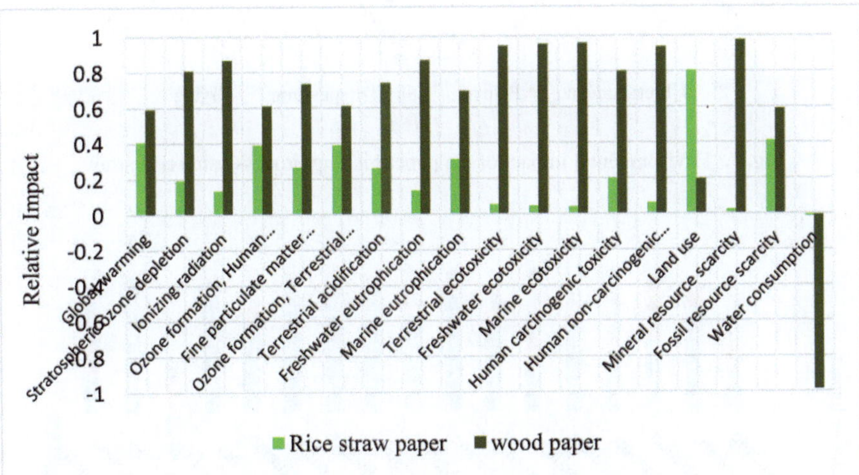

Fig. 6. Relative impact of rice straw & wood paper boards

since both are considered waste. In this context, the high environmental burden of rice straw paper production in the land use impact category is due to the starch used in sheet preparation. The land requirement for cultivating the plants which produce starch contributes to this impact. Both processes yield environmental benefits in the water consumption impact category owing to the recycling of treated wastewater. Since wood pulping generates a higher amount of wastewater compared to rice straw processing

(M'Hamdi et al., 2017), papermaking using wood pulp yields higher benefits in the water consumption impact category.

Figures 7 and 8 clarify the detailed environmental profiles of the paper-making process using rice straw and wood pulp, respectively. According to Fig. 8, the environmental profile of paper making using rice straw is dominated by the pulping and drying stages due to electricity consumption during these stages. Electricity is supplied from the national grid of Sri Lanka, which consists of a mix of renewable energy sources (58%), and fossil fuels: coal (18%) and thermal oil (24%) (Ministry of Power and Energy). In contrast, the pretreatment stage is the most significant life cycle stage in the paper-making process using wood pulp (Fig. 9). The use of sulfuric acid during the pretreatment process contributes to this higher impact.

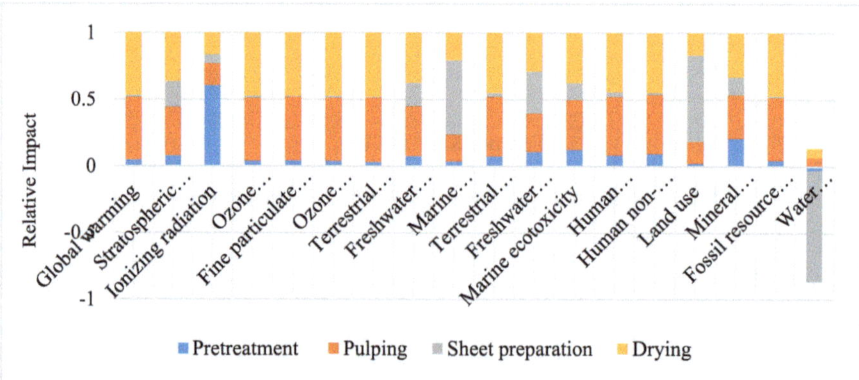

Fig. 7. Environmental impact of the paper making process using rice straw

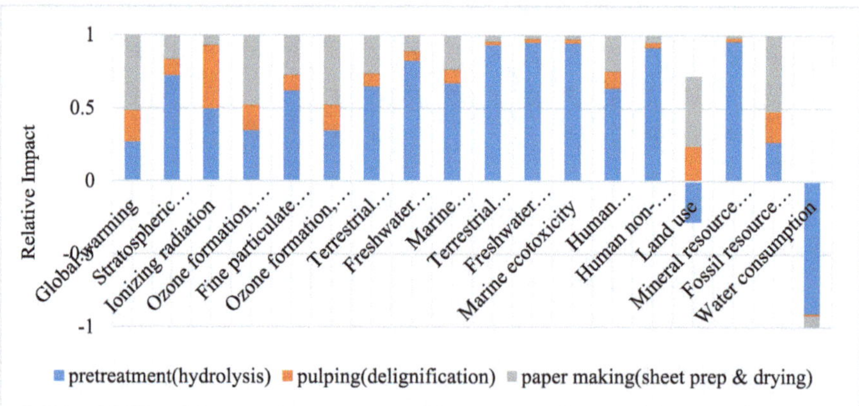

Fig. 8. Environmental impact of the paper making process using wood waste

For the life cycle inventory analysis, a 50% NaOH recovery was assumed in the alkaline recovery stage of the rice straw papermaking process. Therefore, sensitivity analysis was performed to assess its impact when the recovery percentage changes. Recovery percentages of 40%, 50%, and 70% were considered, and the corresponding changes in environmental impact are presented in Fig. 9. Figure 9 shows that the higher the recovery percentage, the lower the environmental burden.

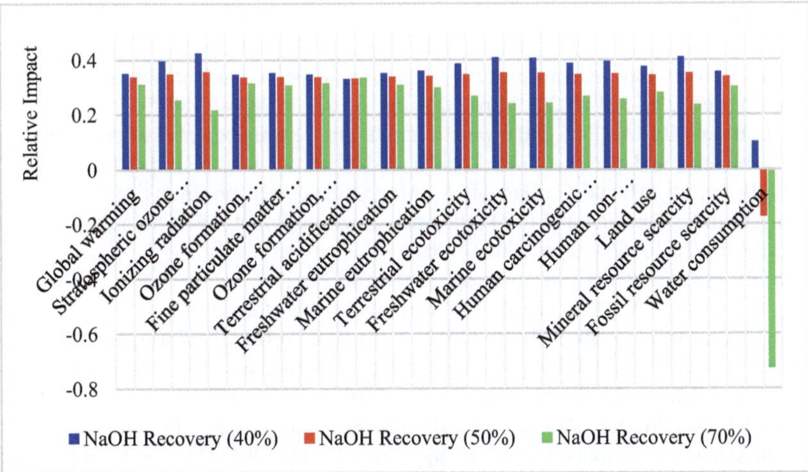

Fig. 9. Sensitivity analysis for NaOH recovery

4 Conclusion

This research focused on optimizing the paper-making process using rice straw, specifically concentrating on sodium hydroxide concentration, cooking time, and cooking temperature as the critical parameters. A NaOH concentration of 1 M, a cooking time of 60 min, and a temperature of 180 °C yielded paperboard with properties meeting commercial standards in Sri Lanka.

Importantly, the life cycle assessment using the ReCiPe midpoint method demonstrated the environmental superiority of the rice straw process compared to traditional wood-based methods. The potential for even greater environmental benefits through increased sodium hydroxide recovery percentage was highlighted, showcasing the sustainability of our approach.

As a transition from lab-scale to industry-scale implementation is necessary to commercialize the product, both experimental and industry-scale data were used in the life cycle inventory analysis. Addressing uncertainties is crucial; hence, a thorough uncertainty analysis is recommended to validate and strengthen our findings in this study. This step will enhance the reliability and applicability of our research, ensuring its impact on promoting environmentally friendly practices in the paper-making industry.

In conclusion, this work not only contributes to sustainable paper production but also underscores the need for ongoing refinement and scrutiny of the methods used to ensure a reliable and impactful shift toward eco-friendly alternatives.

References

Dinh, C.T., Duy, H.N., le Quang, D., Zenitova, L.A., Trung, T.N.: Silica extraction from integrated refinery of rice straw for production of value-added products. IOP Conf. Ser.: Earth Environ. Sci. **337**(1) (2019). https://doi.org/10.1088/1755-1315/337/1/012016

Do, N.H., et al.: The novel method to reduce the silica content in lignin recovered from black liquor originating from rice straw. Sci. Rep. **10**(1) (2020). https://doi.org/10.1038/s41598-020-77867-5

Elhelece, W.A.: Rice straw as a raw material for pulp and paper production. Encyclopedia Renew. Sustain. Mater., 296–304 (2020). https://doi.org/10.1016/b978-0-12-803581-8.10596-x

Francis, B., Basnayake, A.: Paving the way for the future we want potential of converting paddy straw to bio-char and electricity in Sri Lanka (2014). http://www.aprscp.net/11th-APRSCP

Gadde, B., Bonnet, S., Menke, C., Garivait, S.: Air pollutant emissions from rice straw open field burning in India Thailand and the Philippines. Environ. Pollut. **157**(5), 1554–1558 (2009). https://doi.org/10.1016/j.envpol.2009.01.004

Islam, M.: Rice straw in food packaging, pp. 2351–2385 (2023)

Kaur, D., Bhardwaj, N.K., Lohchab, R.K.: Prospects of rice straw as a raw material for paper making. Waste Manage. **60**, 127–139 (2017). https://doi.org/10.1016/j.wasman.2016.08.001

Lorn, S., et al.: Health impact assessment from rice straw production in Cambodia. Appl. Sci. (Switzerland) **12**(20) (2022). https://doi.org/10.3390/app122010276

M'Hamdi, A.I., Kandri, N.I., Zerouale, A., Blumberga, D., Gusca, J.: Life cycle assessment of paper production from treated wood. Energy Procedia **128**, 461–468 (2017). https://doi.org/10.1016/j.egypro.2017.09.031

Ma, X., et al.: Energy and carbon coupled water footprint analysis for straw pulp paper production. J. Clean. Prod. **233**, 23–32 (2019). https://doi.org/10.1016/j.jclepro.2019.06.069

Mishra, A.: J. Geogr. Reg. Plann. Impact Silica Min. Environ. **8**(6), 150–156 (2015). http://www.academicjournals.org/JGRP

Motamedian, H.R., Halilovic, A.E., Kulachenko, A.: Mechanisms of strength and stiffness improvement of paper after PFI refining with a focus on the effect of fines. Cellulose (2019).https://doi.org/10.1007/s10570-019-02349-5

Nong, G., Zhou, Z., Wang, S.: Generation of hydrogen, lignin and sodium hydroxide from pulping black liquor by electrolysis. Energies **9**(1) (2016). https://doi.org/10.3390/en9010013

Rodríguez, A., Moral, A., Serrano, L., Labidi, J., Jiménez, L.: Rice straw pulp obtained by using various methods. Biores. Technol. **99**(8), 2881–2886 (2008). https://doi.org/10.1016/j.biortech.2007.06.003

Rodríguez, A., Sánchez, R., Requejo, A., Ferrer, A.: Feasibility of rice straw as a raw material for the production of soda cellulose pulp. J. Clean. Prod. **18**(10–11), 1084–1091 (2010). https://doi.org/10.1016/j.jclepro.2010.03.011

Suchy, M.: Caustic treatment of the pulp: caustic recovery by reverse caustic treatment of the pulp: caustic recovery by reverse Osmosis Osmosis (2006)

Sun, M., Wang, Y., Shi, L.: Environmental performance of straw-based pulp making: a life cycle perspective. Sci. Total. Environ. **616–617**, 753–762 (2018). https://doi.org/10.1016/j.scitotenv.2017.10.250

Sun, M., Wang, Y., Shi, L., Klemeš, J.J.: Uncovering energy use, carbon emissions and environmental burdens of pulp and paper industry: a systematic review and meta-analysis. Renew. Sustain. Energy Rev. **92**, 823–833 (2018). https://doi.org/10.1016/j.rser.2018.04.036

Wold, S., Esbensen, K., Geladi, P.: Principal component analysis. Chemometr. Intell. Lab. Syst. **2**(1–3) (1987)

Worku, L.A., Bachheti, A., Bachheti, R.K., Rodrigues Reis, C.E., Chandel, A.K.: Agricultural residues as raw materials for pulp and paper production: overview and applications on membrane fabrication. Membranes **13**(2) (2023). https://doi.org/10.3390/membranes13020228

Yavorov, N., Todorova, D., Vrabič Brodnjak, U., Lasheva, V.: Thermal stability of packaging papers treated of silver water (n.d.)

The Impact of Life Cycle Cost-Related Factors to Reduce the Cost of a Green Building in Sri Lanka

C. V. Rajasekara(✉), W. N. Kawmudi, and U. M. Samararathne

Department of Quantity Surveying, Faculty of Built Environment and Spatial Sciences (FBESS), General Sir John Kotelawala Defence University, Southern Campus, Colombo, Sri Lanka
37-qs-0030@kdu.ac.lk

Abstract. As society moves toward greater sustainability, the global building sector is leading the way in adopting green construction. Even though Sri Lanka is now going through an energy and financial crisis, the building sector can grow because of green building ideas. It is advisable to consider a building's Life Cycle Cost (LCC), which comprises cost characteristics related to operations, replacement, maintenance, and demolition. Thus, the primary goal of this research is to determine which component in a green building has the biggest influence on LCC and to offer ways to lessen that aspect's influence to further minimize LCC. A questionnaire was used to poll the relevant authorities, enabling a thorough examination of the research. This employed quantitative analysis to corroborate findings and provide recommendations based on feedback from several construction industry stakeholders. The research shows that the cost data for LCC that is most impacted is the replacement cost. Additionally, to eventually minimize LCC, this study offered solutions for cutting costs in sustainable building. By considering the financial implications of the green construction concept, this research study will be pertinent to Sri Lanka's efforts to address the present economic issues.

Keywords: Green Building · Life Cycle Cost · Operational Cost · Maintenance Cost · Replacement Cost · Demolition Cost

1 Introduction

Green building is a sustainable development approach that aims to reduce global energy consumption and greenhouse gas emissions by rethinking infrastructure, homes, commercial buildings, and industrial facilities. It focuses on preserving the natural environment while enhancing the built environment's ecological impact (Onat et al., 2014). LCC (Life Cycle Cost) is a crucial factor in green building designs, which offer numerous advantages such as environmental conservation, resource conservation, waste reduction, and improved health. These designs also contribute to financial savings, increased worker productivity, and environmental protection, thereby enhancing overall living conditions (Dantata et al., 2005). Green building structures prioritize quality, reducing water, energy, and materials usage, incorporating new building types, and establishing

R. Dissanayake et al. (Eds.): ICRES 2025, PTT, pp. 308–319, 2025.
https://doi.org/10.1007/978-981-96-9551-5_24

life cycle analysis for energy and greenhouse gas emissions reduction (Lu et al., 2016, Anvuur et al., 2011). Sustainability metrics vary globally, with Leadership in Energy and Environmental Design (LEED) being the most reliable standard. Sri Lanka can leverage sustainable development to address energy and economic crises, making cost reduction through sustainable growth a significant topic. (Samari et al., 2013). The economic crisis in Sri Lanka is a major issue, with cost being a key concern. Green Building construction is not popular due to a lack of knowledge, awareness, and effective procurement systems. To overcome this crisis, understanding innovative techniques in the building sector is crucial. Few studies have examined cost, time, and quality comparisons of green building construction, and strategies to reduce LCC. Sri Lanka is addressing the LCC issue by identifying obstacles in green building construction, thereby filling a research gap and meeting industry needs. The problem statement of this study is the identification of the factors of life cycle cost reduction in Sri Lankan green building construction projects. The main objective of this study is to identify the impact of life cycle cost factors to reduce the life cycle cost reduction in Sri Lankan green building construction.

The study's conclusions could only apply to the Sri Lankan context and not necessarily to other nations or areas. This study does not consider the initial cost is also a limitation of this study and as the LCC minimizing factors only maintenance, demolition, operational, and replacement costs are considered.

2 Literature Review

Sustainable development aims to balance the needs of the economy, environment, and culture in the 21st century. It involves growing natural and industrial assets that meet current energy needs without negatively impacting future generations' capacity to meet their own needs. The three most crucial sustainability indicators are environmental, societal, and economic factors (Samari et al., 2013). Sustainable buildings are gaining global attention for reducing construction materials, energy, and waste while increasing building lifespan and considering indirect effects to ensure sustainability in material extraction, construction, and operation (Lemaitre, 2017). Sustainable development requires a balance of economic expansion, population management, equitable income distribution, and comprehensive policies, but is costly and requires global support, particularly in developing nations (Glasby, 2002).

2.1 The Need of Green Building

"Green building" refers to the process of developing and implementing more resource- and health-conscious models for construction, rehabilitation, operation, repairs, and deconstruction (Chan et al., 2009). Green buildings prioritize energy, water, indoor air quality, sustainable materials, solar energy, recycling, and reuse to reduce waste and pollution. They prioritize environmental preservation and user quality, benefiting economic, social, and other aspects of a country's development. Green building design is a suitable method for sustainable development (Kats, 2003). Sustainability in building aims to improve human life and economics without harming the environment. Traditional buildings have higher maintenance and repair costs and overcoming green building concepts is crucial for the construction industry. Education on sustainable development, economic value, and internal health costs can mitigate these issues (Zuo et al., 2012).

2.2 Parameters of Life Cycle Cost

Life cycle costing (LCC) is a process of adding up costs over time, potentially crucial for life cycle management. Despite rising practitioners due to public-private partnerships and build-operate-transfer projects, it remains largely unused in facilities management (Jiang & Tovey, 2010). The Royal Institutes of Chartered Surveyors (RICS) defines initial expenses as site prices, opportunities, finance charges, expert fees, building and infrastructure costs, financial allowances, statutory charges, redevelopment grants, planning gain, and external costs. Additional maintenance expenditures include redecoration, inspection, maintenance, and component replacement. Operational costs include cleaning, utilities, security, and property management. End-of-life expenses include demolition and disposal, and residual value (M.S. Soujanya and H.AE.C Jayasena, 2018).

- **Operational Cost -** Green buildings in Massachusetts save around $750,000 annually over 20 years, compared to traditional buildings, which consume 30% less energy. These savings are based on factors like operating energy supply systems, installing floor heating, water bills, electricity contributions, and district heating grid.
- **Maintenance Cost -** Maintenance costs refer to the expenses incurred in maintaining and ensuring the safety and functionality of a building's structure and components (Dwaikat & Ali, 2018b).
- **Replacement Cost -** The replacement cost category includes the energy system installation costs of heat pumps, windows, and floor heating ventilation systems over the building's lifespan (Marszal & Heidelberg, 2011).
- **Demolition Cost -** Demolition costs involve the removal of a building or structure, often involving heavy machinery or human tools, to create rubble and rubbish.

The LCC (Local Cost) includes initial, running, maintenance, replacements, and destruction costs. Administration, personnel, plant, and trash disposal costs, as well as potential savings from salvage and recycling, contribute to the complex cost structure (Pun et al., 2006).

2.3 Importance of LCC of a Building

The Life Cycle Cost (LCC) is an economic technique for project evaluation that considers all costs linked to owning, maintaining, operating, and disposing of an asset. It includes initial capital expenses, financing expenses, operating expenses, maintenance expenses, and eventual disposal expenses. The LCC is a crucial factor in determining the asset's value over its useful life (Fuller and Petersen, 1995, Pelzeter, 2007). The Life Cycle Cost (LCC) is the total cost of a structure's design, creation, operation, and upkeep, including building waste at the end of its life cycle. It is divided into four categories: design and construction, operation, maintenance, and end-of-life costs (Dwaikat & Ali, 2018b). The four types of costs included in Marszal's paper LCC analysis of a structure are investment, operation, maintenance, replacement, and demolition.

$$LCC = IC + OC + MC + RC + DC \tag{1}$$

Equation 1: Life Cycle Cost Calculation.
IC – Investment cost.
OC – Operational Cost.
MC – Maintenance Cost.
RC – Replacement Cost.
DC – Demolition Cost.

LCC, is a sustainable approach used in green building projects, ensuring high sustainability involvement and long-term economic benefits due to its significant role in the design stage (Taylor, 1981).

3 Methodology

Research techniques may be used to tackle the research challenges methodically, according to Mujumdar (2008). It may be thought of as a science that investigates scientific methodology. In methodology, it looks at the many strategies and justifications that a researcher usually uses to study his study problem (Mujumdar, 2008). This research aims to identify the impact of life cycle cost reduction in green building projects in Sri Lanka. To fulfil this research aim it targets the construction industry experts in Sri Lanka. This research was completely evaluated through the questionnaire survey and interviews with the above-mentioned authorities.

3.1 Data Collection Method

A detailed questionnaire survey (Google Forms) was shared among the construction professionals. Professionals were given questionnaires to complete to get appropriate answers, and various points of view were graded using the "Likert Scale." Google form was distributed to 60 construction industry professionals, and the response rate was 61.7% experienced in the green building industry including 38 Quantity surveyors (QS), 11 Engineers (CE), 6 Architects (AR), 3 Project Managers (PM) and 2 contractors (CO).

3.2 Conceptual Framework

According to the conceptual framework, this research identifies four relationships between the independent variables and dependent variables.

The identified dependent variable is the minimization of the Life Cycle Cost of Green Buildings in Sri Lanka. It is identified as the main variable due to its importance and the main aim that the research is supposed to achieve. Then independent variables were identified as replacement cost, maintenance cost, operational cost, and demolition costs. To analyse the data a hypothesis has developed as follows.

- HA0 (Null factor) States that there is no relationship between management of operational Cost and minimization of LCC of GB in Sri Lanka
- HA1 (Alternative Factor) states that there is a relationship between efficient management of operational cost and minimization of LCC of GB in Sri Lanka
- HB0 (Null factor) States that there is no relationship between the management of Maintenance Costs and minimization of LCC of GB in Sri Lanka

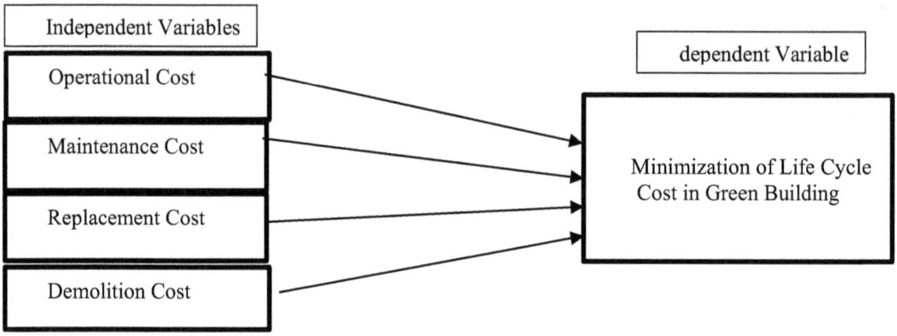

Fig. 1. Conceptual framework. Source: (Marszal & Heiselberg, 2011)

- HB1 (Alternative Factor) States that there is a relationship between efficient management of Maintenance cost and minimization of LCC of GB in Sri Lanka.
- HC0 (Null factor) States that there is no relationship between management of Replacement Cost and minimization of LCC of GB in Sri Lanka
- HC1 (Alternative Factor) States that there is a relationship between efficient management of replacement cost and minimization of LCC of GB in Sri Lanka
- HD0 (Null factor) States that there is no relationship between management of Demolition Cost and minimization of LCC of GB in Sri Lanka
- HD1 (Alternative Factor) States that there is a relationship between efficient management of Demolition cost and minimization of LCC of GB in Sri Lanka

4 Data Analysis Method

The collected data was analyzed through quantitative techniques. The linkages between the created factors linked to cost control and their impact on projects were ascertained with the use of statistical analysis (Devi & Ananthanarayanan, 2017). SPSS software was used to analyze the data and identify the relationship and the impact of the identified independent variables and dependent variables.

1. Correlation analysis: Correlation analysis is used to analyses the relationship between the identified independent variables and dependent variables. This research used the "Pearson Correlation Analysis" to analyze the relationship through the SPSS software. The value of the analysis should be between −1 and +1. The correlation coefficient equals +1 which has a positive relationship and equals -1 which has a negative relationship between the IVs and DV. According to this research, the coefficient value is identified as "P" and the significance value is 0.05.
2. Regression analysis: SPSS software was used to do the regression analysis and using regression analysis identified the impact of identified IVs and DV. This aimed to find out the most impact on cost reduction in the green buildings in Sri Lanka.

Following a study of the literature, the determination of the factors, and the completion of the questionnaire, data was gathered from building professionals. Since this research focuses primarily on identifying among the most important techniques for lowering the life cycle costs of green buildings generally, which have varied effects on each professional at various stages, a variety of approaches was chosen. Therefore, experts in the construction industry quantity surveyors, architects, engineers, project managers, and contractors were covered, and data were obtained for this study. The questionnaire was sent to 65 professionals in the above professions and out of those 60 responses have been received. Figure 2 indicates the respondents that answered the Google questionnaire and gives an idea about the main target sample.

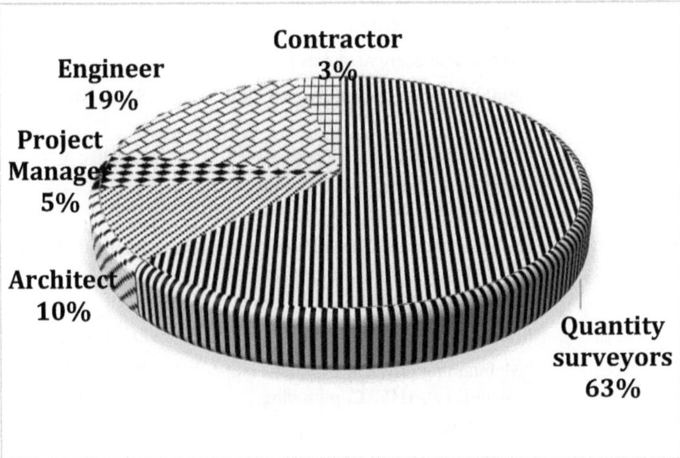

Fig. 2. Respondents' results

Afterward, it was determined if these experts had prior expertise with green building initiatives. According to Fig. 3, 61.7% of those surveyed said they had prior experience constructing green buildings in Sri Lanka.

Fig. 3. Experience in green building construction

We determined the dependent and independent variables using the conceptual framework. The primary metrics inquired about by the participants are listed in the table below (Table 1).

Table 1. Operationalization of variables

Operationalization of Variable	Operationalization	Reference
Operational Cost	System cost (heating, cooling, and lighting systems)	(Bunz et al., 2006)
	Water cost (water efficiency plumbing systems and management of water-saving strategies)	(Illankoon & Lu, 2019)
	Waste reduction cost	(Ahn et al., 2013)
	Renewable energy sources cost	(Marszal & Heiselberg, 2011)
	Building Management System Cost (BMS)	(Smart Market Report, 2018)
	Sustainable practices cost (waste reduction programs, eco-friendly landscaping, cleaning)	(Dwaikat & Ali, 2018b)
Maintenance Cost	Scheduled and Unscheduled Maintenance Activity cost (cleaning, inspections, minor adjustments)	(Illankoon & Lu, 2019)
	Repair cost	(Dwaikat & Ali, 2018b)
	Maintaining physical structures and system costs (HVAC, plumbing, electrical)	(Bunz et al., 2006)
	Maintaining safety equipment	(Radwan et al., 2015)
Replacement cost	Selecting sustainable materials cost (choosing high-quality, long-lasting materials)	(Fowler & Rauch, 2006)
	Replacement of energy efficiency system cost (HVAC, lighting Systems)	(Bunz et al., 2006)
Demolition Cost	Deconstruction cost	(Dantata et al., 2005)
	Waste management cost (sorting, recycling, and disposing)	(Dantata et al., 2005)
	Environmental compliance costs (environmental regulations like air quality standards, noise restrictions, and waste disposal regulations)	(Tan et al., 2011)
	Demolition equipment and labor costs	(Dwaikat & Ali, 2018a)

4.1 Correlation Analysis

To ascertain the relationship between the identified independent variables (IV) and dependent variables (DV), correlation analysis was carried out. The link was examined using Pearson correlation, and SPSS software was employed for data analysis. The dependent variable was used to test each independent variable. When two variables grow, there is a positive correlation; conversely, both variables decline when there is a negative correlation. The significance value (P) is identified as follows identifies then the correlation coefficient will satisfy,

P < 0.05 – The DV has a relationship with identified IVs.
P > 0.05 the DV doesn't have a relationship with IVs.

Table 2. Correlation coefficient analysis of variables (SPSS Generates)

		AVGOC	AVGMC	AVGRC	AVGDC	AVGDV
AVGOC	Pearson Correlation	1	.185	.120	.185	.267*
	Sig. (2-tailed)		.156	.363	.157	.040
	N	60	60	60	60	60
AVGMC	Pearson Correlation	.185	1	.373**	.514**	.379**
	Sig. (2-tailed)	.156		.003	.000	.003
	N	60	60	60	60	60
AVGRC	Pearson Correlation	.120	.373**	1	.451**	.527**
	Sig. (2-tailed)	.363	.003		.000	.000
	N	60	60	60	60	60
AVGDC	Pearson Correlation	.185	.514**	.451**	1	.280*
	Sig. (2-tailed)	.157	.000	.000		.031
	N	60	60	60	60	60
AVGDV	Pearson Correlation	.267*	.379**	.527**	.280*	1
	Sig. (2-tailed)	.040	.003	.000	.031	
	N	60	60	60	60	60

*. Correlation is significant at the 0.05 level (2-tailed).
**. Correlation is significant at the 0.01 level (2-tailed).

Where:
N = No of respondents (60).
AVGOC = Average of Operational Cost Management.
AVGMC = Average of Maintenance Cost Management.
AVGRC = Average of Replacement Cost Management.
AVGDC = Average of Demolition Cost Management.
AVGDV = Average of Dependent Variable (Minimization of Life Cycle Cost in Green Building).

Table 2 shows the correlation results of the questionnaire data of this research. It demonstrates the correlation between the management of operational cost, maintenance cost, replacement cost, demolition cost (IV), and the minimization of LCC in green buildings (DV). The results of the hypothesis are discussed below.

The first variable, operational cost management (OC), has a positive association with a value of 27% and a significance level of 0.040 (P < 0.05), according to Table 2.

This indicates that the alternative hypothesis (HA1), according to which there is a connection between operational cost management and LCC mitigation in green buildings, cannot be rejected.

The second variable, maintenance cost management (MC), shows a positive association with a value of 38% and a significance level of 0.003 ($P < 0.05$), according to Table 2. This indicates that it is not possible to reject the alternative hypothesis (HB1), according to which there is a connection between LCC minimization and maintenance cost management in green buildings.

The management of replacement cost is the third variable. Table 2 shows that this variable has a positive association, with a value of 53% and a significance level of 0.00 (<0.05). This indicates that the alternative hypothesis (HC1), according to which there is a connection between LCC reduction in green buildings and replacement cost management, cannot be rejected.

The management of demolition cost is the fourth variable. Table 2 shows that this variable has a positive association, with a value of 28% and a significant value of 0.031 (<0.05). This indicates that the alternative hypothesis (HD1), according to which there is a connection between the minimizing of LCC in green buildings and the control of demolition costs, cannot be rejected.

4.2 Regression Analysis

After the link has been established, ascertain how each independent variable affects the dependent variable. The impact of the independent variables (maintenance, demolition, operation, and replacement costs) on minimizing the life cycle cost of green buildings was determined using regression analysis. When concentrating on independent and dependent variables, this calls for the use of several simulation and regression techniques for multiple variables. The table below displays the regression analysis summary.

Table 3. Regression coefficient of variables (SPSS Generated)

Model		Unstandardized Coefficients		Standardized Coefficients	t	Sig.
		B	Std. Error	Beta		
1	(Constant)	.692	.627		1.103	.275
	AVGOC	.206	.123	.186	1.670	.101
	AVGMC	.246	.153	.209	1.611	.113
	AVGRC	.382	.103	.458	3.694	.001
	AVGDC	.064	.126	.069	.511	.612

From the above Table 3 shows the regression model as fellows,

$$Y = \alpha + \beta 1X1 + \beta 2X2 + \beta 3X3 + \beta 4X4 + Std.E$$
$$Y = 0.692 + \{(0.206 * AVGOC) + (0.246 * AVGMC) + (0.382AVGRC) + (0.064 * AVGDC)\} + Std.E \quad (2)$$

The table shows a positive impact in the value of Beta for the operational cost as 0.206 at the significance level of 0.101 which showed a 21% variance in the minimization of the LCC in green buildings. More elaborately it states that the minimization of LCC in green buildings increases by 21% with the increase of management of Operational cost.

A positive impact is evident in the management of maintenance costs and the minimization of LCC in green buildings, as the beta value is 0.246 at a significance level of 0.113 which showed a 25% variance in the minimization of LCC in green buildings explained by the management of the maintenance cost. That means the minimization of LCC in green buildings increases by 25% with the increase in the management of maintenance costs.

The table shows a positive impact in the value of Beta for the replacement cost as 0.382 at the significance level of 0.001 which showed a 38% variance in the minimization of the LCC in green buildings. More elaborately it states that the minimization of LCC in green buildings increases by 38% with the increase of management of replacement cost.

A positive impact is evident in the management of demolition costs and the minimization of LCC in green buildings, as the beta value is 0.064 at a significance level of 0.612 which showed a 6% variance in the minimization of LCC in green buildings explained by the management of the demolition cost. That means the minimization of LCC in green buildings increases by 6% with the increase in the management of demolition costs.

5 Conclusion

The main objective of this study was to identify the impact between life cycle cost factors and minimization of LCC in GB construction. Data was collected through the detailed questionnaire survey and analysed through SPSS software. After developing independent and dependent variables, correlation analysis was used to examine the relationship between the variables. The basic definition of the hypothesis was made for this. Table 2 indicates that a positive relationship was only present when the significance value was less than 0.05. The results of the study showed that every independent variable and the dependent variable had a positive relationship.

Based on the previously computed results, which are displayed in Table 3, regression analysis was then performed to determine the impact of the associations. When the variance value is positive, it indicates that the DV increases or decreases in tandem with the IV, and when the variance value is negative, it indicates that the DV decreases or increases in tandem with the IV.

According to the correlation analysis data all IVs and DV has a positive relationship between the variables and according to the regression analysis the minimization of LCC will increase with the increase of all management of IV. And it identified Management of replacement costs in the GB has the highest impact on the minimization of LCC.

5.1 Recommendation

Recommendations to lower the operational cost.

- Install energy-saving lighting, such as LED lights, and regulate the amount of it is used via timers or sensors.

- To produce renewable energy on-site, think about using renewable energy sources like wind turbines or solar panels.
- Install rainwater harvesting systems for collecting and repurpose rainfall for uses other than drinking water, such as flushing toilets and irrigation.
- To recycle sink and shower water for gardening and other uses, install a greywater system.

Recommendations to lower the maintenance cost.

- Make an investment in long-lasting, high-quality building materials that need less upkeep over time.
- Regularly examine building systems (plumbing, electrical, HVAC, etc.) to find problems early and stop serious malfunctions.
- To prevent early failure and maintenance problems, make sure that all building systems, equipment, and materials are installed accurately and in accordance with manufacturer specifications.
- When feasible, use readily replaceable or modular components to reduce disturbance and facilitate maintenance.
- To monitor equipment performance and guarantee prompt service, keep thorough records of all maintenance operations, repairs, and warranty information.

Recommendations to lower the replacement cost.

- Make use of systems with longer lifespans or warranties to reduce the frequency of replacements.
- Make sure building systems are routinely inspected and maintained in a timely manner to spot possible problems early and fix them before they require replacement.
- Create flexible spaces and systems in buildings so that they may readily accept alterations or improvements in the future without needing to be completely replaced.
- Investigate cutting-edge construction technologies that provide longer lifespans and less maintenance needs, which will ultimately result in cheaper replacement prices.

Recommendations to lower the demolition cost.

- Suggestions to reduce the cost of demolition
- Before demolishing, carry out an environmental impact assessment to find any hazardous items that need to be handled carefully and disposed of properly while maintaining legal compliance.
- Employ demolition contractors with expertise and qualifications who are devoted to optimizing material recovery and who are conversant with green demolition procedures.
- Look for ways to use salvaged materials on the spot or give them to projects or organizations that might utilize recycled construction materials.

Acknowledgments. The Faculty of Built Environment & Spatial Sciences (FBESS), Southern Campus of General Sir John Kotelawala Defence University (KDU), deserves special recognition for its assistance and guidance in improving the research, as well as for all the interviewees who provided firsthand information about this study.

References

Ahn, Y.H., Pearce, A.R., Wang, Y., Wang, G.: Drivers and barriers of sustainable design and construction: The perception of green building experience. Int. J. Sustain. Build. Technol. Urban Dev. **4**(1), 35–45 (2013). https://doi.org/10.1080/2093761X.2012.759887

Ali, H.H., Al Nsairat, S.F.: Developing a green building assessment tool for developing countries - Case of Jordan. Build. Environ. **44**(5), 1053–1064 (2009). https://doi.org/10.1016/j.buildenv.2008.07.015

Athapaththu, K.I., Karunasena, G.: Framework for sustainable construction practices in Sri Lanka. Built Environ. Project Asset Manage. **8**(1), 51–63 (2018). https://doi.org/10.1108/BEPAM-11-2016-0060

Awwad, R., El Souki, O., Jabbour, M.: Construction safety practices and challenges in a middle eastern developing country. Saf. Sci. **83**, 1–11 (2016). https://doi.org/10.1016/j.ssci.2015.10.016

Azhar, S., Carlton, W.A., Olsen, D., Ahmad, I.: Building information modeling for sustainable design and LEED ® rating analysis. Autom. Constr. **20**(2), 217–224 (2011). https://doi.org/10.1016/j.autcon.2010.09.019

Azizi, N.Z.M., Abidin, N.Z.: Main elements of soft cost in green buildings. World Academy of Science, Engineering and Technology 72, **6**(12), 992–997 (2012)

Baird, G.: What the Users Think of Sustainable Buildings – a Global Overview. New Zealand Sustainable Building Conference, November, 1–9 (2010)

British Columbia Forest Facts. Building Green & The Benefits of Wood. March (2011).

Bunz, K.R., Henze, G.P., Tiller, D.K.: Survey of sustainable building design practices in North America, Europe, and Asia. J. Archit. Eng. **12**(1), 33–62 (2006). https://doi.org/10.1061/(asce)1076-0431(2006)12:1(33)

Burnett, J.: City buildings-eco-labels and shades of green! Landsc. Urban Plan. **83**(1), 29–38 (2007). https://doi.org/10.1016/j.landurbplan.2007.09.003

Campbell, S.: Green cities, growing cities, just cities?: urban planning and the contradictions of sustainable development. J. Am. Plann. Assoc. **62**(3), 296–312 (1996). https://doi.org/10.1080/01944369608975696

Chan, E.H.W., Qian, Q.K., Lam, P.T.I.: The market for green building in developed Asian cities-the perspectives of building designers. Energy Policy **37**(8), 3061–3070 (2009). https://doi.org/10.1016/j.enpol.2009.03.057

County, L.A.: Green Building Manufacturing Los Angeles County (2006)

Creswell, J.W.: The Selection of a Research Approach. Research Design, 3–23 (2014). 45593:01

Fullbrook, D., Q Jackson, G.F.: Value Case for Sustainable Building in New Zealand: Vol. February 2 (2006). www.mfe.govt.nz

Dagdougui, H., Minciardi, R., Ouammi, A., Robba, M., Sacile, R.: Modeling and optimization of a hybrid system for the energy supply of a "green" building. Energy Convers. Manage. **64**, 351–363 (2012). https://doi.org/10.1016/j.enconman.2012.05.017

Dantata, N., Touran, A., Wang, J.: An analysis of cost and duration for deconstruction and demolition of residential buildings in Massachusetts. Resour. Conserv. Recycl. **44**(1), 1–15 (2005). https://doi.org/10.1016/j.resconrec.2004.09.001

Darko, A., Chan, A.P.C., Owusu-Manu, D.G., Ameyaw, E.E.: Drivers for implementing green building technologies: an international survey of experts. J. Clean. Prod. **145**, 386–394 (2017). https://doi.org/10.1016/j.jclepro.2017.01.043

Delnavaz, M.: Project Managers' Role in Sustainable Building Process. Chalmers, 50 (2012). http://publications.lib.chalmers.se/records/fulltext/161485.pdf

Dovjak, M., Kukec, A., Assessment, A., Factors, H.R.: (n.d.) and Healthy Sustainable Buildings

Dwaikat, L.N., Ali, K.N.: Green buildings life cycle cost analysis and life cycle budget development: practical applications. J. Build. Eng. **18**, 303–311 (2018). https://doi.org/10.1016/j.jobe.2018.03.015

A Study on the Effect of Using Leather Waste in Place of Clay Materials in Molded Rubber Shoe Soles

P. L. L. Arunodhi[1(✉)], C. K. Jayasuriya[1], J. K. Premachandra[2], and A. J. M. Vithanachchi[3]

[1] University of Kelaniya, Kelaniya, Sri Lanka
arunodhiliyanage27@gmail.com
[2] University of Moratuwa, Moratuwa, Sri Lanka
[3] Samson Compounds pvt Ltd., Galle, Sri Lanka

Abstract. Leather is a strong, flexible and durable material which is used during manufacturing of shoe uppers and insole materials. During leather cutting process for shoe upper manufacturing large amount of cutting waste are generated. This solid waste of leather such as trimmings and scraps are usually discarded into landfills and it takes a long time to degrade Santos et al. (2014) Ponsubbiah and Gupta (2019a, b). The idea was to adding valuation to leather waste by replacing with Clay. Raw materials of clay extraction process impact on the imbalance of biodiversity and eco system and moreover can highly effect on unfavourable health concern. The study focusing on use of leather cutting waste as a replacement for clay in the shoe sole compound formulation. The compound mixing involved two roll mill and vulcanization has been followed by compression moulding method. Ground and sieved through 30 mesh (particle size <0.595 mm) Leather cutting waste (Moisture content <3%) were used as a replacement for clay dosage where systematically changed in increments of 0%, 10%, 20%, and 30% by weight percentage. The results revealed increase in leather cutting waste as a replacement for clay led to reduction in physio-mechanical properties such as Hardness, Tensile Strength, Elongation at break Simultaneously specific gravity, 300% Modulus, Tension set, Tear strength, Abrasion Resistance showed an increasing trend. According to the Scanning Electron Microscopy images indication leather waste are consistently incorporated within the rubber matrix. The rubber-leather composite consists with 20% weight of leather cutting waste as a replacement for clay exhibit significantly superior performance. Rubber-leather composite consist with 20% replacement can be applied during manufacturing of moulded sole which has commercial benefit plus greater contribution for saving utilization of clay material while ensuring the environment sustainability.

Keywords: clay fillers · rubber compounding · rubber-leather composite · shoe sole · solid waste

R. Dissanayake et al. (Eds.): ICRES 2025, PTT, pp. 320–331, 2025.
https://doi.org/10.1007/978-981-96-9551-5_25

1 Introduction

Leather is a strong, flexible and durable material which is commonly used in production of shoe uppers, insole materials. During the cutting process of sole upper large amount of cutting waste are generated. This solid waste of leather such as trimmings and scraps are usually discarded into landfills and it takes a long time to degrade. The idea was to adding valuation to leather waste by replacing with clay which is a main filler material used in the shoe sole formulations. Although raw materials of clay extraction process impact on the imbalance of biodiversity and eco system as well as improper use of clay can highly effect on unfavourable health concern.

Leather scraps, composed primarily of collagen fibers, constitute the majority of solid waste generated from the footwear business. (Nguyen Pham Duy Linh and Doan Anh Vu, 2017) Leather particles have been used as reinforcement in polymer composite materials to modify physical and mechanical properties. (Ferreira, Freitas and Almeida, 2011) (M. Vishwas, Sharnappa Joladarashi and Kulkarni, 2017) Clay minerals were obtained from industrial locations and used as basic materials for the rubber clay composite (Tin Tin Aye, N.N.M. 2008). The performance of a shoe sole is determined by the base polymer used, as well as other compounds used and their appropriate concentrations (Nagahara, Kanehisa and Fukunaga, 2018).

The research focusing on use of leather cutting waste as a replacement for clay dosage where systematically changed in increments of leather waste 0%, 10%, 20%, and 30% by weight percentage in moulded rubber shoe sole formula which is currently used in industry. The efforts were made to optimize the industry available formula meet the sustainability purposes. Physio-mechanical properties of rubber leather waste composite such as Hardness, Tensile Strength, Elongation at break, Specific gravity,300% Modulus, Tension set, Tear strength, Abrasion Resistance were tested.

2 Materials and Methods

Rubbers and chemicals were bought from Samson Compounds (Pvt) Ltd. Leather waste were collected which are generated as result of cutting process from D. Samson Industries (PVT) LTD Collected fully homogenized shoe upper cutting leather waste were ground into short fibers and sieved through 30 mesh (Figs. 1, 2 and 3).

Next leather waste was tested for the moisture content weight percentage whose value was <3% were used for compounding. Determination of particle size was carried out using "Fritsch - analysette 22 Nano Tec " particle size analyser, measuring range 0.08 μm to 2100 μm in the wet mode method. A laboratory scale ban bury mixer (Protech industrial eng.co ltd CHSM 2.3, 2.5 Horse Power) and two-roll mill (Protech industrial eng.co ltd CHSM 2.8, 7.5Hourse Power) was used for blending the compounds. Rubber masticated for 4 min in the Banbury mixture. Next resin, pigments, coupling agent, stearic acid, zinc oxide were added and mixed for 3–4 min. Finally, fillers including leather waste and processing oils were added and mixing was carried out for 6 min also temperature inside of the mixer indicated as 148 °C at the end. After compound getting back to the room temperature compound mixture put into the open two roll mill. Vulcanizing agents were added to the compound. The compound was mixing another 10

Fig. 1. Generated Leather waste at Industrial site

Fig. 2. Ground Leather waste in to short fibres

Fig. 3. Sieved Leather waste 30 mesh

min approximately doing 5–8 mill passing. During this compounding process in order to get the uniform compounding and appropriate dispersion of the ingredients in the rubbery matrix proper nip cap and cuts were maintained Following that, the compositions were compressed moulded for 15 min at 150 °C. To ensure complete curing (100% cure), the cure time was set significantly higher than t 95 (time to 95% cure). The t 95 value was calculated by measuring the increase in torque values of none-vulcanized rubber compounds at 180° Different test specimens were obtained after curing as moulded products (Fig. 4 and Table 1).

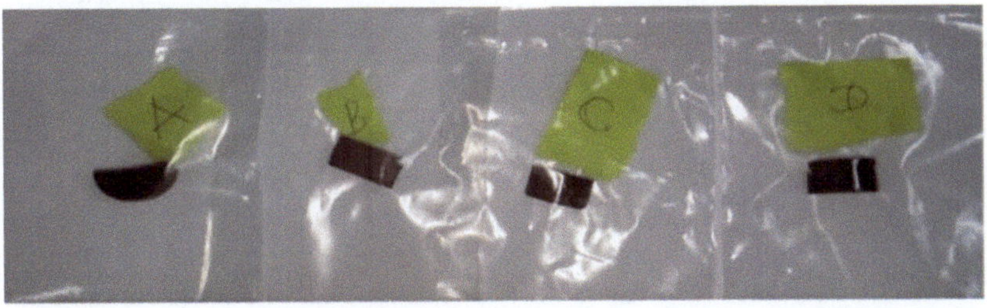

Fig. 4. Test specimens

3 Results and Discussion

According to Fig. 5. Test results were graphically represented and results of sample A (Control) (0%) was taken as reference values. When leather fiber was added, the mobility of the rubber chain decreased, which may cause increasing hardness. Although disruption of Polymer Chains with the addition of filler particles can disrupt the packing and alignment of polymer chains within the matrix will lead reduces the material's resistance to indentation. The composite may experience greater leather waste concentration and density as a result of agglomeration, while specific gravity decrement may be due to leather waste added could have absorbed water molecules. As the leather waste content rises further leather bundles made of fiber-to-fiber bonding aggregate may be making the composite weaker when an outside force is applied may cause for the overall decrement of tensile strength. The overall toughness and elongation at break of the composite material are reduced as leather waste content rises because the proportion of tough, deformable polymer matrix drops relative to the rigid leather waste particles. The increase in the 300% modulus in-dictates that these leather waste successfully reinforce the polymer matrix, adding to the material's stiffness and resistance to deformation. Result of filler-filler interaction which raise tension set by preventing matrix from reverting to its original form after tensile stress. Increased leather waste content contributes to the matrix's ability to resist crack growth and strengthen tears. Resistance to abrasive wear is the results of this increased filler-matrix interaction but weak filler-matrix inter-faces decrease abrasion resistance strengthen tears. Resistance to abrasive wear is the results of this increased filler-matrix interaction but weak filler-matrix interfaces decrease abrasion resistance (Tables 2, 3 and Figs. 6, 7, 8, 9, 10, 11, 12, 13 and 14).

Integrating leather waste fillers into polymer matrices has demonstrated significant promise in improving the performance attributes of composite materials. In-depth examination shows that these leather waste additives are evenly distributed within the polymer matrix, demonstrating slight clumping. This consistent distribution is essential, as it guarantees that the composite's mechanical properties are maximized, avoiding the creation

Table 1. Sample formulation

Ingredient	Quantity in Phr			
	A	B	C	D
SBR	17.5	17.5	17.5	17.5
HSR	12.5	12.5	12.5	12.5
RSS	70.0	70.0	70.0	70.0
Processing oils	12.5	12.5	12.5	12.5
Silica	25.0	25.0	25.0	25.0
Clay	30	27	24	21
Leather Waste	-	3.0	6.0	9.0
Pigments	0.5	0.5	0.5	0.5
Antioxidant	1.0	1.0	1.0	1.0
Stearic Acid	2.0	2.0	2.0	2.0
Zinc Oxide	5.0	5.0	5.0	5.0
Pigment 1	1.9	1.9	1.9	1.9
Pigment 2	0.5	0.5	0.5	0.5
Processing Aids 1	2.5	2.5	2.5	2.5
Processing Aids 2	2.0	2.0	2.0	2.0
Resin	3.0	3.0	3.0	3.0
Wax	1.0	1.0	1.0	1.0
Coupling Agent	1.25	1.25	1.25	1.25
Sulfur	2.5	2.5	2.5	2.5
Accelerators 1	0.18	0.18	0.18	0.18
Accelerators 2	0.5	0.5	0.5	0.5
Accelerators 3	2.7	2.7	2.7	2.7
Accelerators 4	0.3	0.3	0.3	0.3

of weak spots that might undermine the overall structural integrity. (SEM) images offer strong proof of this occurrence, Myint et al. (2008) distinctly showing that there are no noticeable indications of fractures or cracks in the polymer matrix encasing the leather waste filler particles. The lack of structural damage is important, as it suggests a robust and stable interaction between the polymer and the leather waste, allowing for efficient load transfer under mechanical stress. This strong bonding is crucial for preserving the composite's integrity, especially in dynamic situations where flexibility and resilience are necessary. Additionally, the absence of flaws like cracks or breaks not only improves the composite's durability but also aids in its longevity, rendering it appropriate for a range of applications.

Table 2. Test Results of prepared samples

Properties	unit	Test standard method	A(control) 0%	B 10%	C 20%	D 30%
Hardness	IRHD	ISO 48-2010	64.7	65.5	71.2	64.4
Specific gravity			1.162	1.174	1.151	1.176
Tensile strength	MPa	ISO 37-2005	14.3	13.6	11.6	12.7
Elongation at break	%	ISO 37-2005	674	560	458.1	479
300% Modulus	MPa	ISO 37-2005	2.9	5.2	6.5	6.6
Tension set	%	ISO 2285-2007	21	22	26	28
Tear strength	N/mm	ISO 34-12010	37.4	38.2	48.4	49
Abrasion resistance	mm^3	DIN 53516	295	321	224	334

Table 3. Test Results of Leather waste

Particle size	μm	10%	135.12	20%	602.24	90%	1002.08
Moisture content	%	Sample 1	2.78	Sample 2	2.62	Sample 3	2.52

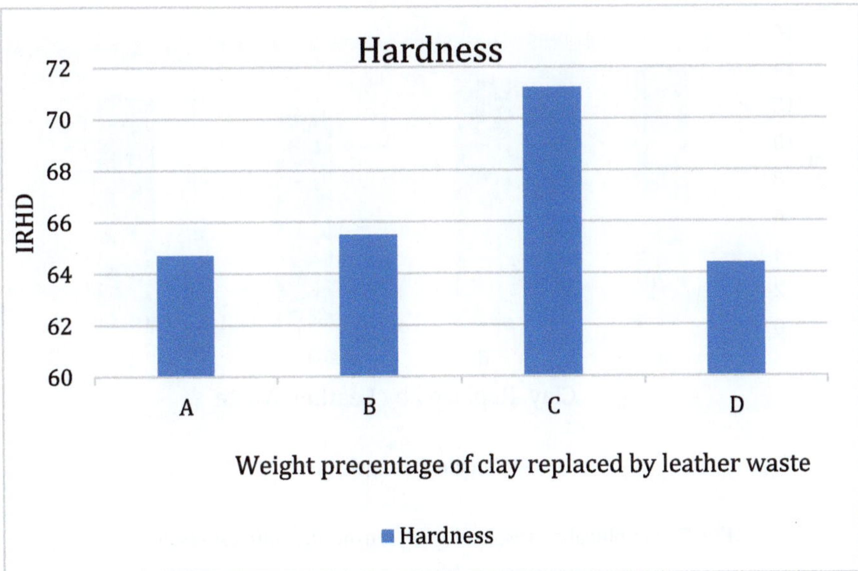

Fig. 5. Graphical representation of Hardness test results

These results highlight the twofold advantages of using leather waste fillers: they contribute to sustainability by recycling waste products while also improving the mechanical

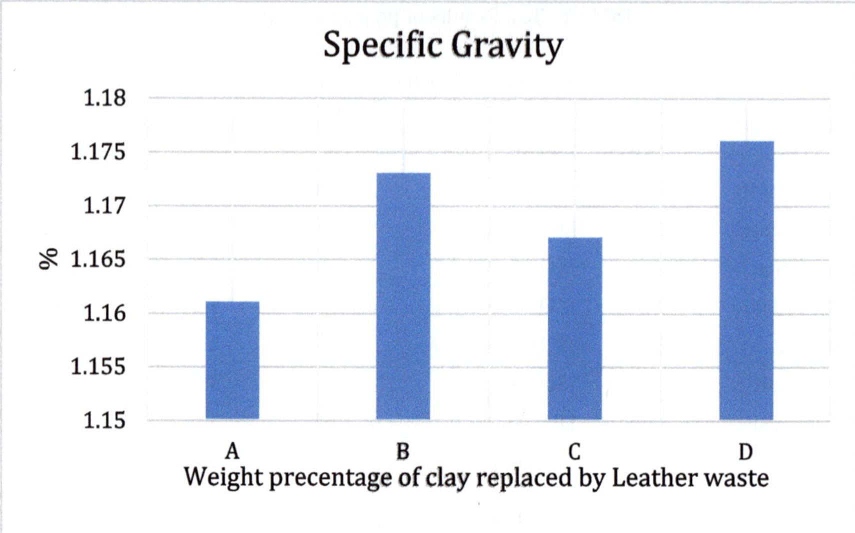

Fig. 6. Graphical representation of specific gravity test results

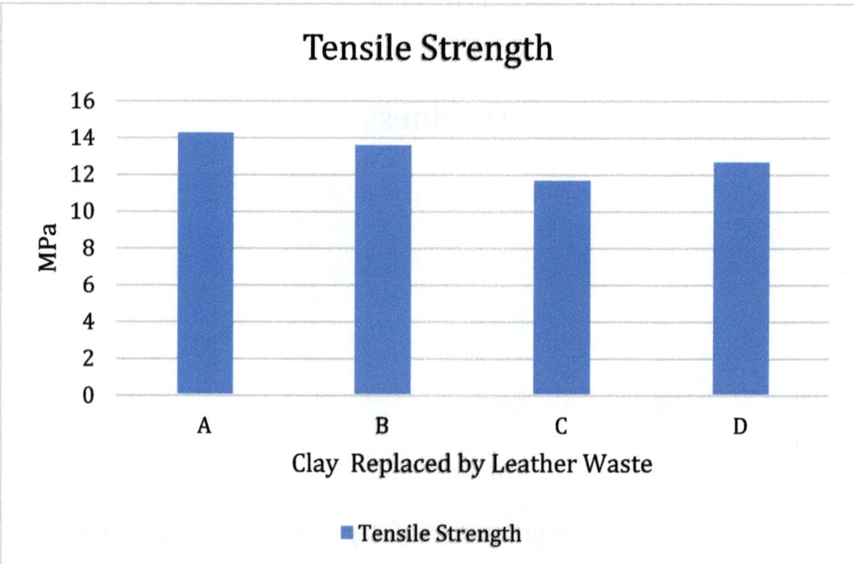

Fig. 7. Graphical representation of Tensile strength test results

properties of the composites produced. This ground breaking method creates fresh opportunities for the advancement of high-performance, eco-friendly materials across diverse sectors, including footwear and automotive uses, emphasizing the promise of a more sustainable future in materials science.

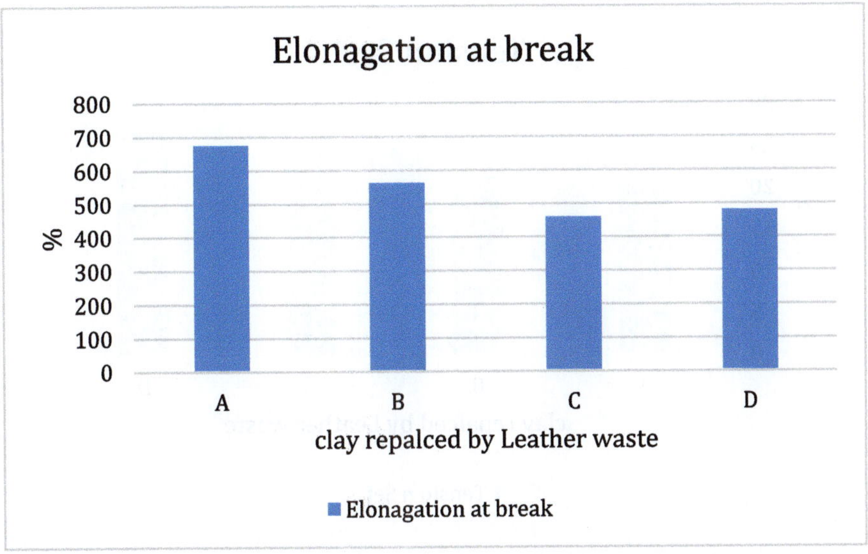

Fig. 8. Graphical representation of Elongation at break test results

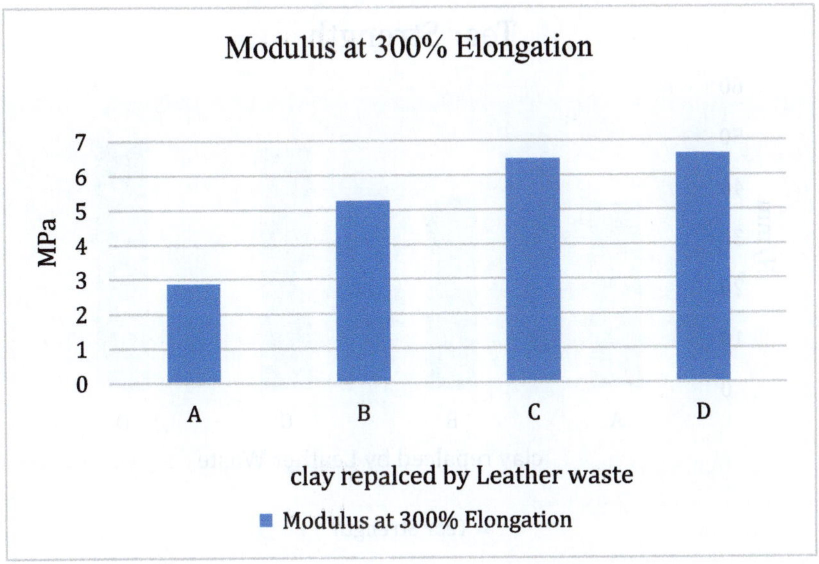

Fig. 9. Graphical representation of Modulus at 300% Elongation test results

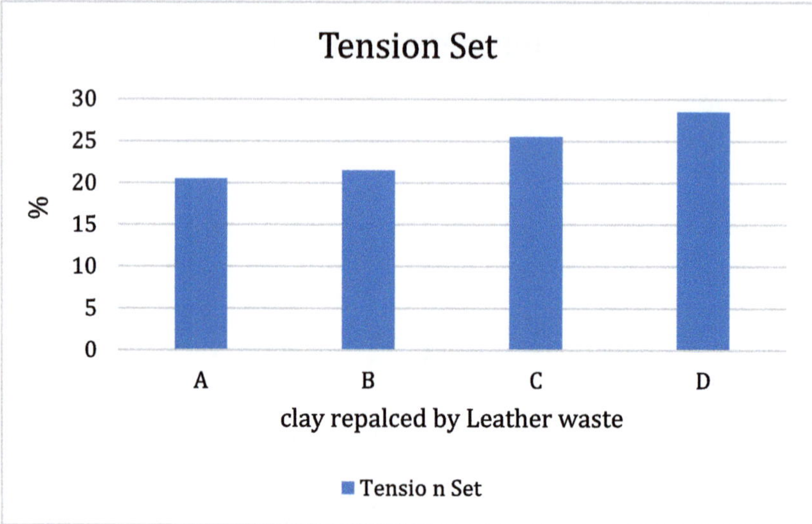

Fig. 10. Graphical representation of Tension set test results

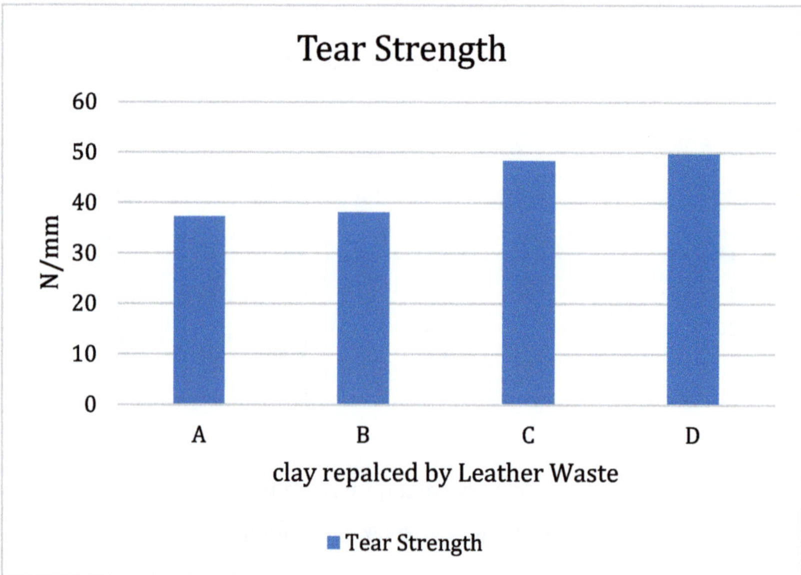

Fig. 11. Graphical representation of Tear strength test results

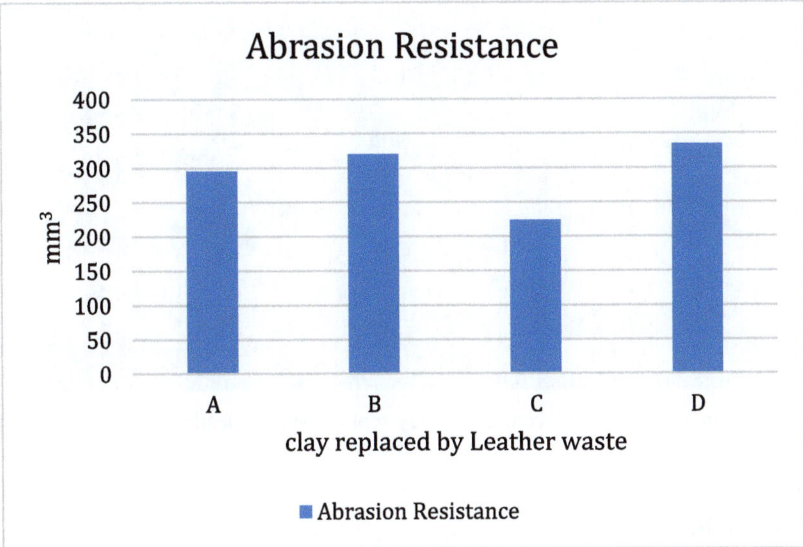

Fig. 12. Graphical representation of Abrasion resistance test results

Fig. 13. SEM image of 0% percentage of clay replacement by leather waste (Magnification 500x)

Fig. 14. SEM image of 20% percentage of clay replacement by leather waste (Magnification 500x)

4 Conclusion

The innovative use of a composite material that replaces 20% of leather waste with clay in footwear soles offers substantial benefits in both performance and sustainability, marking a significant advancement in the shoe industry. This approach enhances the stability and durability of shoes, making them particularly well-suited for activities that demand a solid base and resilience under strain, Thuy Hang et al. (2020) such as sports and outdoor pursuits. The incorporation of clay contributes to a remarkable 300% modulus, which allows the soles to endure considerable deformation without compromising their structural integrity an essential feature for footwear subjected to dynamic movements. Furthermore, the greater tension set achieved through this composite material provides critical support for rapid direction changes, enhancing the overall performance and safety of the footwear. The abrasion-resistant properties of the soles ensure longevity, particularly on harsh surfaces like concrete and gravel, reducing the need for frequent re-placements and thereby minimizing waste. Beyond its performance advantages, this method addresses pressing environmental concerns by effectively reducing leather waste and utilizing eco-friendly materials, aligning with the increasing consumer demand for sustainable products. The positive reception of this composite material suggests promising potential for future innovations in footwear design that prioritize both high performance and ecological responsibility, ultimately paving the way for a more sustainable shoe industry that meets the needs of modern consumers while fostering a more responsible approach to manufacturing.

Acknowledgements. I express my thanks to my colleagues' University of Kelaniya for their collaboration and insightful feedback, which greatly enhanced the quality of my work. Special appreciation goes to Samson compounds Pvt Ltd and D Samson Industries for providing the necessary resources and facilities that facilitated my research, which made this project possible. I would like to thank my family and friends for their unwavering encouragement and belief in me during this journey.

References

Ferreira, M.J., Almeida, M.F., Freitas, F.: Formulation and characterization of leather and rubber wastes composites. Polym. Eng. Sci. **51**(7), 1418–1427 (2011). https://doi.org/10.1002/pen.21643

Hang, L.T., Viet, D.Q., Linh, N.P.D., Doan, V.A., Dang, H.L.T., Dao, V.D., Tuan, P.A.: Utilization of leather waste fibers in polymer matrix composites based on acrylonitrile-butadiene rubber. Polymers **13**(1), 117 (2020). https://doi.org/10.3390/polym13010117

Myint, N., Myo Naing, K., Wynn, N.: Performance study of the natural rubber composite with clay minerals. MM 1000119 J. Myan. Acad. Arts Sc, VI (1) (2008). https://inis.iaea.org/collection/NCLCollectionStore/_Public/41/086/41086844.pdf

Nagahara, R., Kanehisa, H., Fukunaga, T.: Influence of shoe sole bending stiffness on sprinting performance. J. Sports Med. Phys. Fitness **58**(12) (2018). https://doi.org/10.23736/s0022-4707.17.07834-3

Santos, R.J., Agostini, D.L.S., Cabrera, F.C., Budemberg, E.R., Job, A.E.: Recycling leather waste: preparing and studying the microstructure, mechanical, and rheological properties of leather waste/rubber composite. Polym. Compos. **36**(12), 2275–2281 (2014). https://doi.org/10.1002/pc.23140

Ponsubbiah, S., Gupta, S.: Characterisation of shoe soling material prepared by using rubber and leather solid waste. Int. J. Eng. Adv. Technol. **9**(2), 1539–1545 (2019a). https://doi.org/10.35940/ijeat.a1818.129219

Ponsubbiah, S., Gupta, S.: Preparation of low cost shoe soling material from leather waste. Int. J. Res. Advent Technol. **7**(3) (2019b)

Hang, L.T., Linh, N.P.D., Vu, D.A.: The effect of leather fibers on vulcanization behavior of natural rubber. Vietnam J. Sci. Technol. **55**(1B), 85 (2018). https://doi.org/10.15625/2525-2518/55/1B/12095

Isolation and Characterization of Polyhydroxyalkanoate Bio Plastics Producing Bacteria from Putrefied Rice

Ahmed Ahsan, Charitha Basnayaka, Maheshi Somasiri, Zumaira Nazeer, Nirath Thilini, and Eustace Y. Fernando(✉)

Department of Biological Sciences, Faculty of Applied Sciences, Rajarata University of Sri Lanka, Mihintale 50300, Sri Lanka
eustace6192@as.rjt.ac.lk

Abstract. Polyhydroxyalkanoates (PHAs) are internal microbial polyesters that are produced when there is an abundance of carbon sources and a scarcity of nutrients like nitrogen. They function as energy storage carbon reservoirs and provide prokaryotes with stress tolerance. These biopolymers are viable substitutes for carbon-based synthetic polymers (plastics) and have been previously isolated from bacteria growing on putrefied rice. The objective of this study was to isolate and characterize PHA and PHA-producing bacteria from the natural environment and to synthesize PHA-type biopolymer through a laboratory-scale procedure. Boiled and putrefied rice samples were utilized to collect, culture and isolate bacteria for the current study. These bacteria were then screened for PHA production capability. Sudan III dye staining was conducted to confirm the presence of PHA accumulation in the bacteria. When provided with a complex growth medium, bacteria growing in the medium exhibited diauxic growth. Molecular identification based-on 16srRNA marker gene confirmed the identity of the isolated PHA producing bacterium as *Bacillus tropicus* strain MCCC 1A01406. Polyhydroxyalkanoate produced from a cultured bacterial sample was characterized using Raman spectroscopy. Results from Raman analysis confirmed that the PHA produced was a copolymer between 3-hydroxybutyrate(3HB) and 3-hydroxyvalerate(3HV) units. Characteristic Raman markers confirm the presence of this copolymer which simultaneously occurred at the spectral positions 436 cm^{-1}, 837 cm^{-1} and 1726 cm^{-1} in the Raman spectra. The diauxic growth together with the production of a copolymer confirms that the bacteria used in the present study were feeding on more than one carbon substrate. This study clearly demonstrated that free-living environmental organisms can be isolated and conveniently used for the purpose of polyhydroxybutyrate (PHB) bioplastics production.

Keywords: *Bacillus tropicus* · PHA · PHA copolymer · microbial polyesters · Raman spectroscopy

1 Introduction

Plastic pollution persists as a global environmental challenge, with its adverse impacts on biotic and abiotic components being documented for over half a century. The environmental challenges posed by plastics can be reduced by incorporating biopolymers

R. Dissanayake et al. (Eds.): ICRES 2025, PTT, pp. 332–346, 2025.
https://doi.org/10.1007/978-981-96-9551-5_26

and bio-based fibers derived from bioresources and renewable waste. The emergence of bio-based plastics represents a crucial advancement, promoting the bioeconomy and decreasing dependence on fossil fuel-based materials (Nanda et al., 2022). It is clear that bioplastics have been developed using starch, cellulose, and protein as alternatives (XZhao et al., 2023). Additionally, microbial polyesters, produced by widely available bacteria, offer another viable option (de Sousa et al., 2024). The family of biodegradable polyesters known as polyhydroxyalkanoates (PHAs) is one of the natural polymers that is formed by a wide range of microorganisms for the storage of intracellular carbon and energy as part of their survival mechanism (Li, Yang & Loh, 2016). They are synthesized in the presence of excess carbon and limiting nutritional components like N, P, S, O, or Mg (Yup Lee et al., 1999). These polyhydroxyalkanoates are usually found as large inclusion bodies within bacterial cells (Vicente et al., 2023). Bacteria of several different genera can make PHAs including *Bacillus* sp, Pseudomonas sp, and *Ralstonia* sp, (Kim et al. 1994, Preusting, Hazenberg &Witholt, 1993, Obruca et al., 2011). While PHAs are completely biodegradable in soil, they are robust enough for regular use and closely mimic the physical and elastomeric properties of their synthetic and petrochemical plastics counterparts. Many soil microorganisms degrade PHA at a rapid rate (duration of around 90–270 days), converting it to CO_2 and H_2O by their own secreted PHA hydrolases (Shahid et al., 2021). Thus, they can be considered biopolymers, which are potential alternatives for the petrochemical and synthetic plastic materials that we use frequently in our lives. The quantification of individual PHA types in complex polymer mixtures produced by bacteria is essential not only for answering microbiological questions but also for biotechnological and ecological problems. For instance, it is hypothesized that a high ratio of PHA to phospholipid fatty acids is a sign of an imbalanced or stressed environment for the microorganisms (Koller et al., 2011).

Polymers of PHA can be categorized based on the length of the carbon chain in the monomeric form. Polyhydroxybutyrate (PHB) is the most common type of PHA and is produced by many different bacteria (McAdam et al., 2020). It has a high melting point and is brittle, making it difficult to process (McAdam et al., 2020). Polyhydroxyvalerate (PHV) is a type of PHA which has a lower melting point than PHB and is more flexible, making it easier to process (Kim et al., 2023). It is produced by some bacteria, including *Pseudomonas* sp. and *Alcaligenes* sp. (McLeod et al., 2020). Polyhydroxybutyrate-co-valerate (PHBV) is a copolymer of PHB and PHV and has properties that are intermediate between the two homopolymers. It is produced by various bacteria, including *Ralstonia eutropha* (Parikh et al., 1998). Polyhydroxyhexanoate (PHHx) is a type of PHA which has a longer side chain than PHB and is produced by some bacteria, including *Pseudomonas putida* (Sikkema et al., 2023). Polyhydroxyoctanoate (PHO) is a type of PHA has an even longer side chain than PHHx and is produced by some bacteria, including *Aeromonas hydrophila* (Riaz, Rhee & Park, 2021).

Acetyl Co-A units act as the raw material for PHB biosynthesis. All three enzymes of the PHB biosynthesis are organized into an operon (consisting of genes encoding for enzymes phaA, phaB and phaC), where the synthesis of PHB can either be up or down regulated depending on cellular stress conditions. (de Vogel et al., 2021). Polyhydroxyalkanoate hydrolases (phaY) are enzymes that hydrolyse PHAs into smaller molecules like hydroxy alkanoic acids, which the bacteria can use as a source of carbon and energy.

Hydrolase-based degradation is a critical step in the recycling of PHA polymers and the maintenance of a stable metabolism in bacterial cells. This method offers a sustainable and environmentally responsible way to generate these valuable biopolymers, making it significant for the manufacture of PHA monomers for industrial applications. These hydrolases are specific enzymes that recognize and bind to the ester bonds connecting the PHA polymer chain's monomers. They then catalyze the hydrolysis of these bonds, releasing the individual monomers. After leaving the bacterial cell membrane, the liberated monomers enter the metabolic pathways for generating energy or biosynthesis (Zhou et al., 2023).

Apart from this, PHA is a biopolymer that can effectively replace synthetic and conventional polymers. It has a wide range of applications in medicine, from soft tissue engineering to creating cardiac stents (Pulingam et al., 2022) to hard tissue engineering for bone tissue regeneration (Gregory et al., 2022). Additionally, it can be utilized in tissue regeneration matrices for the healing of burn wounds (Sahana and Rekha, 2018) and injured nerves (Pulingam et al., 2022). Moreover, PHA microspheres can also be used for medication delivery purposes, particularly for the targeted release of pharmaceuticals like cancer chemotherapy drugs (Shrivastav, Kim and Kim, 2013).

Expanding the production of polyhydroxyalkanoate (PHA) bioplastics in Sri Lanka presents a sustainable alternative to conventional plastics and offers significant economic benefits. Environmental isolates of novel PHA-producing bacteria are crucial for large-scale biopolymer production. Studies have demonstrated that bacteria isolated from industrial waste environments, such as cardboard manufacturing wastewater and pulp industry sludge, can efficiently produce PHAs, highlighting the potential of utilizing local microbial resources for bioplastic production (Bhuwal et al., 2013). Currently, there is a paucity of research on PHA-related bioplastics in Sri Lanka, which hampers the country's potential economic growth in this sector. Investigating the adoptability of bioplastics as sustainable materials in the Sri Lankan construction industry could pave the way for new industries and employment opportunities, thereby bolstering the local economy (Muhammed et al., 2023). The global challenge of finding economically viable and environmentally friendly alternatives to traditional plastics is intensifying. Conventional plastics degrade slowly, persisting for hundreds of years and causing environmental pollution and harm to wildlife. In contrast, PHAs are biodegradable and can be produced from renewable resources, making them a more sustainable choice. However, to fully harness the potential of PHA-related bioplastics, further research is needed to optimize production processes and develop cost-effective, scalable manufacturing techniques (Mitra et al., 2020). Utilizing environmental isolates of PHA-producing bacteria and investing in research to optimize production processes are essential steps toward establishing a sustainable and economically beneficial PHA bioplastic production industry, especially in developing nations such as Sri Lanka.

The primary aim of this study was to isolate, identify & characterize PHA producing bacteria by obtaining putrefied rice sample in the environment and utilize them in the production of PHA type biopolymer via a laboratory-scale process. It is a proof of a concept study to isolate PHA producing organisms from the environment and to produce the polymer only at laboratory scale. Therefore, the objectives of this study are to isolate and Identify PHA producing bacteria from putrefied rice sample characterize PHA producing bacteria using molecular and biochemical methods and laboratory-scale production of PHA type biopolymer from PHA producing bacteria isolated from putrefied rice samples.

2 Materials and Methods

2.1 Isolation and Culturing of Bacteria Producing PHA

Bacteria producing polyhydroxyalkanoate (microbial polyesters) were isolated from putrefied cooked "Nadu" rice, which was obtained from a canteen in the faculty of applied sciences at Rajarata University was tightly sealed inside a container and kept for a few days (Fig. 1(A)). The bacteria obtained were cultured in a special agar medium (Fig. 1(B)) dedicated to bacteria producing PHA, known as the "Kannan and Rehacek medium prepared by adding 2 g of glucose, 0.25 g of yeast extract, 0.3 g of potassium chloride, 0.5 g of ammonium sulfate, 0.1 g of tryptone soya broth, and 2.5 g of agar per 100 ml (Valappil et al., 2008). After the appearence of bacterial colonies in the agar medium (after 2 days), one of those colonies was transferred to a 100 ml broth containing "Kannan and Rehacek" medium components. Bacteria were cultured in this broth for 2 days. Approximately 40 ml of the cultured broth were then transferred to 1 L of freshly prepared Kannan and Rehacek solutions poured into flasks (Fig. 1(C)), and the bacteria were allowed to be cultured at room temperature for around 62–72 h.

2.2 Biomass Measurement of PHA Producing Bacteria

The cultured broth was transferred into several 50-ml Falcon tubes and centrifuged at 5000 RPM for around 9 min. The cell pellets settled at the bottom of the centrifuge tube were collected, and the supernatant was discarded. Collected wet biomass inside Falcon tubes were prepared for drying (Fig. 1(D)). The wet biomass was oven dried until a constant weight at a temperature of 105 °C. The dried pellets were collected and prepared for solvent extraction of PHA (Fig. 1(D)). The mass of the pellets was calculated by taking the mass difference before and after placing the aluminium foil.

2.3 Extraction of PHA from the Biomass

The dried cell pellets were ground until they became a fine powder by using a mortar and pestle (Fig. 1(E and F)). This powder was then mixed with 30% H_2O_2 solution and pure chloroform at a ratio of 1:1 and shaken for about 30 min on a shaker platform. A separation funnel was used to separate the dead cell mass from the chloroform layer. The chloroform solution at the bottom, which contained the PHA, was filtered out into a 250ml beaker (Fig. 1(G)).

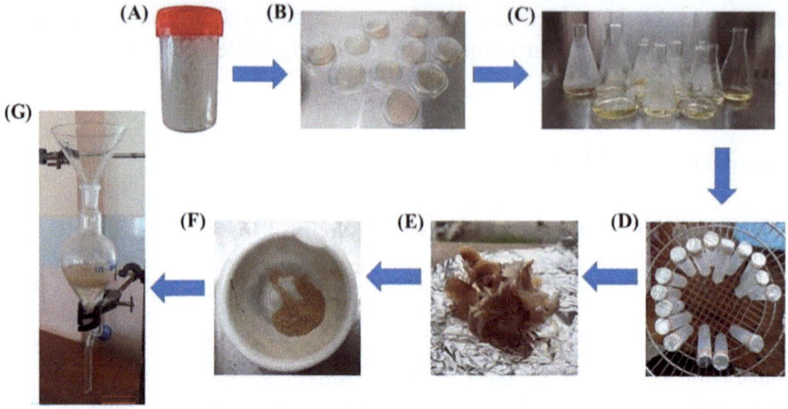

Fig. 1. Extraction procedure of PHA: (A) Putrefied source rice sample containing bacteria kept in a closed tightly packed container (B) Bacteria streaked into Kannan & Rehacek medium in culture plates (C) Flasks used to harvest bacteria obtained from single colony in Kannan & Rehacek broth (D) Centrifuged harvested bacterial cell pellets in falcon tubes wrapped with Aluminum foils (E) Dried bacterial cell pellets (F) Ground cell pellets (F) Separation of Chloroform layer using a separating funnel.

2.4 Solvent Precipitation and Separation of PHA

Four volumes of pure ice-cold methanol were slowly poured into the beaker containing PHA and chloroform. The beaker was then carefully covered on top using aluminum foil and kept in a freezer operating at −20 °C overnight. The PHA precipitations were immediately visible in the cold solvent mixture, and PHA flakes could be separated by filtration using Whatman grade 42 filter paper (Sigma Aldrich, UK). The separated PHA was air-dried and stored at room temperature for further analysis and characterization.

2.5 Microscopic Observation of PHA-Producing Bacteria

The bacteria producing PHA were detected using a lipophilic stain, Sudan III. Cells were grown for 48 h in Kannan and Rehacek medium containing Sudan III: 0.5 g $(NH_4)_2SO_4$, 0.25g Difco yeast extract, 0.3g KCl, 0.1g Tryptone Soya Broth, 2g D-glucose, 2.5g bacteriological agar, and 9 ml Sudan III made by mixing 0.5% solution (w/v) in ethanol. Smears of the bacteria producing PHA were prepared on a glass slide and heat-fixed. Stained samples were observed through Optikam B5 digital camera attached to a Carl Zeiss Primo Star microscope (Carl Zeiss, Germany).

2.6 Bacterial Growth Measurement

Bacteria grown for 48 h in a Kannan and Rehacek medium were inoculated into a 75-ml broth containing the Kannan and Rehacek components. Samples drawn from the culture liquid at regular time intervals were heat fixed on glass slides and the cells were Gram-stained (with Crystal Violet for 60 s, Gram's Idoine for 60 s, alcohol (95%) for 20 s, and Safranin for 60 s) for observation under a Nikon Eclipse 100 LED MV R Microscope.

Bacterial density was measured at each interval using the "Breed Method" (Vincent, 1918).

$$\text{bacterial density/mL} = \frac{ND^2}{nVd^2}$$

where:

N = total number of bacterial cells counted
D = diameter of suspension (10 mm)
n = number of fields counted
V = volume of suspensions (50 L)
d = diameter of field under high power objective (0.46 mm)

2.7 Molecular Microbiological Characterization and Phylogenetic Analysis of PHA Producing Bacteria

Isolated PHA producing bacterium from Kannan & Rehacek medium was subjected to molecular characterization by using 16s ribosomal RNA (16s rRNA) marker gene (Dissanayake et al., 2021). The total genomic DNA from single colonies of isolated bacterium from the Kannan and Rehacek medium were purified as per the manufacturer's instructions using a WizardTM genomic DNA extraction kit (Promega Corporation, USA). The 16s rRNA gene was amplified from the extracted genomic DNA using the primer pair 27F ((5′ - AGA GTT TGA TCM TGG CTC AG - 3′) and 907R (5′ - CCG TCA ATT CCT TTR AGT TT - 3′). The PCR master mix (Promega, USA) was used as per the manufacturer instructions to amplify the 16s rRNA marker gene, in a PCR thermocycler (Thermo-Fisher, UK) under the following set of conditions; initial denaturation at 95 °C for 4 min, followed by 30 cycles of 95 °C for 0.5 min, 58 °C for 1 min, 72 °C for 0.5 min, and finally at 72 °C for 7 min. PCR products were confirmed and verified on 1% agarose gels before being sequenced using the Sanger DNA sequencing technology at Eton Bioscience Sequencing Services, New Jersey, USA. The consensus sequences from the bidirectional raw reads were assembled using the program BioEdit, and the nucleotide sequences acquired from the bidirectional sequences were verified for chimeras using the software DECIPHER. Using the NCBI BLAST nucleotide search tool, fully assembled and quality-checked consensus sequences were compared with the 16s rRNA gene entries in the databases NCBI GenBank 16S rRNA nucleotide sequence repository and Ribosomal Database Project (RDP). Phylogenetic analysis of the sequences was conducted as described in earlier studies (Dissanayake et al., 2021 and Fernando et al., 2013) by using Mega 7.0.026 molecular evolutionary genetics analysis software with maximum likelihood method (1000 bootstrap replicates). *Escherichia coli* reference 16S rRNA sequence was used as the out-group of the rooted phylogenetic tree.

2.8 Raman Micro-Spectroscopic Characterization of PHA

PHA extracts from experiments were dissolved in Chloroform and mounted directly on Raman spectroscopy-grade CaF_2 slides (Crysran, UK). Samples mounted were evaporated to dryness to leave an ultra-thin film of PHA biopolymer on the CaF_2 Raman windows, as described previously (Fernando et al., 2019). Briefly, Raman microspectroscopy was conducted using a Reneshaw™ InVia (UK) Raman microscope and the Raman excitation laser beam had a wavelength of 514 nm. The diffraction grating of the Raman instrument used during all analyses was a 1200 grooves/mm. The incident Raman laser beam was attenuated down to 5% of the total power using a neutral density (ND) filter. The excitation laser of the Raman system was focused onto the biofilm sample using a 40 X air objective with a numerical aperture of 1.0. The return Raman signal was collected using the same objective and the Raman spectra were recorded on a CCD detector cooled to - 70 °C. Raman spectra from the biofilm samples were collected within the spectral range of 200 cm^{-1} to 1700 cm^{-1}. The Raman system was calibrated to the first-order scattering signal of a standard silicon wafer at 520.7 cm^{-1}. All spectra were recorded using the in-built Wire 5.0 software of the Raman system (Renishaw Instruments, UK).

3 Results and Discussion

3.1 Characterization of PHA Producing Bacteria

3.1.1 Yield Coefficient of PHA Producing Bacteria

The calculated yield coefficient for PHA ($Y_{X/S}$) of the cell pellets obtained by centrifugation was 0.0662, indicating the conversion efficiency of substrate into the intended form of biomass. Maximum yield results in optimal PHA production and usually it's also dependant upon downstream processing steps such as concentrating cells through centrifugation, cross-flow filtration or flocculation (Zinn et al., 2001). However, in our experiment, cells were not cultured for more than 72 h to prevent the loss of already produced PHA. When bacterial cells exhaust the available carbon substrate, typically within 72 h, they begin utilizing the stored PHA as an energy source. Therefore it is essential to harvest the bacterial cells during the stationary phase of growth.

3.1.2 Colony Morphology and Microscopic Observation of PHA Producing Bacteria

The bacterial colonies appear white in colour and gummy in texture, with a well-rounded shape (Fig. 2(A)). The colonies become fully non-transparent after three days of inoculation. Figure 2(B) contrasts how these colonies appear when Sudan III is used on them to observe them through microscopic imaging. Sudan III is a nonpolar, fat-soluble dye that interacts hydrophobically with the lipid chains of hydrocarbons to color them red and thus can be used for the identification of PHA inside the bacterial cells (Staśczak et al., 2022).

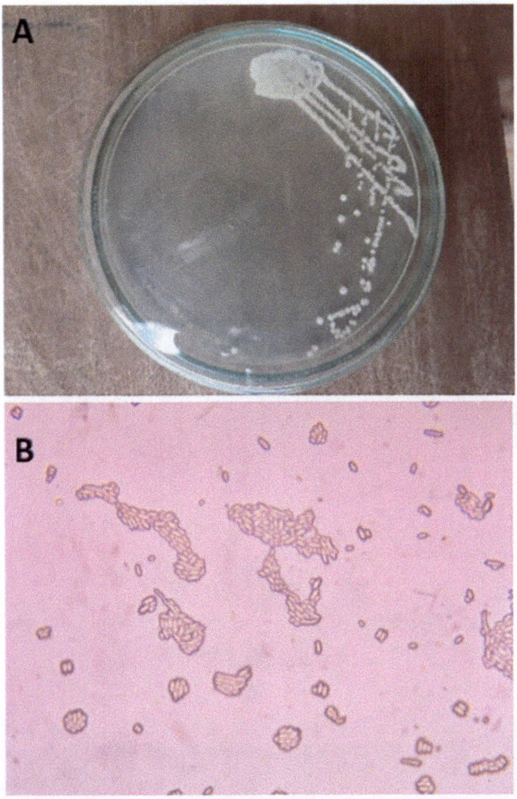

Fig. 2. (A) Unstained colonies appear opaque and off white in colour (B) Results from Sudan III staining indicates that the lipophilic dye has been absorbed by the bacteria (Red) forming marked contours around cells, thus confirming the presence of PHA within.

3.1.3 Cell Growth Measurement

When a single carbon source is used, a monoauxic growth curve can be observed. However, in our study, a diauxic growth curve was observed due to the selective and sequential use of two different carbon sources by the bacteria. Tryptone soy broth (TSB) and yeast extract are nutrient-rich sources in the Kannan and Rehacek medium, and one of the components in these sources could've acted as a secondary carbon source in addition to glucose. The use of two different carbon sources by bacteria during PHA production has also been noted in previous studies. The bacterium *Methylomonas extorquens* displays diauxic growth by first converting methanol before using glycerol as a carbon substrate when non-purified crude glycerol phase (CGP) medium is used (Koller et al., 2017). Wild-type E. coli has been known to utilize glucose, followed by the slow release of Carbon Catabolite Repression (CCR), resulting in the slow utilization of xylose, during PHA production (Li et al., 2007). However, even though a clear secondary source of carbon in addition to glucose was not used in this experiment, the diauxic growth (Fig. 3(A)) may be due to the utilization of proteins present in yeast extract as a secondary carbon

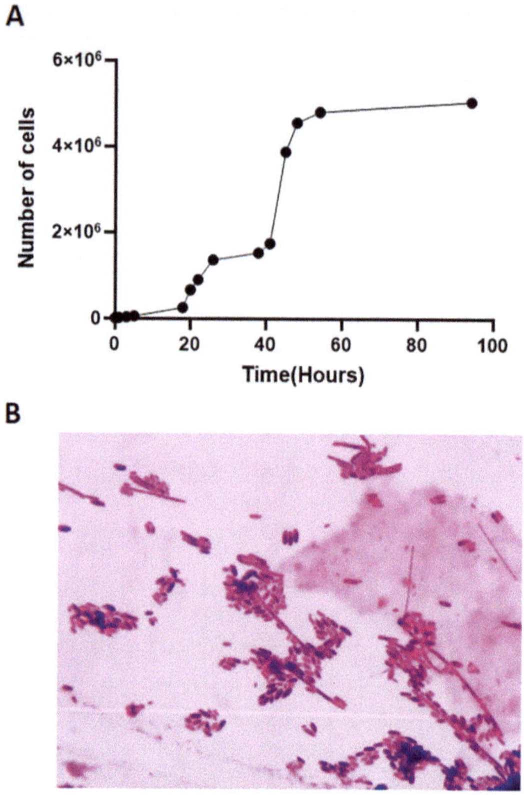

Fig. 3. (A) The growth curve observed from an unknown bacterium indicates diauxic growth, thus confirming the bacteria sequentially uses different substrates supplemented in the growth medium. (B) Bacterial aggregates under gram stain.

containing substrate for growth. A study conducted in Queensland, Australia, demonstrates how low-cost mixed microbial PHA production was successful when protein-rich waste streams were used as a substrate (Oliveira et al., 2018). Gram staining was used to identify the PHA producing bacteria and make the enumeration process easier. As it appears in Fig. 3(B), the bacterial aggregates appeared clearly purple and therefore they were confirmed as gram positive.

3.1.4 Molecular Characterization of Isolated PHA Producing Bacteria

The closest match to the bacterial isolate obtained using the Kannan and Rehacek medium with a percentage identity of 99.86% was a *Bacillus tropicus* strain MCCC 1A01406, with the accession number NR_157736 in the NCBI database. *Bacillus* species are well known for PHA production. *Bacillus* sp. Involved in PHA copolymer production with co-metabolism of different substrates have also been previously recorded. *Bacillus thuringiensis* EGU45 in order to create PHA co-polymers, may co-metabolize a variety of different and structurally unrelated Carbon sources, including Crude Glycerol and Glucose (Ray & Kalia, 2017). This also explains the diauxic growth pattern observed in Fig. 3(A). The synthesis of the PHA polymers can happen through multiple pathways involving different *Pha* genes. The pathway following glycolysis often requires three genes, *PhaA* (acetyl-CoA acetyl transferase, *PhaB* (acetoacetyl-CoA reductase), and *PhaC* (PHA synthase) (de Vogel et al., 2021). However, the involvement of two different pathways, one containing the substrate propionyl-CoA, and the other containing acetoacetyl-CoA can lead to synthesis of the PHA Co-polymer poly(3-hydroxybutyrate-co-3-hydroxyvalerate) (Parveez, Kadir & Bahariah, 2008).

3.1.5 Phylogenetic Analysis of the PHA Producing Isolate

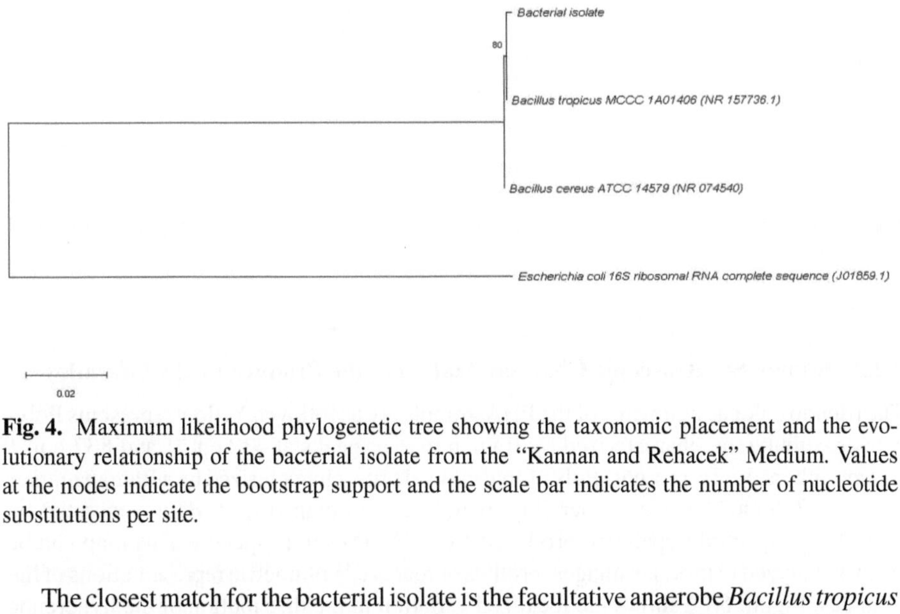

Fig. 4. Maximum likelihood phylogenetic tree showing the taxonomic placement and the evolutionary relationship of the bacterial isolate from the "Kannan and Rehacek" Medium. Values at the nodes indicate the bootstrap support and the scale bar indicates the number of nucleotide substitutions per site.

The closest match for the bacterial isolate is the facultative anaerobe *Bacillus tropicus* strain MCCC 1A01406 (Fig. 4). The bacterium *Bacillus cereus* ATCC 14579, which was previously well known for PHA production, shares a percentage identity of 99.79%, in the NCBI 16s ribosomal RNA marker gene database, with the bacterial isolate found in this study. Therefore, it can be concluded that the PHA producing property of the aforesaid *Bacillus tropics* MCC 1A01406 was hitherto unknown and have been conclusively demonstrated by the outcomes of this study.

3.2 Characterization of PHA

3.2.1 PHA Precipitation

Observation with the naked eye indicated that PHA biomass to be floating on the ice-cold methanol as a white-colored substance as shown in Fig. 5. The increased visibility of PHA after prolonged exposure to temperatures below 0 °C may be attributed to the ductile deformation of the PHA granule core (Lauzier & Marchessault, 1992) and the swelling of PHA granules, similar to the effects observed in an acetone-water solution (Lundgren & MmaCH, 1968). This method of PHA precipitation works well when no visible precipitation is observed immediately after using ice cold Methanol or when there is no possibility to get dried pellets for observation.

Fig. 5. PHA precipitation observed as white colour floating masses within the solvent in one of the experiment trials.

3.2.2 Raman Spectroscopic Characterization of the Produced PHA Granules

The pseudo color assignments of the PHA granule are as follows: Yellow represents Poly-3-hydroxybutyrate, characterized by Raman marker peaks at 432 cm^{-1} and 837 cm^{-1} wavenumbers. Red represents Poly-3-hydroxyvalerate, characterized by a Raman marker peak at 1726 cm^{-1} wavenumber (Fig. 6(A)). Raman map (Fig. 6(B) is generated by collecting a spectral hypercube produced from 11,000 single spectra. This map can be further analyzed to produce images, profiles or rendered volumetric representations of the granules. Figure 6(C) shows the localized variation in the distribution of the hypercube as a 3D plot. The peak with the highest intensity in Fig. 6(A) corresponds to the "Raman substrate" CaF_2 used in the instrument. Commercial glass microscope slides don't work as substrates because they have a very strong fluorescent background when excited in the near infrared (Fullwood et al., 2013). Evidence for the presence of hydroxy-3-butyrate and hydroxy-3-valarate molecules indicates that the polymer produced by the bacteria is co-polymer, poly-3-(hydroxybutyrate-co-hydroxyvalerate). The ratio of 3HB to 3HV in P(HB-co-HV) is dependent on the applied concentration of the substrates. In a previous study, the mass ratio of 3HV in P(HB-co-HV) was increased by a significant amount when the applied co-substrate, valerate concentration was increased (Koller et al., 2008). PHB & other Co-polymers of PHB, including P(HB-co-HV) have many diverse applications such as tissue and skin implants, 3D printing, drugs and drug coatings, nutritional supplements and chemical constituents of medical devices and implants (S. Das & Samantaray, 2017).

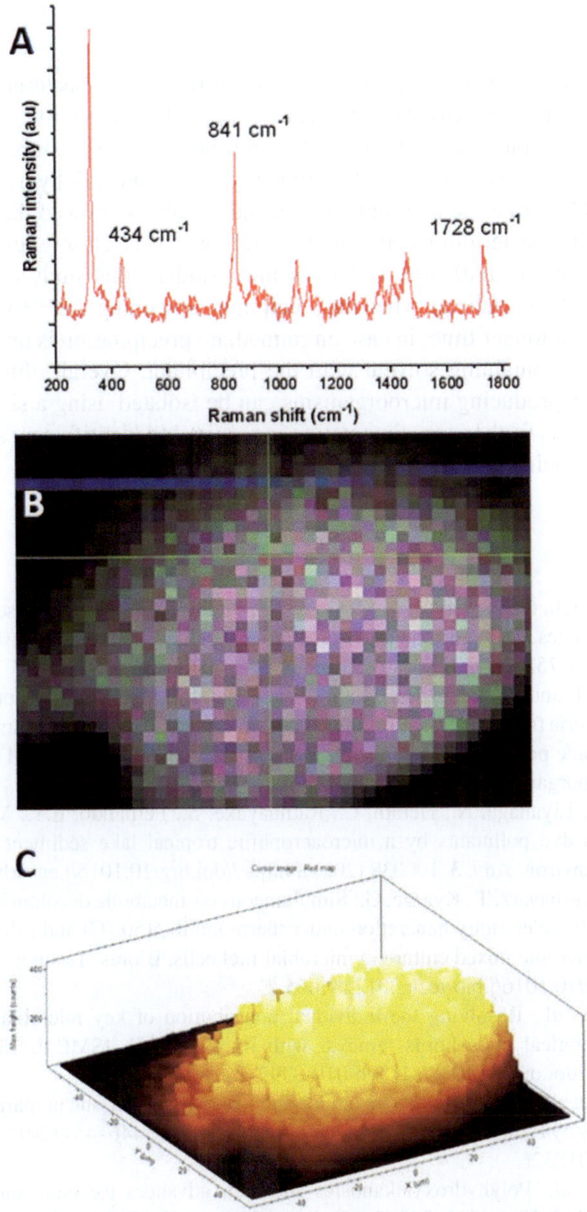

Fig. 6. Raman spectroscopic data for the PHA produced. (A) Representative Raman spectrum from a PHA droplet from this study indicates all of the signature peaks that are associated with the co-polymer poly-3-hydroxybutyrate – co – poly-3-hydroxyvalarate occurring at wavenumbers 434 cm^{-1}, 837 cm^{-1} and 1726 cm^{-1} respectively. (B) Raman polymer map of one the PHA granules, where green areas represent Poly-3-hydroxyvalarate (PHV) and purple areas represent Poly-3- hydroxybutyrate (PHB). (C) PHA granule 3D plotted using MatLab software.

4 Conclusion

This study demonstrates that PHA produced from free-living bacteria in the environment, isolated from putrefied rice samples can be used in PHB bioplastics production. The RAMAN spectroscopic results of this study confirmed that the produced PHA biopolymer is a co-polymer between Poly-3-hydroxybutyrate and Poly-3-hydroxyvalarate. The use of 16s rRNA marker gene based phylogenetic analysis showed that the PHA bioplastics producing bacterium isolated in this study was very closely related to *Bacillus tropicus* strain MCCC 1A01406. Unlike in most studies, this study proves that PHA precipitation can be visualized, when the precipitant containing the PHA is kept at lower temperatures for a longer time, in case an immediate precipitation is not observed after mixing the PHA containing solvent with the precipitant. Overall, this concept study proves that PHA producing microorganisms can be isolated using a simple laboratory procedure, and thus further experimentation is required to identify new species that can contribute to replacing petrochemical based plastics.

References

Bhuwal, A.K., Singh, G., Aggarwal, N.K., Goyal, V., Yadav, A.: Isolation and screening of polyhydroxyalkanoates producing bacteria from pulp, paper, and cardboard industry wastes. Int. J. Biomater. **2013**, 752821 (2013). https://doi.org/10.1155/2013/752821

de Vogel, F.A., Schlundt, C., Stote, R.E., Ratto, J.A., Amaral-Zettler, L.A.: Comparative genomics of marine bacteria from a historically defined plastic biodegradation consortium with the capacity to biodegrade polyhydroxyalkanoates. Microorganisms **9**(1), 1–26 (2021). https://doi.org/10.3390/microorganisms9010186

Dissanayake, M., Liyanage, N., Herath, C., Rathnayake, S., Fernando, E.Y.: Mineralization of persistent azo dye pollutants by a microaerophilic tropical lake sediment mixed bacterial consortium. Environ. Adv. **3**, 100038 (2021). https://doi.org/10.1016/j.envadv.2021.100038

Fernando, E.Y., Keshavarz, T., Kyazze, G.: Simultaneous co-metabolic decolourisation of azo dye mixtures and bio-electricity generation under thermophillic (50 °C) and saline conditions by an adapted anaerobic mixed culture in microbial fuel cells. Biores. Technol. **127**, 1–8 (2013). https://doi.org/10.1016/j.biortech.2012.09.065

Fernando, E.Y., et al.: Resolving the individual contribution of key microbial populations to enhanced biological phosphorus removal with Raman–FISH. ISME J. **13**(8), 1933–1946 (2019). https://doi.org/10.1038/s41396-019-0399-7

Fullwood, L.M., et al.: Effect of substrate choice and tissue type on tissue preparation for spectral histopathology by Raman microspectroscopy. Analyst **139**(2), 446–454 (2013). https://doi.org/10.1039/c3an01832f

Gregory, D.A., et al.: Polyhydroxyalkanoates and their advances for biomedical applications. Trends Mol. Med. **28**(4), 331–342 (2022). https://doi.org/10.1016/j.molmed.2022.01.007

Kim, B.S., Lee, S.C., Lee, S.Y., Chang, H.N., Chang, Y.K., Woo, S.I.: Production of poly(3-hydroxybutyric acid) by fed-batch culture of alcaligenes eutrophus with glucose concentration control. Biotechnol. Bioeng. **43**, 892–898 (1994). https://doi.org/10.1002/bit.260430908

Kim, J., et al.: Biodegradation studies of polyhydroxybutyrate and polyhydroxybutyrate-co-polyhydroxyvalerate films in soil. Int. J. Mol. Sci. **24**(8), 7638 (2023). https://doi.org/10.3390/ijms24087638

Koller, M., et al.: Polyhydroxyalkanoate production from whey by Pseudomonas hydrogenovora. Biores. Technol. **99**(11), 4854–4863 (2008). https://doi.org/10.1016/j.biortech.2007.09.049

Koller, M., Gasser, I., Schmid, F., Berg, G.: Linking ecology with economy: Insights into polyhydroxyalkanoate-producing microorganisms. Eng. Life Sci. **11**(3), 222–237 (2011). https://doi.org/10.1002/elsc.201000190

Koller, M., Maršálek, L., de Sousa Dias, M.M., Braunegg, G.: Producing microbial polyhydroxyalkanoate (PHA) biopolyesters in a sustainable manner. New Biotechnol. **37**, 24–38 (2017). https://doi.org/10.1016/j.nbt.2016.05.001

Lauzier, C., Marchessault, R.H., Smith, P., Chanzy, H.: Structural study of isolated poly (β-hydroxybutyrate) granules. Polymer **33**(4), 823–827 (1992)

Li, R., Chen, Q., Wang, P.G., Qi, Q.: A novel-designed Escherichia coli for the production of various polyhydroxyalkanoates from inexpensive substrate mixture. Appl. Microbiol. Biotechnol. **75**(5), 1103–1109 (2007). https://doi.org/10.1007/s00253-007-0903-2

Li, Z., Yang, J., Loh, X.J.: Polyhydroxyalkanoates: opening doors for a sustainable future. NPG Asia Mater. **8**(4), e265–e265 (2016). https://doi.org/10.1038/am.2016.48

Lundgren, D.G., MmaCH, R.H.: Morphology of Poly-/~-hydroxybutyrate Granules. J. Mol. Biol. **35** (1968)

McAdam, B., Fournet, M.B., McDonald, P., Mojicevic, M.: Production of polyhydroxybutyrate (PHB) and factors impacting its chemical and mechanical characteristics. Polymers **12**(12), 2908 (2020). https://doi.org/10.3390/polym12122908

Mitra, R., Xu, T., Xiang, H., Han, J.: Current developments on polyhydroxyalkanoates synthesis by using halophiles as a promising cell factory. Microb. Cell Fact. **19**(1), 86 (2020). https://doi.org/10.1186/s12934-020-01342-z

Muhammed, N.S., Gallage, S.D., Eranga, B.A.I., Madushanka, T.H.: Adoptability of bioplastic as a sustainable material in Sri Lankan building construction industry. In: World Construction Symposium, vol. 1, pp. 86–97 (2023). https://doi.org/10.31705/WCS.2023.8

Nanda, S., Patra, B.R., Patel, R., Bakos, J., Dalai, A.K.: Innovations in applications and prospects of bioplastics and biopolymers: a review. Environ. Chem. Lett. **20**(1), 379–395 (2022). https://doi.org/10.1007/s10311-021-01334-4

Obruca, S., Marova, I., Melusova, S., Mravcova, L.: Production of polyhydroxyalkanoates from cheese whey employing Bacillus megaterium CCM 2037. Ann. Microbiol. **61**(4), 947–953 (2011). https://doi.org/10.1007/s13213-011-0218-5

Oliveira, C.S.S., Silva, M.O.D., Silva, C.E., Carvalho, G., Reis, M.A.M.: Assessment of protein-rich cheese whey waste stream as a nutrients source for low-cost mixed microbial PHA production. Appl. Sci. **8**(10), 1817 (2018). https://doi.org/10.3390/app8101817

Parikh, M., Gross, R.A., McCarthy, S.P.: The influence of injection molding conditions on biodegradable polymers. J. Injection Molding Technol. **2**(1), 30–36 (1998)

Parveez, A., Kadir, G.: Transformation of PHB and PHBV genes driven by Maize Ubiquitin Promoter into Oil Palm for the Production of Biodegradable Plastics Malaysian Palm Oil Board. J. Oil Palm Res. (2008). https://www.researchgate.net/publication/233927836

Preusting, H., Hazenberg, W., Witholt, B.: Continuous production of poly(3-hydroxyalkanoates) by Pseudomonas oleovorans in a high-ceil-density, two-liquid-phase chemostat (1993)

Pulingam, T., Appaturi, J.N., Parumasivam, T., Ahmad, A., Sudesh, K.: Biomedical applications of polyhydroxyalkanoate in tissue engineering. Polymers **14**(11), 2141 (2022). https://doi.org/10.3390/polym14112141

Ray, S., Kalia, V.C.: Co-metabolism of substrates by Bacillus thuringiensis regulates polyhydroxyalkanoate co-polymer composition. Biores. Technol. **224**, 743–747 (2017). https://doi.org/10.1016/j.biortech.2016.11.089

Riaz, S., Rhee, K.Y., Park, S.J.: Polyhydroxyalkanoates (Phas): biopolymers for biofuel and biorefineries. Polymers **13**(2), 253 (2021). https://doi.org/10.3390/polym13020253

Maity, S., Das, S., Samantaray, D.P.: Effect of Vitamin on accumulation of PHB by Zobellella species under submerged fermentation process. Int. J. Curr. Microbiol. Appl. Sci. **6**(2), 1310–1316 (2017). https://doi.org/10.20546/ijcmas.2017.602.148

Sahana, T.G., Rekha, P.D.: Biopolymers: applications in wound healing and skin tissue engineering. Mol. Biol. Rep. **45**(6), 2857–2867 (2018). https://doi.org/10.1007/s11033-018-4296-3

Shahid, S., Razzaq, S., Farooq, R., Nazli, Z.: Polyhydroxyalkanoates: next generation natural biomolecules and a solution for the world's future economy. Int. J. Biol. Macromol. **166**, 297–321 (2021). https://doi.org/10.1016/j.ijbiomac.2020.10.187

Shrivastav, A., Kim, H.-Y., Kim, Y.-R.: Advances in the applications of polyhydroxyalkanoate nanoparticles for novel drug delivery system. BioMed Res. Int. **2013**, 1–12 (2013). https://doi.org/10.1155/2013/581684

Dirk Sikkema, W., Cal, A.J., Hathwaik, U.I., Orts, W.J., Lee, C.C.: Polyhydroxyalkanoate production in Pseudomonas putida from alkanoic acids of varying lengths. PLoS ONE **18**(7), e0284377 (2023). https://doi.org/10.1371/journal.pone.0284377

Staśczak, A., et al.: Optical evaluation of effects of energy substrates on PHB accumulation for bioplastic production. Energies **15**(22), 8390 (2022). https://doi.org/10.3390/en15228390

Valappil, S.P., Rai, R., Bucke, C., Roy, I.: Polyhydroxyalkanoate biosynthesis in Bacillus cereus SPV under varied limiting conditions and an insight into the biosynthetic genes involved. J. Appl. Microbiol. **104**(6), 1624–1635 (2008). https://doi.org/10.1111/j.1365-2672.2007.03678.x

Vicente, D., Proença, D.N., Morais, P.V.: The role of bacterial polyhydroalkanoate (PHA) in a sustainable future: a review on the biological diversity. Int. J. Environ. Res. Public Health **20**(4), 2959 (2023). https://doi.org/10.3390/ijerph20042959

Vincent, C.: The direct or breed method for counting bacteria in tomato catsup, pulp or paste. J. Bacteriol. **3**(2) (1918). https://journals.asm.org/journal/jb

XZhao, X., et al.: Sustainable bioplastics derived from renewable natural resources for food packaging. Matter **6**(1), 97–127 (2023). https://doi.org/10.1016/j.matt.2022.11.006

Lee, S.Y., Choi, J.-i, Wong, H.H.: Recent advances in polyhydroxyalkanoate production by bacterial fermentation: mini-review. Int. J. Biol. Macromol. **25**(1–3), 31–36 (1999). https://doi.org/10.1016/S0141-8130(99)00012-4

Zhou, W., Bergsma, S., Colpa, D.I., Euverink, G.-J., Krooneman, J.: Polyhydroxyalkanoates (PHAs) synthesis and degradation by microbes and applications towards a circular economy. J. Environ. Manag. **341**, 118033 (2023). https://doi.org/10.1016/j.jenvman.2023.118033

Zinn, M., Witholt, B., Egli, T.: Occurrence, synthesis and medical application of bacterial polyhydroxyalkanoate. Adv. Drug Deliv. Rev. **53**, 5–21 (2001). https://doi.org/10.1016/S0169-409X(01)00218-6

Life Cycle Assessment Comparison Between Conventional Concrete and Rice Husk Ash (RHA) Blended Concrete

G. D. R. I. Harischandra(✉), K. J. L. Fernando, P. H. Samarajeewa, and H. A. D. S. Buddika

Department of Civil Engineering, University of Peradeniya, Peradeniya 20400, Sri Lanka
ravishkairoshana@gmail.com

Abstract. Concrete is one of the most popular building materials used in the construction industry in the world. Cement is one of the primary materials used in concrete. Cement production results in enormous carbon dioxide emissions, leading to serious sustainability issues that badly affect the environment. As a solution, supplementary cementitious materials (SCMs) come into action. From various types of SCMs, agro-waste-based SCMs are becoming a promising solution. The potential of rice husk ash (RHA) as a supplementary cementitious material (SCM) to improve the environmental performance of concrete is examined in this study. RHA is a by-product of milling rice, and it can be used as a partial substitute for cement due to its pozzolanic qualities and high silica content. A Life Cycle Assessment (LCA) was carried out to evaluate the environmental effects of utilizing RHA in concrete, focusing on key indicators such as Global Warming Potential (GWP), energy consumption, and resource depletion. The results demonstrate that incorporating RHA significantly reduces the carbon footprint of concrete. Through the literature review, it can be seen that RHA can be used as an SCM in concrete production to improve its mechanical and durability properties, which is confirmed by the concrete experiment conducted during this study. This study highlights the importance of adopting agro-waste-based SCMs like RHA to promote sustainable construction practices and reduce the environmental burden of the construction industry.

Keywords: Sustainability · Rice Husk Ash (RHA) · Supplementary Cementitious Material (SCM) · Life Cycle Assessment (LCA) · Global Warming Potential (GWP) · Ordinary Portland Cement (OPC)

1 Introduction

The Construction industry faces increasing pressure to adopt sustainable practices as concerns over environmental impact continue to grow. Concrete, the most widely used building material, mainly relies on cement, which contributes a high percentage of global carbon dioxide emissions, and has raised serious sustainability concerns. Addressing these challenges requires innovative approaches that reduce the environmental impact

R. Dissanayake et al. (Eds.): ICRES 2025, PTT, pp. 347–360, 2025.
https://doi.org/10.1007/978-981-96-9551-5_27

of concrete without compromising its performance. One promising solution for this major sustainability issue is the use of Supplementary Cementitious Materials (SCMs), which enhance concrete properties while reducing reliance on traditional cement.

Among the various SCMs, Rice Husk Ash (RHA) emerges as a highly promising alternative, valued for its exceptional silica richness and remarkable pozzolanic reactivity, making it an effective and sustainable cement replacement Uduweriya and Silva (2010). The silica in RHA can be used to enhance the strength and durability of the concrete, improving its long-term performance and pozzolanic properties, enabling the formation of additional binding compounds that contribute to the strengthening of the concrete.

To fully understand the sustainability of RHA as an SCM, going through a Life Cycle Assessment (LCA) is essential. LCA provides a systematic evaluation of the environmental impacts associated with RHA concrete, from raw material extraction and processing to its use and end-of-life disposal. By assessing key indicators such as carbon footprint, energy consumption, resource depletion, and waste generation. This study aims to quantify the environmental benefits of incorporating RHA into concrete through an LCA so we can ensure both performance efficiency and environmental sustainability by using RHA as an SCM. The findings in this study strengthen the use of RHA as a sustainable cement alternative and also highlight the critical role of LCA, ultimately advancing the transition towards a much greener, more resilient infrastructure in the future.

2 Literature Review

Life Cycle Assessment is an essential tool for evaluating the environmental impacts of construction materials in the industry. According to Gursel, the cement industry alone contributes 7–10% of global CO_2 emissions, making it one of the highest sources of greenhouse gases. This happens due to emissions in the production of calcination of limestone and the energy-intensive clinker production in the cement industry. (Gursel, Maryman and Ostertag, 2016). Given these circumstances, the introduction of SCMs like fly ash, silica fume, and Rice Husk Ash (RHA) has been widely explored to mitigate the traditional carbon footprint of concrete (Henry and Lynam, 2020).

Studies have shown that conventional concrete production generates high energy, resource depletion, and significant CO_2 emissions Mosaberpanah and Umar (2020). As shown in Fig. 1, the Global Warming Potential (GWP) of 100% Ordinary Portland Cement (OPC) concrete can reach 544 kg CO_2-eq/m^3 of concrete. In comparison, replacing up to 30% of cement with RHA has been found to significantly reduce CO_2 emissions while maintaining or enhancing durability (Gursel, Maryman and Ostertag, 2016).

RHA is an agro-waste material rich in amorphous silica Virajini et al. (2014), making it a highly reactive pozzolan that can partially replace cement. Research confirms that RHA improves the mechanical strength, durability, and sustainability of concrete while significantly reducing its carbon footprint (Henry and Lynam, 2020). Also, the use of RHA in concrete aligns with circular economy principles by utilizing an agricultural by-product that would otherwise contribute to landfill waste (Fernando et al., 2021).

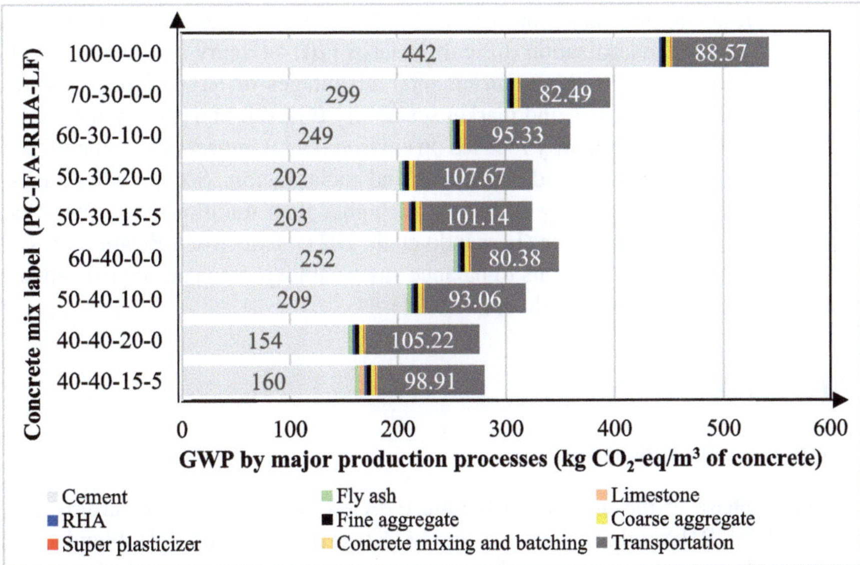

Fig. 1. Comparison of LCA results per unit volume of concrete mixes (Gursel, Maryman and Ostertag, 2016)

There are several LCA studies conducted to show the key advantages of using RHA blended concrete over conventional OPC concrete. Regarding the Global Warming Potential (GWP), Henry and Lynam found that RHA has a negative embodied energy of -26MJ/kg, making it an energy-saving material compared to OPC (Henry and Lynam, 2020). Research has shown conventional OPC-based concrete exhibits high CO_2 emissions due to energy-intensive clinker production; in contrast, RHA blended concrete reduces emissions by offsetting a portion of cement (Fernando et al., 2021). Similarly, Gursel, demonstrated that ternary and quaternary concrete mixes incorporating RHA and fly ash resulted in lower GWP while simultaneously improving durability (Gursel, Maryman and Ostertag, 2016).

In terms of Energy consumption, RHA production requires lower energy inputs than OPC, particularly when processed under controlled combustion conditions (Henry and Lynam, 2020). However, some LCA studies emphasize the need to optimize the transportation and grinding energy to maximize sustainability benefits (Gursel, Maryman and Ostertag, 2016). Additionally, RHA enhances the long-term compressive strength of concrete, as observed in the findings of Gursel, making it a viable alternative to OPC. Regarding durability, research has confirmed that concrete incorporating up to 15% RHA can achieve comparable and superior strength and durability properties relative to conventional OPC concrete (Fernando et al., 2021). Furthermore, RHA blended concrete exhibits chloride penetration resistance up to 10 times greater than conventional OPC-based concrete, making it particularly suitable for marine and highly corrosive environments (Gursel, Maryman and Ostertag, 2016).

Despite its advantages, there are several challenges to the widespread adoption of RHA in concrete. One of the major challenges we can observe is the variability in

RHA properties. The pozzolanic activity of RHA depends on combustion conditions, though making the standardization difficult Helarisi (2014)(Henry and Lynam, 2020). While LCA studies confirm the environmental advantages of RHA, further research is needed on cost-effectiveness and market scalability (Gursel, Maryman and Ostertag, 2016). LCA studies have primarily focused on environmental impacts, but a comprehensive assessment should also include economic and social factors. Some studies suggest that the initial cost of processing RHA can be higher than traditional cement, which may impact its market acceptance (Fernando et al., 2021). Effective utilization of RHA requires optimized mix proportions to balance sustainability, mechanical performance, and economic feasibility (Fernando et al., 2021).

3 Methodology

3.1 Experimental Procedure

In the initial phase of the research, a literature review was conducted regarding RHA used as an SCM, RHA utilized concrete, and LCA was conducted on RHA blended concrete. Through literature, it was found that the optimum percentage of replacing cement by RHA in concrete without compromising the performance of concrete is between 15% and 20% Ahmed, et al. (2020), Amran et al. (2021)(Replacement percentage of cement). Considering the information obtained through the literature, 15% was considered the optimum replacement percentage for conducting the LCA Helarisi (2014), Chopra et al. (2015). For the LCA to be precise and to obtain the inputs for the processes, the proportions of materials used in concrete should be obtained. In order to obtain the proportions of materials used in conventional concrete and RHA blended concre te, mix designs were developed and tested to verify using a trial mix and considering the 28-day compressive strength result. An experiment was also conducted to analyze the compressive strength and workability of concrete containing 15% RHA. LCA Analysis was done according to the procedure in ISO 14040 and 14044 and using the openLCA 2.2.0 open-source software. Following the standards, the LCA was conducted in 4 steps, namely, goal and scope definition, inventory analysis, impact assessment, and interpretation. The LCA was done for concrete containing RHA and for conventional concrete. Finally, the results of the LCA comparison of conventional concrete and RHA blended concrete were taken into consideration for the conclusions.

3.2 Development of a Mix Design

In this study, the LCA is conducted with the aim of comparing the environmental impacts of concrete containing RHA and conventional concrete without RHA. This assessment will help determine the sustainability advantages of using RHA as an SCM. To conduct the LCA, the proportions of materials required in producing concrete were obtained by developing a mix design. A compressive strength of 55MPa, a water-to-binder ratio of 0.45, and a slump range of 60–180 mm were considered in developing the concrete mix design. Also, a high water-reducing admixture (Superplasticizer - Chryso Optima SL 183) of 1% cement content was incorporated to maintain the workability of the mix

Zerbino et al. (2011). During the confirmation of this mix design, the trial mix resulted in a 28-day compressive strength of 58.8 MPa and a slump of 120 mm. Table 1 shows the mix proportion for the concrete mix without RHA, and after replacing 15% of OPC with RHA, the mix proportions are shown in Table 2. The concrete was cast and materials were prepared before the casting according to the standard procedures Habeeb and Mahmud (2010), ASTM C (2012).

Table 1. Concrete mix proportions without RHA

Constituents	Quantity (kg/m^3)
Cement (OPC)	400.0
Water	177.6
Fine aggregate	673.2
Coarse aggregate	1196.8
Superplasticizer (l)	4.0

Table 2. Concrete mix proportions with RHA

Constituents	Quantity (kg/m^3)
Cement (OPC) + RHA	340.0
RHA	60.0
Water	177.6
Fine aggregate	673.2
Coarse aggregate	1196.8
Superplasticizer (l)	6.0

3.3 Concrete Experiments

Using the mix design mentioned in Tables 1 and 2, experiments were done for concrete containing 0% (Control) and 15% RHA replacement percentages. Superplasticizer was used to increase or maintain the workability of the concrete mixes Jittin et al. (2020). The aim of this experiment was to confirm the literature by analyzing the results obtained by the concrete tests Ahmed, et al. (2020), Amran et al. (2021).

3.4 Life Cycle Assessment (LCA)

According to the procedure in ISO 14040 and 14044, the LCA was done in 4 stages, namely, goal and scope definition, inventory analysis, impact assessment, and interpretation. OpenLCA 2.2.0 open-source software was used to model the LCA. The LCA conducted during this study has followed a cradle-to-gate approach. It focuses on the environmental impacts from raw material extraction (cradle) to the production of concrete (gate). It has not included the use phase or end-of-life disposal of the concrete (grave). For the LCA analysis, Ecoinvent v3.10 Cutoff Unit-Processes Superseded Regionalisation database was used. This database included data from all over the globe. Since the region of Sri Lanka was not found in this database, the region was considered as RoW (Rest of the World).

Goal and Scope Definition

The goal of this LCA is to analyze the effectiveness of utilizing RHA as an SCM in concrete in the construction industry from an environmental perspective by comparing the environmental impacts of conventional concrete production and RHA-blended concrete production. As the functional unit of this LCA, 1 cubic meter (1 m^3) of concrete is considered.

RHA is produced as a waste material from many operations, such as a by-product of boilers generating electrical and thermal energy, and burning rice husks in rice mills. During this LCA, RHA is considered a waste material that has been produced by one of those processes. Emissions during these processes, such as during energy generation, are not considered in this study due to a lack of data in the literature and the database used in the software (Ecoinvent v3.10 Cutoff Unit-Processes Superseded Regionalisation database). However, RHA produced should go through a grinding process before being used as an SCM. So, the emission factor for the grinding process was taken from the literature as 0.0027 kg CO_2-eq. (Fernando et al., 2021) Some assumptions used in this study are, that transportation distances are considered average values specific to the study region which is provided in the Ecoinvent v3.10 Cutoff Unit-Processes Superseded Regionalisation database, energy sources utilized in the software are consistent across processes unless specified, 15% is considered the optimum RHA replacement for cement from the literature review and the experimental results obtained, and the emission factor for the grinding process of RHA is taken by referring the literature as 0.0027 kg CO_2-eq. As a limitation of this study, the lack of data in the openLCA 2.2.0 software regarding RHA and only focusing on environmental impacts while not considering economic feasibility or social implications, which are important for large-scale adoption, can be identified.

Inventory and Boundary Analysis

For the LCA to be conducted, processes should be defined in the openLCA 2.2.0 software. The conventional concrete production and RHA blended concrete production were considered as the two processes. In order to model the inputs and outputs of the processes, the mix designs mentioned in Tables 1 and 2 were used. In the software, various flows are predefined and are used to model the processes. Those predefined flows used in the processes have data on the extraction of raw materials, average transportation distances for the considered region, and energy and emissions of the processes. These are modelled in the Ecoinvent v3.10 Cutoff Unit-Processes Superseded Regionalisation database. The processes modelled in the software are used to produce 1 m^3 of concrete.

During this LCA, extraction and processing of raw materials (i.e., cement, aggregates, water, superplasticizer), production of RHA, including grinding, transportation of materials (average distances specific to the study region are assumed - region used - RoW), and concrete production are included. However, due to the lack of data available in the database and the scope of this study, some processes are excluded during the analysis. Emissions from the combustion process that generates RHA (e.g., emissions from boilers or rice mills), fuel consumption for transportation of RHA to concrete production facilities (Due to the high variation of travel distances), use phase of the concrete (e.g., maintenance, repair), and end-of-life disposal or recycling of concrete are excluded. The flow chart of processes in RHA-blended concrete production is shown in Fig. 2.

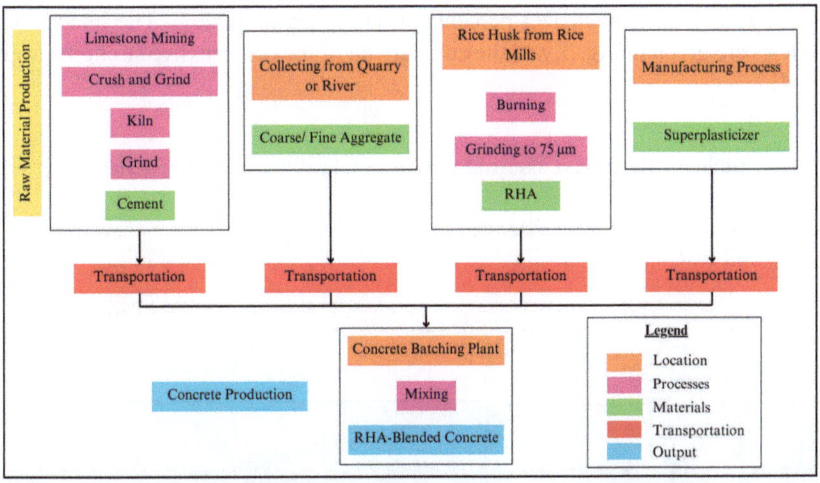

Fig. 2. Process flow chart in RHA-blended concrete production

Impact Assessment

The impact assessment was done using the ReCiPe impact assessment method, and it is one of the most popular impact assessment techniques in LCA. The environmental impacts of conventional concrete and RHA blended concrete were assessed using this method. This method contains a wide range of environmental impact categories using inventory data available, such as emissions and resource extractions. The ReCiPe impact assessment method can analyze both midpoint and endpoint impact categories. Impacts are measured closer to the actual stages where environmental influences occur in midpoint assessment, and long-term impacts or end consequences are measured in endpoint assessment. In this study, the ReCiPe midpoint impact assessment method is used since it is good for comparisons and gives detailed insights into specific environmental impacts.

Interpretation

The impact on the environmental footprint of the two products (conventional concrete and RHA blended concrete) can be compared by evaluating the results provided by the openLCA 2.2.0 software. In the results provided by the software, many impact categories can be found, including Global Warming Potential (GWP), energy consumption, water use, acidification potential, eutrophication potential, resource depletion, etc.

4 Results and Discussion

4.1 Concrete Experiment Results

Concrete experiment results were obtained using several trials to obtain the target strength of the mix design and to achieve considerable workability of concrete. After two trials, it was decided that a superplasticizer content equal to 1% of cement content can be incorporated for 10% RHA mix and 1.5% of cement content for 15% RHA mix. The slump, flow, and superplasticizer percentages are given in Table 3. The images of the slump tests are given in Fig. 3, and the comparison of the compressive strengths is given in Fig. 4.

Table 3. The impact assessment results of the different project variants

RHA Content	Superplasticizer Content	Slump (mm)	Flow (mm)
0% Control	1%	166	510
15%	1.5%	177	480

Fig. 3. Slump of 0% (Left) and 15% (Right) RHA blended concrete

4.2 Life Cycle Assessment Results

In order to obtain the results, two processes were defined in the software, namely, conventional concrete production and RHA-blended concrete production. Included in those processes were the flows representing the components used in producing concrete. The flows used for this analysis were Portland cement, coarse aggregates, fine aggregates, tap water, superplasticizer, and RHA. The flow types and reference units/providers used in the LCA model are shown in Table 4.

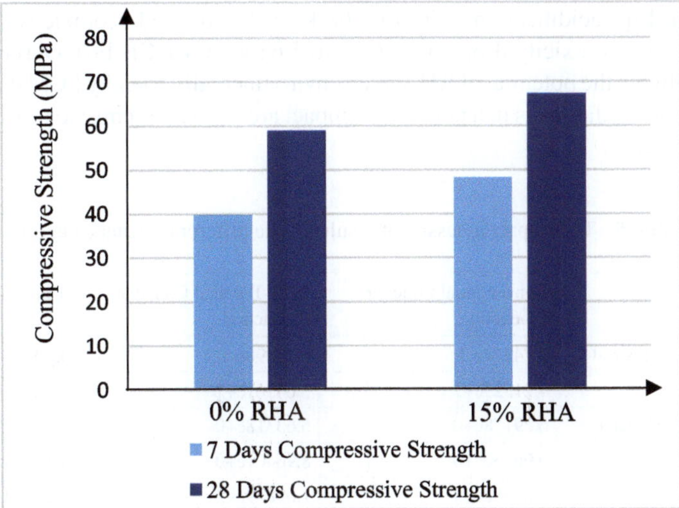

Fig. 4. Compressive strength results of concrete experiments

Table 4. Flow types (Inputs) and reference units

Flow (Input)	Provider
Cement, Portland	Market for cement, Portland I cement, Portland I Cutoff, U - RoW
Coarse Aggregates	Market for gravel, crushed I gravel, crushed I Cutoff, U - RoW
Fine Aggregate	Market for sand I sand I Cutoff, U - RoW
Tap Water	Market for tap water I tap water I Cutoff, U - RoW
Superplasticizer	Market for plasticizer, for concrete, based on sulfonated melamine formaldehyde I plasticizer, for concrete, based on sulfonated melamine formaldehyde I Cutoff, U - GLO
RHA	Considered as a waste material from boilers in energy production

To obtain the results, the Ecoinvent v3.10 Cutoff Unit-Processes Superseded Regionalisation database and ReCiPe 2016 v1.03, midpoint (H) impact assessment method was used. The results were used for the comparison of the impact of conventional and RHA-blended concrete.

Impact Assessment Results

When comparing the results of the impact assessment of conventional concrete and RHA blended concrete, it shows that utilizing RHA in concrete has considerable environmental benefits. It can be seen that the environmental implications of RHA-blended concrete are reduced across all indices. Table 5 shows the impact assessment results of the different project variants of the selected LCIA method of the project. By utilizing RHA-blended concrete, reductions can be observed in energy resource use (51.46 to

45.58 kg oil-Eq), acidification (0.78 to 0.68 kg SO_2-Eq), and ecotoxicity (for example, freshwater ecotoxicity drops from 6.29 to 5.63 kg 1,4-DCB-Eq). Observing these results highlights the potential of RHA as an environmentally friendly SCM that can help lessen environmental loads in a number of impact areas, such as emissions, toxicity, and resource depletion.

Table 5. The impact assessment results of the different impact categories

Indicator	Conventional Concrete Production	RHA-Blended Concrete Production	Unit
Acidification: terrestrial	7.74874e-1	6.79998e-1	kg SO_2-Eq
Climate change	4.21523e+2	3.64361e+2	kg CO_2-Eq
Ecotoxicity: freshwater	6.29178e+0	5.63318e+0	kg 1,4-DCB-Eq
Ecotoxicity: marine	9.96450e+0	8.96401e+0	kg 1,4-DCB-Eq
Ecotoxicity: terrestrial	1.84388e+3	1.69829e+3	kg 1,4-DCB-Eq
Energy resources: non-renewable, fossil	5.14575e+1	4.55751e+1	kg oil-Eq
Eutrophication: freshwater	4.32681e-2	3.81036e-2	kg P-Eq
Eutrophication: marine	3.55128e-3	3.15112e-3	kg N-Eq
Human toxicity: carcinogenic	2.80523e+1	2.56344e+1	kg 1,4-DCB-Eq
Human toxicity: non-carcinogenic	1.79458e+2	1.59156e+2	kg 1,4-DCB-Eq
Ionising radiation	4.30747e+0	3.93656e+0	kBq Co-60-Eq
Land use	6.58423e+0	6.17007e+0	m^2*a crop-Eq
Material resources: metals/minerals	3.30405e+0	2.95436e+0	kg Cu-Eq
Ozone depletion	3.53754e-5	3.18517e-5	kg CFC-11-Eq
Particulate matter formation	3.28853e-1	2.89644e-1	kg PM2.5-Eq
Photochemical oxidant formation: human health	9.36141e-1	8.20325e-1	kg NOx-Eq
Photochemical oxidant formation: terrestrial ecosystems	9.58204e-1	8.40080e-1	kg NOx-Eq
Water use	1.07076e+0	9.96110e-1	m^3

Relative Results

The comparisons are shown in Fig. 5, which shows the relative indicator results of the respective project variants. For each indicator, the maximum result is set to 100% and the results of the other variants are displayed in relation to this result. The impact of climate change is shown in Fig. 6 in kg CO_2-eq per functional unit.

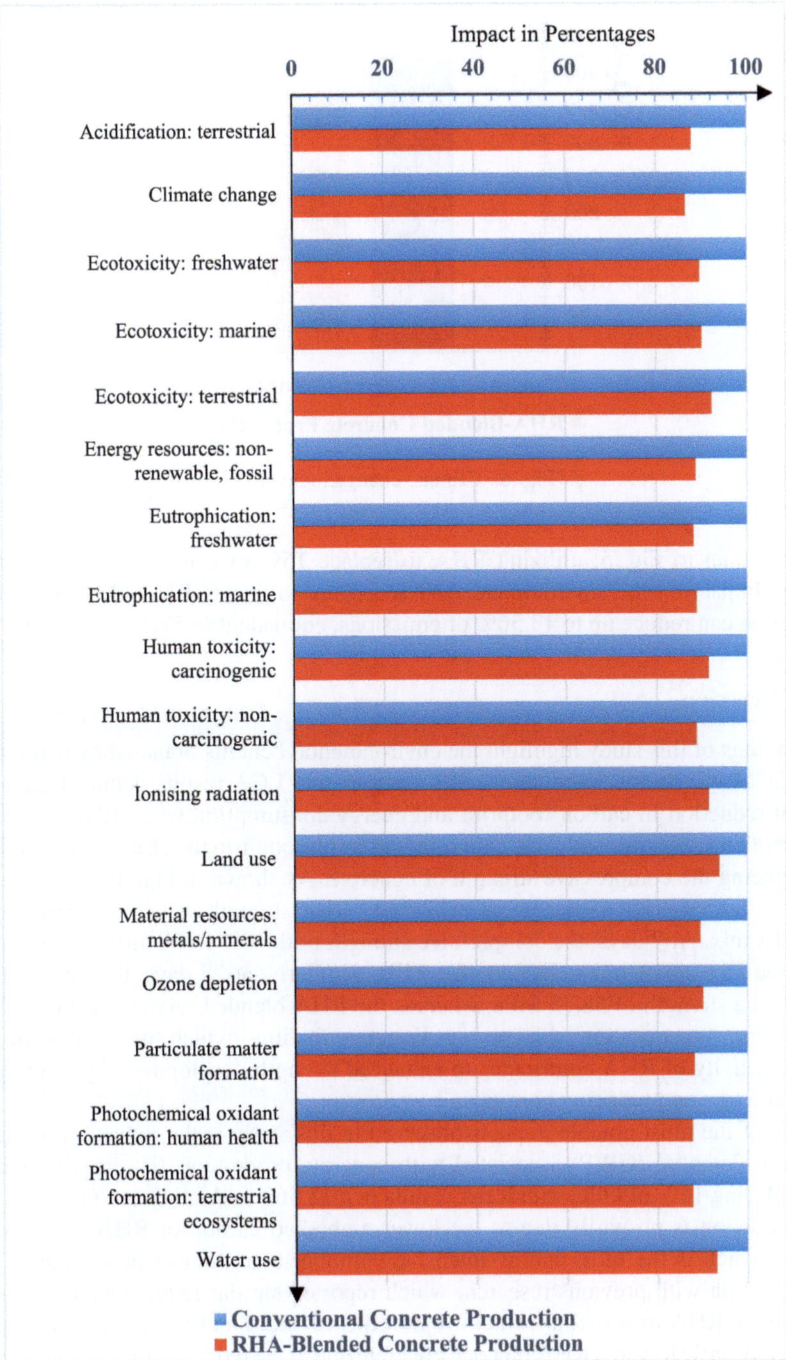

Fig. 5. Relative LCIA comparison

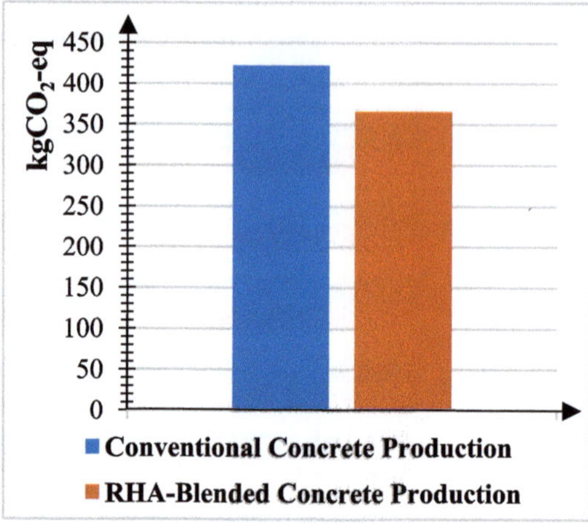

Fig. 6. Climate change indicator

According to Fig. 5, utilizing RHA to replace 15% of cement in concrete has a positive impact on the environment. And according to Fig. 6, RHA-blended concrete production can reduce up to 13.56% of emissions, equivalent to 57.162 kg of CO_2.

4.3 Discussion

The findings of this study highlight the environmental benefits obtained by using RHA as an SCM in concrete production. The comparative LCA results demonstrate a significant reduction in carbon footprint and energy consumption when RHA replaces a portion of OPC. Additionally, the experimental results confirm the effectiveness of RHA in enhancing the compressive strength of concrete. As shown in Fig. 4, concrete specimens with 15% RHA replacement exhibited higher strength values compared to the 0% RHA mix. At 7 days, the compressive strength of the 0% RHA mix was 39.7 MPa, while the 15% RHA mix achieved 48.1 MPa. Similarly, at 28 days, the mix with 0% RHA had a strength of 58.75 MPa, whereas the RHA blended mix reached 67.1 MPa. These improvements align with previous literature findings, which suggest that the pozzolanic activity of RHA contributes to enhanced strength development by refining the microstructure and reducing porosity.

One of the most notable impacts observed in this study is the reduction of Global Warming Potential (GWP) associated with concrete production. The results indicate that replacing 15% of OPC with RHA results in a 13.56% reduction in CO_2 emissions. This reduction is primarily due to the lower embodied carbon of RHA compared to clinker, which is the most energy-intensive component of cement production. These findings align with previous research, which reported similar reductions in emissions when using RHA as a partial cement replacement. Additionally, energy consumption was significantly lower in RHA blended concrete due to the reduced clinker content. This demonstrates that incorporating SCMs like RHA leads to a decrease in energy-intensive clinker production. Furthermore, the use of RHA contributes to resource conservation by utilizing an agricultural by-product that would otherwise be discarded as waste.

5 Conclusion

From the concrete experiments, it can be seen that the 7-day and 28-day compressive strength of 15% RHA-containing concrete is higher than 0% RHA-containing concrete (control). This confirms the literature and the selection of 15% as the optimum RHA replacement percentage. From the LCA, it can be seen that RHA as an SCM significantly enhances the sustainability of the concrete. We can confirm the results that, by replacing 15% of cement with RHA, greenhouse gas emissions are reduced by 13.56%, lowering the overall carbon footprint of concrete production. Additionally, the results highlight improvements across multiple environmental impact categories, including resource depletion, energy consumption, and climate change mitigation. These findings help to confirm the viability of agro-waste-based SCMs in reducing the ecological burden of conventional cement while maintaining structural performance.

Although this study primarily focuses on environmental impacts, the long-term durability of RHA blended concrete should be further evaluated through experimental investigations. Despite its numerous benefits, several challenges must be addressed for the widespread adoption of RHA in concrete production. One major challenge is the availability of RHA properties due to differences in combustion conditions. Standardizing RHA production techniques will be essential to ensure consistent performance. Additionally, transportation distances for RHA procurement could impact overall emissions, and further optimization in logistics would be beneficial. to build upon the findings of this study, future research should focus on large-scale experiential studies to validate the durability and mechanical performance of RHA blended concrete, explore the economic feasibility of RHA integration in construction projects, and optimize transportation and grinding processes to further minimize environmental impacts.

However, the LCA results presented in this study focus primarily on environmental impacts, while other crucial factors such as economic feasibility and social implications should also be considered before large-scale adoption. Factors such as production costs, supply chain logistics, and stakeholder acceptance need further exploration to ensure a complete understanding of a sustainable assessment. Moving forward, optimizing RHA processing techniques, re-evaluating cost-effectiveness, and developing supportive policies will be essential to facilitate its widespread implementation throughout the world. However, it can be seen that using RHA in promoting sustainable construction practices contributes to a much greener and more resilient built environment.

References

Virajini, R.M.S., Perera, V.P.S., Rajendra, J.C.N.: Physical properties and content of silica in different varieties of rice husk found physical properties and content of silica in different varieties of rice husk found in Sri Lanka. In: Proceedings of the Technical Sessions, Institute of Physics - Sri Lanka, vol. 30, pp. 105–110 (2014)

Uduweriya, R.B.Y., De Silva, S.: Effect of mechanical properties of concrete containing rice husk ash. In: Proceedings of the 8th Academic Sessions, University of Ruhuna, vol. 8, pp. 87–94 (2010)

Ahmed, A., et al.: Rice husk ash as a cement replacement in high strength sustainable concrete. In: Materials Science Forum, vol. 1007, pp. 90–98 (2020). https://doi.org/10.4028/www.scientific.net/MSF.1007.90

Habeeb, G.A., Mahmud, H.B.: Study on properties of rice husk ash and its use as cement replacement material. Mater. Res. **13**(2), 185–190 (2010). https://doi.org/10.1590/S1516-14392010000200011

Zerbino, R., Giaccio, G., Isaia, G.C.: Concrete incorporating rice-husk ash without processing. Constr. Build. Mater. **25**(1), 371–378 (2011). https://doi.org/10.1016/j.conbuildmat.2010.06.016

ASTM C1202. Standard Test Method for Electrical Indication of Concrete's Ability to Resist Chloride Ion Penetration. , American Society for Testing and Materials, (C), pp. 1–8 (2012). https://doi.org/10.1520/C1202-22E01.2

Helarisi, A.: Effects of biomass-based pozzolanic material on chloride ion penetration in concrete. IESL Trans. (2014)

Amran, M., et al.: Rice husk ash-based concrete composites: a critical review of their properties and applications. Crystals. MDPI AG, pp. 1–33 (2021). https://doi.org/10.3390/cryst11020168

Mosaberpanah, M.A., Umar, S.A.: Utilizing rice husk ash as supplement to cementitious materials on performance of ultra high-performance concrete: – a review. Mater. Today Sustain. (2020). https://doi.org/10.1016/j.mtsust.2019.100030

Fernando, S., et al.: Life cycle assessment and cost analysis of fly ash–rice husk ash blended alkali-activated concrete. J. Environ. Manag. **295**(June), 113140 (2021). https://doi.org/10.1016/j.jenvman.2021.113140

Jittin, V., Bahurudeen, A., Ajinkya, S.D.: Utilisation of rice husk ash for cleaner production of different construction products. J. Clean. Prod. **263**, 121578 (2020). https://doi.org/10.1016/j.jclepro.2020.121578

Gursel, A.P., Maryman, H., Ostertag, C.: A life-cycle approach to environmental, mechanical, and durability properties of "green" concrete mixes with rice husk ash. J. Clean. Prod. **112**, 823–836 (2016). https://doi.org/10.1016/j.jclepro.2015.06.029

Chopra, D., Siddique, R., Kunal: Strength, permeability and microstructure of self-compacting concrete containing rice husk ash. Biosyst. Eng. **130**, 72–80 (2015). https://doi.org/10.1016/j.biosystemseng.2014.12.005

Henry, C.S., Lynam, J.G.: Embodied energy of rice husk ash for sustainable cement production. Case Stud. Chem. Environ. Eng. **2**, 100004 (2020). https://doi.org/10.1016/j.cscee.2020.100004

Author Index

R. Dissanayake et al. (Eds.): ICRES 2025, PTT, pp. 361–362, 2025.
https://doi.org/10.1007/978-981-96-9551-5

MIX
Papier aus verantwortungsvollen Quellen
Paper from responsible sources
FSC® C105338

If you have any concerns about our products,
you can contact us on
ProductSafety@springernature.com

In case Publisher is established outside the EU,
the EU authorized representative is:
Springer Nature Customer Service Center GmbH
Europaplatz 3, 69115 Heidelberg, Germany

Printed by Libri Plureos GmbH
in Hamburg, Germany